The Pacific Northwest
Geographical Perspectives

James G. Ashbaugh, Editor
Portland State University

Contributors

John A. Alwin
James G. Ashbaugh
Ronald R. Boyce
Teresa L. Bulman
Sheldon Bluestein
Claude W. Curran
Keith S. Hadley
John Hall
Steven R. Kale
Larry King
Eugene P. Kiver
Keith W. Muckleston
William Rabiega
James W. Scott
Alex Sifford
Dale F. Stradling
Bettina von Hagen
Robert Whelan
William Woodward

KENDALL/HUNT PUBLISHING COMPANY
4050 Westmark Drive   Dubuque, Iowa 52002

Maps pages 127, 317, 352, 353 and 356 by Rick Vrana.

Copyright © 1997 by Kendall/Hunt Publishing Company

Library of Congress Catalog Card Number: 97-72001

ISBN 0-7872-3606-3

All rights reserved. No part of this publication may be reproduced, stored in a retrieval system, or transmitted, in any form or by any means, electronic, mechanical, photocopying, recording, or otherwise, without the prior written permission of the copyright owner.

Printed in the United States of America
10 9 8 7 6 5 4 3 2 1

Contents

Preface v
Acknowledgments vii
Introduction ix

Chapter 1 Historical Geography 1
   James W. Scott *Western Washington University*

The Physical Environment 39

Chapter 2 Landforms 41
   Eugene P. Kiver and Dale F. Stradling *Eastern Washington University*
Chapter 3 Climate 77
   Claude W. Curran *Southern Oregon University*
Chapter 4 Vegetation of the Pacific Northwest 99
   Keith S. Hadley *Western Oregon University*
Chapter 5 Soils of the Pacific Northwest 117
   Keith S. Hadley *Western Oregon University*

Natural Resources 139

Chapter 6 Fishing 141
   James G. Ashbaugh *Portland State University*
Chapter 7 Forests 185
   Teresa L. Bulman and Bettina von Hagen *Portland State University*
Chapter 8 Regional Water Resources 213
   Keith W. Muckleston *Oregon State University*
Chapter 9 Energy Resources of the Pacific Northwest 259
   Alex Sifford *Sifford Energy Services*
Chapter 10 Mining 291
   Robert Whalan *Oregon Dept. of Geology and Mineral Industries*

Economic Geography 307

Chapter 11 The Economy 309
   Steven R. Kale *Oregon Department of Transportation*
Chapter 12 Manufacturing in the Northwest 345
   William A. Rabiega with John Hall *Portland State University*
Chapter 13 Agriculture 363
   James W. Scott *Western Washington University*

Chapter 14 Tourism and Recreation 401
  Larry King *Portland Community College*
Chapter 15 Urban Systems 427
  Ronald R. Boyce and William Woodward *Seattle Pacific University*

EXTERNAL RELATIONS 447

Chapter 16 The Northwest and the Pacific Rim 449
  John A. Alwin *Central Washington University*

PREFACE

It has been apparent for the last thirty-five years that a text dealing with the contemporary geography of the Pacific Northwest was needed. Even as regional geography was being de-emphasized in colleges and universities there was growing student interest in learning about the region in which they live.

The problem for instructors was truly daunting. Most were specialists in fields such as climate, geomorphology, urban, economic and historical geography. Few claimed the Pacific Northwest as a regional specialty.

Many relied on the excellent *Atlas of the Pacific Northwest*. However, even a superb atlas is a reference source, not a text. Articles from professional journals were available but often were too specialized to provide faculty and students with an analysis in depth of the many facets of the regional geography.

In an attempt to solve this problem a group of geographers were asked to write a chapter on their specialty for a book on the Pacific Northwest. These individual studies provide geographic perspectives on the Pacific Northwest. They each stand by themselves. Each instructor can use the book as it best serves their goals and objectives. Suggested readings are provided for each chapter. Chapters on soils, transportation, manufacturing, minerals, service industries and the relationship of the Pacific Northwest to British Columbia are currently being written and will be added to future editions.

ACKNOWLEDGMENTS

Without the dedication and work of the chapter authors and cartographer, this book would not have been possible. Despite many other competing obligations they were convinced of the importance of the project and committed themselves to its successful completion. They all have my heartfelt thanks.

I also wish to thank my wife Dorothy, son Stuart and daughter Megan for their help and encouragement which made my task much easier.

INTRODUCTION

Over the last 40 years, a new Pacific Northwest geography—one based on both internal change and external relationships of the region—has evolved. Where once trade with the Far East was dominated by exports of wheat and logs, it has now expanded to include potatoes and other agricultural products, as well as a wide variety of imported manufactured products, including automobiles from Japan and Korea. Hydroelectric installations, especially on the Columbia River, no longer produce huge surpluses of power. Boeing, one of many United States manufacturers of air frames, has become a world leader in the production of jet air carriers. The wine industry, long renowned as the producer of cheap fortified fruit wines, has expanded to wines of varietal grapes prized in international competition. Large publicly owned and once pristine forests have been decimated by excessive cutting, damaging the ecology of the region. The once plentiful salmon resource is now endangered because of alteration of environment and resultant loss of habitat. And the Pacific Northwest and British Columbia now jointly cooperate, under treaty, on flood control and hydroelectric development on the Columbia River. Almost 40 years has passed since Otis Freeman and Howard Martin edited their excellent book on the Pacific Northwest. And although the region has changed tremendously, it remains tremendously valuable as a historical resource on the Pacific Northwest.

People in their ever-increasing numbers have changed the face of the land. The conversion of agricultural land to urban use has removed farm land from production and resulted in land-use legislation. And population growth continues to reshape the region. In fact, these changes have had a profound effect on many different aspects of the Pacific Northwest: from the physical environment, including landforms, climate, and vegetation, to the human-made one of agriculture, the timber industry, fishing, manufacturing, water resources, energy, economic development, and urban growth.

For many years, the need for an updated book on the Pacific Northwest has been discussed. One that would present the Pacific Northwest of today from several perspectives was decided on. We make no claim for comprehensiveness and plan that future editions will include additional perspectives as important as the ones in this book.

Just what is the Pacific Northwest? What geographic area does it include? *Pacific Northwest Magazine* believes northern California and southern British Columbia should be counted as part of the region. Freeman and Martin considered Alaska and western Montana as part of the Pacific Northwest. *The Atlas of the Pacific Northwest,* in its earlier editions, included western Montana. The area is now usually limited to Washington, Oregon, and Idaho. This book includes Washington, Oregon, Idaho, and the 17 western counties of Montana.

HISTORICAL GEOGRAPHY

Chapter 1

An Overview of the Historical Geography of the Pacific Northwest

James W. Scott
Western Washington University

For more than 10,000 years, humans have occupied portions of the Pacific Northwest and made use of its abundant natural resources. Until a couple of centuries ago, however, these groups were exclusively Amer-Indian in racial composition. Only since the 1770s have Europeans, Africans, Asians, or Australasians been traveling to the region, trading there, exploiting its resources, building communities, laying out roads and railroads, and leaving their imprint on the face of the land.

Understandably, the effect of these settlers and sojourners, both ancient and more recent, has varied enormously. For hundreds, even thousands of years, effects were on a relatively modest scale; whereas in recent centuries, impacts have increased in magnitude. Today, people's shaping of the land extends to every corner of the region, from the greatest depths of the adjacent seas to the highest elevations of the land, even into the atmosphere and the troposphere beyond. No portion of the region remains pristine, untouched by human hand.

This chapter identifies and examines the activities and achievements of the successive waves of newcomers, and assesses the impact each group has made on the land and its various resources.

NATIVE AMERICAN MIGRATIONS AND SETTLEMENT

Why, when, and by which routes the earliest Native American groups arrived in the region remain matters of conjecture and contention. No longer in doubt, however, are the conclusions that these were people with origins in some part of eastern Asia, and who at some time thousands of years ago traversed the land bridge that joined Siberia and Alaska—across the present Bering Strait—between 25,000 and 14,000 years before the present (B.P.) and for at least two shorter periods thereafter.

Why these hunter-gatherers moved eastward to Alaska, and then south to the Americas in the next thousand or so years is unclear. One can assume both "push" and "pull" factors were at work. A dwindling food supply—the result perhaps of population pressure, some climatic fluctuation, or a reduction in available food supplies—would be the most powerful "push" factor. The expectation of more abundant wild herds and other food elsewhere was undoubtedly the dominant "pull" factor. Once begun, the progression eastward into the Yukon Valley and beyond proved both inevitable and inexorable as the larger mammals—in particular the wooly mammoth and the mastodon—were drastically reduced in numbers and even exterminated in some areas. Backtracking for most was likely impossible, if not unthinkable, and the only alternative was to move on until all regions of North and South America had been explored and all resources and possibilities had been assessed.

Fixing the precise dates of these movements poses many vexing problems. Despite decades of field research, the dating of the earliest arrivals remains in dispute. Dates of 35,000 to 40,000 B.P., and even earlier ones, have been proposed by reputable scholars, although with slender evidence to back their speculations. However, the remains of sizeable numbers of sites occupied or utilized by early hunters in the years between 13,500 and 11,000 B.P. suggest that the principal migrations were well under way during these years. Further excavations will doubtless add to this growing body of knowledge and perhaps will help modify, if not radically change, many of today's tentative conclusions.

While specific routes of migrating groups likely varied to some extent from year to year—as we know happened on both the Oregon and the Californian Trails—it is now generally agreed by anthropologists and prehistorians that in North America migration routes were concentrated along a fairly broad corridor just east of the Rocky Mountains. A few groups probably moved southward along the Pacific shore, which at that time extended a good deal further west than it does now, because of a considerably lower sea level—as much as 400 feet lower than it is today. However, various obstacles made such movements hazardous and difficult, including large valley glaciers moving seaward from the adjacent Coast Range, and numerous rocky peninsulas and precipitous coastal cliffs. Further, little firm evidence exists that the earliest groups to use this coastal route were equipped with boats or canoes, which were to become the preferred mode of travel in later centuries.

Thus, while a few Pacific Northwest Indian groups might have arrived via the coastal route, it is generally agreed that the majority of the region's earliest Native Americans came from the northeast and east, having moved southward along the main migratory corridor east of the Rockies. Evidence of human occupance in the Pacific Northwest in the period prior to 10,000 B.P. remains scanty, although there is ample proof that a number of groups were moving through the region at that time in search of game. For example, one can point to the 1977 discovery at the Manis mastodon site, near to Sequim on the Olympic Peninsula, of a human artifact (a spearpoint) found embedded in a mastodon's rib. This artifact has been dated to approximately 12,000 years ago.

Over the many centuries, Indian settlements grew in quantity and, especially near the coast, in size, as successful ways of life evolved following the general retreat of the continental ice-sheets northward and the contraction of mountain glaciers to higher elevations, events which occurred about 9,000 years ago. In recent years, archeological excavation of hundreds of Indian sites across the Pacific Northwest has provided ample evidence of a succession of permanent and seasonal settlements in places where one would expect to find them located: at level coastal sites and at similar ones along the banks of the lower reaches of the region's many rivers. Population distribution and density depended largely on the nature of the subsistence economy. In general, the permanent villages of the fishing cultures that developed along the coastal strip of Oregon, Washington, and British Columbia, and on nearby islands were larger than those of groups living on the interior plateaus and intermontane valleys further east, where a mixed hunting-gathering economy prevailed, and where carrying capacities were correspondingly much lower.

Among the coastal Indian groups, the main diet consisted of usually abundant salmon and shellfish supplemented by berries, camas roots, mammals, and birds. Their permanent settlements at the time of the European voyages in the late eighteenth century were large cedar dwellings, along with some smaller storage sheds. Social customs were usually complex, and political and economic organization were advanced and markedly hierarchical. Population density in the settled coastal strip reached averages as high as 25 to the square mile, while the total population for the southern portion of the Northwest coast has been estimated at 72,300 at the time of the first European explorations.[1]

By contrast, the plateau Indians of the interior areas of the Pacific Northwest lived a much sparser existence, depending on berries, roots and whatever game and small mammals could be shot or trapped. Settlements were smaller, with groups often being no more than extended families. Usually these settlements were seasonal rather than permanent, and their dwellings were less durable than those found along the coast. The advent of the horse, however, greatly increased the hunting capabilities of such groups as the Cayuse and Nez Perce, and this inevitably led to a more nomadic lifestyle. Many groups became less dependent on river sites, supplementing traditional sources of food with larger mammals such as the bison.[2]

By the mid-eighteenth century, Indian marts were flourishing, as perhaps they had been for some centuries. Seasonal gatherings, these marts brought together tribes from coastal and interior regions at such sites as The Dalles, Celilo Falls, and Kettle Falls: areas where great numbers of salmon were speared, netted, and dried, and where items from all over the Pacific Northwest could be exchanged. Alexander Ross, one of Astor's Pacific Fur Company, writes that although the "constant inhabitants" of the Narrows (that is, The Dalles) "do not exceed 100 persons" during the salmon season the number could rise to more than three thousand.[3] Similar reports were left by many of Ross's contemporaries.[4]

While developing suitable lifestyles, Indian groups inevitably made changes in their physical environment. Not only were major additions made to the land in the form of cedar houses and sheds in villages; but along most streams and rivers, weirs were constructed and the discharge of these was seasonally modified. More significantly, sizeable areas of forest were cleared to provide sites for villages, as well as places for hunting or the cultivation of camas roots. Elsewhere, in areas where the hunting of game was a major activity, grasses and shrubs were seasonally set on fire, in part to herd the animals to places where they could more easily be killed, but perhaps in larger part to encourage the growth of new grasses that would attract such animals. Many paths and trails criss-crossed the forest to provide access to hunting and other food sites, changing the mix of trees and shrubs, not to mention the accelerated erosion that could also occur.

Inexorably, the face of the Pacific Northwest was modified and occasionally radically transformed. While the changes made appear to have been minuscule compared to the dramatic changes of recent times, there is no doubt that some of these did leave permanent scars and helped encourage even greater changes when the land was seized by Euro-American settlers after 1800. For instance, most of the so-called Indian prairies, which were avidly eyed by the newcomers, were the result of years, perhaps centuries, of Indian burning.[5]

THE EURO-AMERICAN EXPLORATIONS

Although the arrival of explorers from Europe and the eastern United States during the later eighteenth century might have been preceded by groups from east Asia, knowledge about these groups is so inconclusive that further discussion here is unwarranted. Equally inconclusive, and to virtually every professional scholar of exploration totally unproven, is the purported voyage in 1592 of Juan de Fuca, a Greek sailor supposedly in the service of Spain.

More reliable is the evidence that fifty years earlier Juan Cabrillo and his pilot Bartolomé Ferralo sailed north along the Pacific coast to some point between the 42nd and 44th parallel; to at least the Oregon-Californian border and perhaps beyond. In the 1570s, some decades later, the English sailor Sir Francis Drake also reached northern California and perhaps traveled as far north as the coast of southern Oregon. However, neither the Cabrillo/Ferralo nor the Drake voyages led to any immediate results that in any way disrupted the lives of the native American groups occupying the region.

Not so the voyages that followed—one after another in ever-quickening succession in the 1770s and 1780s. Reacting to rumors of Russian expansion eastward from the Kamchatka Peninsula of Siberia, in 1774 Spanish authorities in New Spain

(Mexico) sent Juan Perez on a voyage north, with instructions to claim for Spain all lands sighted to as far north as the 60th parallel.[6] Perez reached only as far as 54°N before turning back, without having made any effective contact with the native peoples of the Northwest coast. A year later Bruno Heceta and Juan Francisco de Bodega y Quadra sailed north and achieved the first recorded landing on the Washington coast near Point Grenville—a disastrous one, all those sent ashore being lost. In addition, Heceta on his voyage north became the first European to make a sighting of the Columbia River. This, however, Heceta mistook for the fable Strait of Anian. Two years later, in 1778, Captain James Cook, the most renowned of all British explorers and then on his third voyage to the Pacific, sailed along the Oregon and Washington coasts but failed to sight or even detect either the Columbia River or the Strait of Juan de Fuca. Nevertheless, Cook did complete a number of superbly executed charts of harbors along the Northwest coast, including one of Nootka Sound, and in Alaska he eventually concluded that the long-sought Northwest Passage, purported to connect the Pacific and the Atlantic Oceans, did not exist.[7] Cook's principal legacy to the Pacific Northwest was both unforeseen and unplanned: the opening up of the Pacific Northwest to traders from many nations following the auction in Canton, China, of sea otter pelts obtained by members of his crew from the Indians of Vancouver Island.

In 1785 the British trader Captain James Hanna inaugurated the sea otter trade, which in a scant two decades resulted in the virtual extinction of the species along the Northwest coast. It also led to the permanent presence of Europeans and Americans that was to destroy or seriously endanger every native American culture in the Pacific Northwest, as well as transform the land and its resources almost beyond recognition in the ensuing century. The inevitable political clash between Spain and the traders came in 1789 with the seizure of two ships belonging to an English trader, Captain John Meares. Within a year, however, the Nootka Sound Convention had been agreed to by Spain and Great Britain. This gave the "foreigners" the right to fish, trade, and even to establish settlements in the region. To insure that the Convention was being implemented effectively, Captain George Vancouver was despatched by the British Government in 1791. He was also instructed to complete the exploratory survey and mapping begun by James Cook more than a decade earlier. Vancouver's maritime explorations included detailed surveys of Puget Sound and the first circumnavigation of Vancouver Island. Scores of geographical features and landmarks were named. Many still carry the names Vancouver bestowed on them, including Mt. Rainier, Mt. Hood, Mt. St. Helens, Mt. Baker, Puget Sound, Admirality Inlet, Hood Canal, Port Townsend, Whidbey Island, Bellingham Bay, and Point Roberts.[8]

Between 1789 and 1792, Spanish seamen were busy exploring the region and preparing maps and charts. They included Manuel Quimper, who made the first systematic survey of the north coast of the Strait of Juan de Fuca; Francisco de Eliza, who explored Haro Strait and the San Juan Islands; and Dionisio Alcala Galiano and Cayetano Valdes, who explored both shores of the Strait of Juan de Fuca. To the south,

Figure 1. Galiano-Valdés Map.

the Columbia River, which had first been sighted by Heceta in 1776, was entered by the American trader Captain Robert Gray in the spring of 1792. Shortly after that William Broughton, a member of the Vancouver Expedition and commander of the vessel "Chatham," an armed tender, crossed the bar of the river and traveled approximately a hundred miles upstream.

Consequently, by the end of 1792 virtually every portion of the coastline of the Pacific Northwest and all its adjacent islands, from the California border to Alaska, had been visited by one or more European or American ships, while the whole coastline had been mapped with remarkable accuracy by Captain George Vancouver. Also at this time, Spain decided to withdraw from the region for good, leaving the area to the British and the Americans to exploit and to eventually divide what came to be called the Oregon Country between them.[9]

Throughout this early phase of coastal exploration the interior had remained "terra incognita" to Europeans and Americans alike, but this was soon to change. Alexander Mackenzie, a trader in the employ of the North West Company, made his historic crossing of the continent and on July 20, 1793, reached the "western ocean" near Bella Coola, British Columbia. Although no immediate results came of Mackenzie's daring exploit, more than a decade later the interior Northwest began to be systematically explored. However, in 1798 David Thompson, another North West Company fur trader, began his series of geographical surveys that eventually unlocked the secrets of the exceedingly complex hydrography of the northern Rocky Mountains and the regions east and west. Beginning with the upper Mississippi River basin within a few years Thompson extended his surveys into the Rocky Mountains and eventually to the Columbia River basin. It was not until 1811, however, that he traveled downstream to the mouth of the river, only to find that John Jacob Astor's Pacific Fur Company had already established its first post near the mouth of the river at present day Astoria, effectively reinforcing American claims to the region.

Making his way west across the Rockies was yet another North West Company trader: Simon Fraser's efforts between 1806 and 1808 to tap new fur-bearing regions and discover an easy route west provided ample evidence of the rugged terrain west of the Continental Divide, as well as the discovery of the Fraser and Thompson Rivers. But the Fraser River was not to become a viable trade route until many decades later, when the Canadian Pacific Railroad was completed.

More impressive by far was the transcontinental expedition of Captains Meriwether Lewis and William Clark between 1804 and 1806. Despatched by President Thomas Jefferson, ostensibly for scientific purposes, the information obtained by the two commanders was to have profound economic and political consequences in the following quarter century.[10]

Leaving Saint Louis on May 14, 1804, the expedition traveled up the Missouri River and wintered with the Mandan Indians in North Dakota. In April 1805 they moved into the foothills to the west and eventually into the heart of the Rocky Mountains, exploring and mapping many of its river valleys while seeking suitable routes through the mountains. Crossing Lemhi Pass in the Bitterroot Mountains, they reached the Clearwater River, a tributary of the Snake River, before traveling to the Columbia, arriving at its mouth on November 7, 1805. Not surprisingly, given the time of year, it was a fog-enshrouded scene that met them. But later in the day, the fog and clouds cleared and, in Clark's words, "we enjoyed the delightful prospect of the ocean—that ocean, the object of all our labors, the reward of all our anxieties."[11] They spent a somewhat miserable winter in the damp and cold of the hastily constructed Fort Clatsop close to present day Astoria.

On their return journey, having made numerous brief trips north and south of the Columbia, particularly along the Oregon coast, the party retraced its steps to Lolo Pass where the Rocky Mountains were crossed in late June. At this point the expedition was divided into two groups, and for the next six weeks or so each group explored

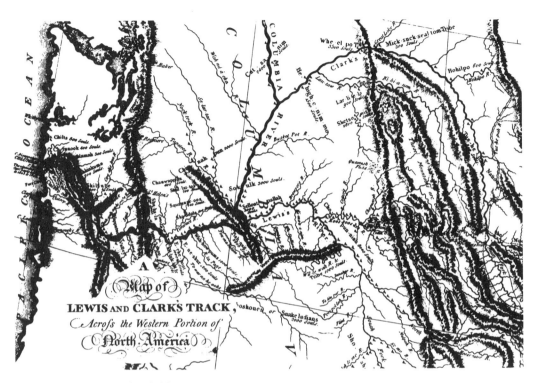

Figure 2. Lewis and Clark Map.

portions of the complex terrain and river systems of western Montana. The two parties met again at the mouth of the Yellowstone River on August 12, 1806, and six weeks later, on September 23, 1806, the expedition was safely back in Saint Louis.

A staggering amount of information had been collected during the two years about terrain, climate, flora and fauna, other natural resources of note, and much else of scientific and economic significance. They had also assembled information on many of the region's tribes. Arguably, no expedition in the history of exploration has achieved so much at so little loss of life—one man—or at so little cost to its backers—while gaining so much beneficial information. Information that was soon to attract the attention of trader and traveler, scientist and politician, and just about anyone else interested in new lands and strange places.

Exploration of the region, of course, did not cease with the Lewis and Clark Expedition. It took decades before all the pieces fitted into place. Among those who had a hand in this somewhat protracted activity were fur traders and curious travelers, government scientists and military officers, missionaries and would-be settlers. Notable among this miscellaneous group of explorers are: Peter Skene Ogden and John

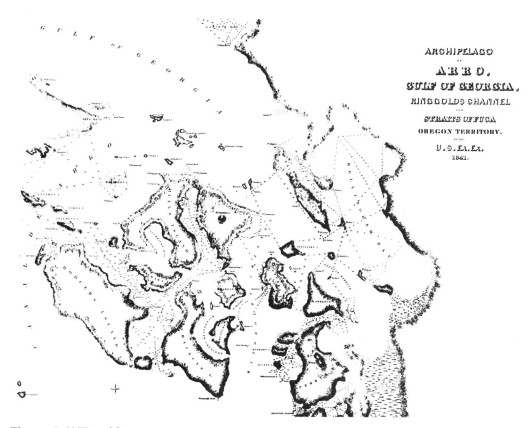

Figure 3. Wilkes Map.

Work of the Hudson's Bay Company; mountain men such as Jedidiah Smith, Jim Bridger, and the Sublette Brothers; botanists David Douglas, Thomas Nuttall and John Kirk Townsend; missionaries Samuel Parker and Father de Smet; Lieutenant Charles Wilkes of the United States Navy and members of his 1838–42 expedition, and John C. Fremont, "pathfinder" and politician extraordinaire.[12]

THE FUR TRADE ERA

Commercial exploitation of the Pacific Northwest's natural resources began, as noted, in 1785 with the establishment of the *sea otter trade*. Concentrated along the coast of British Columbia—particularly on Vancouver Island and in the Queen Charlotte Islands—this earliest of the region's *staple industries* was a short-lived one, lasting only to about 1810. At that time, however, the interior parts of the region began to be exploited for beaver and other fur-bearing animals, an enterprise that saw its greatest success between 1821 and the early 1840s, when the arrival of increasing numbers of settlers brought most fur-trading activities to a halt, except in the more remote areas of the Northwest such as northern and northeastern British Columbia.

Started by the British in 1785, the sea otter trade soon attracted vessels from other countries in the next few years, most notably from the eastern seaboard of the United States. British vessels outnumbered American by about three to one during the first ten years of the trade, but in the remaining fifteen years, American vessels were six to seven times as numerous as the British. Sea otter pelts traded by Native American groups at Nootka Sound and other places on Vancouver Island, as well as in the Queen Charlottes, were exchanged for a variety of items including knives, axes, buttons, blankets, and —most unfortunately—alcohol. The traders' most lucrative market was Canton, but they also developed other outlets in China as well as in New England and Europe. An estimated half million pelts were harvested and sold during that quarter century, with as many as 15,000 pelts sold in Canton in a single season. Overharvesting of animals became increasingly apparent in the years after 1800, and by 1810 the sea otter was an endangered—if not totally extinct—species in most of the Pacific Northwest.*

At no time during the sea otter hunting era did a single company or individual merchant come to dominate the industry: not so, however, the land-based fur trade that developed after 1810. American and British companies had long had their sights turned toward the region. And from bases east of the Rockies, from Saint Louis, Montreal and Hudson Bay, both American and British traders began their move toward the Pacific Northwest. Already established, and busily exploiting fur-bearing districts east of the Rockies, John Jacob Astor's American Fur Company was but one of a number of American outfits interested in the resource. In 1810 Astor formed a new company, the Pacific Fur Company to exploit the riches west of the Rockies and to develop a trade route linking Saint Louis and the mouth of the Columbia. Two expeditions were organized: a maritime expedition on the "Tonquin" that would sail around Cape Horn, and an overland expedition that would follow in large part the route already laid out by Lewis and Clark. Misfortunes of one sort or another dogged the company almost from the start.[13]

In March 1811 the "Tonquin" reached the mouth of the Columbia River, and having traversed the dangerous bar across it, a site was chosen and the building of Fort Astoria begun. The "Tonquin" then sailed north in search of furs. But in the late summer of 1811, the ship came to grief off the coast of Vancouver Island when it was attacked by Indians who scuttled the vessel, either drowning or murdering all but one member of the crew. Meanwhile, the overland expedition, led by William Price Hunt, was having incredible difficulty finding a suitable route through the Rockies, and it

---

* Only in the past two decades have populations of sea otters been successfully replaced along the Olympic coast of Washington.

was not until February 1812 that the party arrived at Fort Astoria, just as the War of 1812 was about to erupt. Rivalry in the Columbia Basin between the Monreal-based North West Company and the Astor company had already resulted in the establishment of trading posts close to one another at the confluence of the Okanogan and Columbia Rivers and on the Spokane River. It was to no avail, however, for the Pacific Fur Company: Its resident partners were forced to sign an agreement with the North West Company to withdraw from the region, rather than face the threatened consequences of removal by Royal Navy ships. By mid-1813 active American participation in the fur trade of the Oregon Country was at an end in any meaningful way.

With the North West Company in sole possession of the region for the next half dozen years, it might have been expected that great changes would occur; in fact, very little happened. The company, it transpired, was more concerned with engaging its British rival, the Hudson's Bay Company, in districts east of the Rocky Mountains and north of the 49th parallel. A forced merger of the two companies was mandated by the British government, and in 1821 the Hudson's Bay Company took control of the whole region west of the Rockies from the 42nd parallel north to Russian Alaska.

Reorganization of the Hudson's Bay Company, Britain's oldest chartered company,* by its North American governor Sir George Simpson led to the appointment of Dr. John McLoughlin as superintendent and chief factor, and to the choice of a site on the north bank of the Columbia one hundred miles upstream from Astoria.[14] Begun in 1824, Fort Vancouver immediately became the economic and social hub of the Pacific Northwest, as it was to remain for the next twenty years. Astoria, now renamed Fort George, was placed on inactive status, and new posts were established east and west of the Cascades, among them Fort Boise, Fort Colvile, Fort Langley, Fort Nez Perce (also called Fort Walla Walla), Fort Nisqually, and Spokane House. Overland communication was maintained with posts on Hudson Bay, and regular maritime service was inaugurated between the Columbia and Hawaii, as well as with London.

The forts themselves became increasingly self-sufficient as agriculture was established. In the 1840s, for example, Fort Vancouver had some 6,000 acres, both tilled land and pasture, in cultivation. By that time also Fort Nisqually and Cowlitz Farm had been established to help provide additional products for its newly formed and fully owned subsidiary, the Puget Sound Agricultural Company, to trade with the Russians in Alaska.[15] Saw mills and dairies at or close to Fort Vancouver and fish-packing plants at Fort Langley and Fort Vancouver provided other commercial products for the Alaskan and Hawaiian trade, as well as much needed items for company use.

The fur brigades, sent out annually by the company from Fort Vancouver, scoured the interior portions of the Oregon Country for beaver and other pelts, doing whatever they could to deter, but not fully prevent, the incursion of American traders from east of the Rockies. There the Rocky Mountain Fur Company, the American Fur Company,

---

* Founded in 1668, the Hudson's Bay Company was grated its charter by King Charles II in 1670.

and scores of mountain men—individualists to the core—continued to harvest the by then rapidly-diminishing populations of beaver and other fur-bearing animals, as well as seriously impinging on the life styles of the Indian groups they came into contact with. This, it should be noted, almost always had unfortunate consequences for the Indians, either then or later.

Political considerations had in part led to the choice of the site of Fort Vancouver as headquarters for the Hudson's Bay Company west of the Rockies, and it was political considerations that led the Company in 1843 to choose a site on Vancouver Island as a suitable substitute should the imminent division of the Oregon Country result in the Americans acquiring control of both banks of the Columbia, as indeed did happen.[16] When the division of the Oregon Country was agreed upon by the Americans and the British in 1846, by the Treaty of Washington, the Company was forced to move its center of operations to Fort Victoria on Vancouver Island. Following long and tedious negotiations that resulted in compensation being paid by the American government for losses incurred by the company, Fort Vancouver was abandoned in 1859. By this time the fur trade was in full retreat. Company profits had plummeted, and new economic ventures were being undertaken that would eventually result in the Hudson's Bay Company becoming one of Canada's most successful and highly diversified enterprises.

THE MISSIONARY ERA AND PIONEER SETTLEMENT

The period from 1834 to 1847, the so-called missionary era, was little more than a prelude to the inflow of large numbers of permanent white settlers, which began as a trickle in the late 1830s and became a substantial stream by the middle of the following decade.

Three groups of Christian missionaries, two of them Protestant, and the third Roman Catholic, vied to bring Christianity and civilization to the Indians of the Pacific Northwest.[17] First on the scene in 1834 were the Methodists, led by Jason Lee and his nephew Daniel Lee. On the advice of Dr. John McLoughlin of the Hudson's Bay Company, which was keen to keep the newcomers in areas south and east of the Columbia, the first mission was established a few miles south of Champoeg in the Williamette Valley. Other missions were built later at The Dalles and at Clatsop near the mouth of the Columbia. All attempts to Christianize and civilize the Indians failed, despite the arrival in 1838 of reinforcements of clergy and laymen. It was these clergy and laymen who became the nucleus of the first American settlers in the Willamette Valley.[18]

Equally unsuccessful were the efforts of the missionaries of the American Board of Commissioners for Foreign Missions, a Congregational/Presbyterian consortium. In 1835 the Reverend Samuel Parker was despatched to the Northwest to inaugurate this mission, but he returned immediately on reaching Fort Vancouver. And it was Marcus Whitman, a medical missionary, who established the first American Board mission at Waiilatpu, just to the west of present day Walla Walla in 1836. His companion, Henry H. Spalding, moved somewhat further east to establish a mission at Lapwai on the Clearwater River. Later missions were established at Kamiah, upstream from Lapwai on the Clearwater, and at Tshimikain a few miles north of the Spokane River. Attempts to Christianize the Cayuse, Nez Perce, and Spokane Indians and to persuade them to settle down and farm achieved virtually no success, while the increasing numbers of settlers arriving in the region from 1843 on helped spread measles, cholera, and typhoid among the native groups. Whitman's lack of success in treating those Indians afflicted by diseases, in contrast to his much greater success with the Whites, helped increase cross-cultural tension. This culminated in November 1847 in the massacre of the Whitmans—Marcus and Narcissa—and a dozen others at the Waiilatpu Mission by Cayuse leaders.[19]

Most successful of the three missionary groups were the Roman Catholics, who began to serve the region in 1838. Despite opposition from Sir George Simpson and others in the Hudson's Bay Company, two priests—Father Blanchet and Father Demers—were allowed to travel west with a transcontinental party of the Company in 1838. Soon after, they established their first mission near Cowlitz Farm. In the following year, Father Pierre Jean de Smet traveled from Saint Louis to the Flathead country of Montana to begin a major missionary effort that quickly spread Christianity throughout the interior valleys of the region. This led in the next few years to the establishment of scores of missions and the erection of many magnificent churches—such as that at Cataldo—that still stand today.

The Whitman Massacre may have brought to an end the short-lived missionary era and the withdrawal of all Protestant missionaries from the region east of the Cascades, but the Catholic fathers—known among the Indians as the Black Robes—who were intent only on working with the Indians, remained throughout the Indian wars that followed the capture of the Indian chiefs responsible for the massacre. In so doing, a strong Catholic tradition became rooted among a high percentage of the Indian tribes of the Pacific Northwest.[20]

Westward expansion, already far advanced in the Midwest by the 1830s, was encouraged to move still further west into the Oregon Country by the writings of propagandists such as Hall J. Kelley and by the more sober first-hand accounts of people like the Reverend Samuel Parker and the lectures and sermons of the Reverend Jason Lee. Marcus Whitman's journey east in 1842 and his return in 1843 did not "save Oregon for the United States," as later apologists for the slain missionary were to claim. Nevertheless, Whitman certainly proved helpful to the more than nine-hundred people who traveled over the Oregon Trail during the summer and fall of 1843. Thereafter the trail carried thousands of would-be settlers until the first transcontinental railroad was completed in 1869. Many who made the crossing during those years found it relatively easy, and few troubles were recorded as coming from Indian attacks, a lack of fresh pasture or feed, or the outbreak of such diseases as cholera and typhoid. In some years, undoubtedly, the latter were dangers that beset many of the companies that traveled the trail, but as John Unruh so convincingly demonstrates in his work *The Plains Across* these dangers were far from typical.[21]

By 1845 there were more than two thousand White settlers in the Pacific Northwest, nearly all of them in the Willamette Valley. The census taken that year by the Provisional Government recorded 405 heads of families and a total population of 2,110 Euro-Americans. When the 1849 Territorial Census was taken numbers had risen to 8,779—all but 183 of them citizens of the United States.[22]

By 1849, of course, the Oregon Country had been divided between the United States and Great Britain; a territorial government had replaced the ad hoc provisional government; and a start had been made on legislation that would help unscramble the problems of land ownership and enable the newcomers to obtain title to the lands they occupied.

Acting with some urgency to organize the region and deal with various problems that stemmed, at least in part, from prior enactments of the Provisional Government, the Congress established the Oregon Territory in 1849. Particularly difficult to address was the matter of land acquisition. Federal law specifically forbade the acquisition of land titles until the government had secured cession of lands from the region's Indian tribes. But whereas the federal government had traditionally allowed settlers to preempt 160 acres prior to completion of all legal requirements, including the survey of lands involved, the Provisional Government allowed settlers to preempt 640 acres. To resolve this problem, the Oregon Donation Land Act was passed in 1850. Settlers who were domiciled in the territory prior to December 1, 1850, were allowed to claim 320 acres if single or 640 acres if married and both parties were residents. Settlers arriving after that date and before the end of December 1855, when the act was to expire, were permitted to claim 160 acres if single or 320 acres if married. An unusual feature of the act was that it allowed women, and not only men, to file claims in their own name. During the next few years close to 7,500 claims, covering approximately 2,500,000 acres, were filed in what is now the state of Oregon, and just over 1,000, covering approximately 300,000 acres, were filed in what is now the state of Washington. Survey of these claims and completion of all legal requirements proved to be a slow process, and it was not until after 1862 that any public land was offered for sale in the Pacific Northwest.[23]

Nearly all donation claims were located in the Willamette Valley, the Lower Columbia Valley, and the Puget Sound area. There were, however, a few donation claims filed in regions east of the Cascades, as for example in the Walla Walla Valley and in the vicinity of Fort Colvile. Not surprisingly, lands chosen by the settlers were among the most accessible and potentially the most fertile. During the first few decades of settlement, these areas became the bases of both Oregon and Washington economies.

Later land laws included the Homestead Act of 1862, the Desert Land Act of 1877, the Enlarged Homestead Act of 1909, and the Grazing Land Act of 1916. All had some influence on the process of settlement and development of the Pacific Northwest.

Meanwhile, the territorial governors of Oregon and Washington—the latter had become a territory in 1853—were negotiating treaties with the tribes of their respective regions. Before that task was completed and the treaties negotiated were approved by the Senate, a series of Indian wars began. A somewhat temporary nuisance to the settlers, but a downright disaster for the Indians, the Indian wars led eventually to the enforced cession of nearly all Indian lands east and west of the Cascades. Most of the troubles had been brought under effective control, if not a totally satisfactory solution, by the late 1850s. The Indians for the most part—there were a few exceptions—were consigned to the newly-established reservations, and the region east of the Cascades was thrown open to permanent White settlement. Prior to this, the discovery of gold in Washington, Oregon, and Idaho had already attracted a sizable number of prospectors and miners, but it was not until the advent of the railroads in the 1870s that systematic settlement and economic development of the interior parts of the Pacific Northwest were effectively begun.

THE PIONEER ECONOMY TO 1880

Although all the resource industries—agriculture, fishing, logging and mining—were developed in the Pacific Northwest during this time, it was agriculture, that "first of all the arts" as Jefferson had called it, which provided the mainstay of the pioneer economy. The produce of both tilled land and pasture, however, was destined largely for home consumption or the provisioning of some nearby town, logging camp, or mine working. Transportation systems were insufficiently developed and the forms of transportation available so slow that most agricultural goods—especially the more perishable items—were unable to be carried to any far-distant market in saleable condition. There were exceptions: wool, hides and grain were shipped to regions outside the Pacific Northwest after the Californian Gold Rush of 1849, but for the most part the "tyranny of distance," as it has been called, prevailed. Agriculture throughout the pioneer period tended to be of the mixed subsistence sort, with a variety of crops being grown and an assortment of animals kept on most farms.

As most farms during the first twenty or so years of the pioneer period were located in the wetter areas west of the Cascades, the land had first to be cleared of brush and trees before farming could begin. This was a slow, tedious, and time-consuming task that might take a number of years to finish. Some of the earliest settled lands did include extensive Indian prairies, as in the Willamette Valley and around Puget Sound, but they were the exception rather than the rule. What is abundantly clear is that the 160 acres quarter-section provided for by the Homestead Act, not to mention the even more generous allotments of the Donation Land Act, were more than enough for the family farm of the period, and a large number of successful farms then, as now, were of smaller acreage.

In those areas first settled east of the Cascades, it was the bottom lands of the Walla Walla, Snake, Spokane and other rivers that were first chosen for development, but here also agriculture initially was of the mixed sort, various crops and animals being raised on virtually all of them. Things began to change when areas away from the rivers in the Palouse Hills, the Big Bend Country of the Middle Columbia, and various parts of eastern Oregon, as well as lands east and north of the Snake River in Idaho, began to be utilized as rangelands. Used for the raising of both cattle and sheep, competition for this still predominantly public land increased rapidly in the 1870s and 1880s. Range wars and disputes between cattlemen and sheepmen were bitter and frequent. And it was not until effective government regulations were introduced, and railroad grant lands began to be sold or leased to the hitherto landless ranchers, that these began to diminish.[24] Improvements in overland transportation, and the introduction of new technologies, including the spread of irrigation, the increasing use of the wind pump, and the installation of barbed wire became the harbingers of a new commercial era as well as marking the passing of the pioneer period of agriculture.

Logging and lumbering, like agriculture, were already well-established if small-scale activities in western Oregon, western Washington, and British Columbia when the demand for timber increased dramatically following the 1849 gold rush in California. Spurred by the growth of San Francisco, Sacramento and other Bay Area towns, new lumber mills, many underwritten with Californian capital, were established throughout the Pacific Northwest, particularly around Puget Sound. Among them were the Pope & Talbot mill at Port Gamble, which is still in operation, the Yesler mill in Seattle, and mills at Port Blakely, Port Madison, Port Ludlow, Seabeck, Utsalady, and Whatcom. With forests stretching to tidewater in many places, transportation of the logs to the nearby mills, and later shipment of sawn timber to markets both near and far, posed few problems, although the reliability of machinery and other equipment was often another matter. However, as the forests receded inland and upriver, and from river bank to hillside and eventually to mountain, the problems of transportation increased and production costs mounted dramatically. Maintaining a constant flow of timber was sometimes as difficult as finding a secure market. Both problems plagued the lumber industry throughout the pioneer period while technological changes introduced elsewhere and brought to mills on the west coast during the period hastened the demise of many under-capitalized early mills, including the Yesler mill in Seattle and mills at Port Orchard and Utsalady.[25]

Mineral exploitation in the Pacific Northwest, as in so many other regions of the American West, began with the discovery of gold, in 1851. Soon after that a growing demand for fuel in the Bay Area led in 1853 to the shipment of coal mined on the shores of Bellingham Bay to San Francisco. For the next quarter century, small quantities of coal continued to be mined there, while other coalfields—notably those at Namaimo on Vancouver Island and others in King County and Pierce County, Washington—were brought into production. The economic impact of coal mining during the pioneer period, however, was modest; no great investments were recorded and no great fortunes were made by those involved in the coal mining industry.

Much more significant in terms of its widespread economic and political impact was metalliferous mining—especially in the early years of *placer mining* of gold. The news, whether well-founded or false, that gold had been discovered in a region was enough to start a gold rush, and during the second half of the nineteenth century every western state, as well as British Columbia and the Yukon, experienced one or more of these. Oregon's first gold rush was in the southwestern counties of Josephine and Jackson beginning in 1851. Ten years later two eastern Oregon counties, Baker and Grant, were involved in another gold rush. Washington's first rush began in 1855 near Fort Colvile and spread east along the Colville Valley and northwards toward the Canadian border. Others came later in the Similkameen Valley and in the Cascades. In 1856 gold was discovered on Gold Creek, Montana, although it was not until the discovery of gold at Bannack in 1862 and at Virginia City in 1863 that western Montana became a mecca for prospectors and miners. In the meantime gold had been discovered at Pierce (City) on the auspiciously named Oro Fino Creek, and this set off yet another gold rush. North of the border, the Fraser gold rush of 1858 was followed by an even more significant one—the Cariboo gold rush—a couple of years later.

Placer gold deposits were soon exhausted in most of the early gold operations, and although in a few places *hydraulic mining* helped uncover other water-deposited gold dust and gold nuggets in adjacent hills, it was the discovery of gold-bearing quartz veins that led to the greatest changes. Highly capitalized *hard rock mining* quickly replaced placer mining, in which even the poorest had been able to participate; as a consequence, the independent miner became a paid employee. Along with hard rock mining came other changes that were to have as many political as economic impacts. Investments in underground operations and the installation of surface stamps and smelters were accompanied by demands for better transportation systems, the growth of company towns, and a glut of influence-peddling that became an integral part of territorial government, particularly in Idaho and Montana. The problem of disaffected miners plagued the Coeur d'Alene and other mining districts for decades to come. By this time, of course, the pioneer era was over, with silver, lead, zinc, and copper replacing gold as the principal metals sought and mined.[26]

Although the fishing industry had been developed in the 1820s as an integral part of its export trade by the Hudson's Bay Company at Fort Langley and Fort Vancouver, it was a slow starter in the pioneer economy of the Pacific Northwest. Little use was made of the abundant salmon and other fish stocks by the first settlers, although small quantities of dried and salted salmon continued to be exported to the east coast and elsewhere. Oysters were being collected from Willapa Bay for shipment to the Bay Area as early as 1854.[27]

With the introduction of new and effective techniques, as well as the establishment of the first cannery in the west on the Sacramento River in 1864 by the Hume brothers, great changes began to occur. In 1867 the Humes moved their operation to the Columbia River and within a few years the Columbia River salmon fishery was a thriving one involving more than a hundred fishing vessels of various sorts and employing many hundreds of workers in the canneries, some of them Chinese. The Puget Sound fishery came into its own a decade later when the first cannery was established at Mukilteo in 1875. By 1880 the Pacific Northwest coast from the Californian border to Alaska was dotted with numerous small canneries, each supplied by local fishermen from the many bays and islands as well as from the lower reaches of the region's many rivers.[28]

Joining the canneries in the manufacturing sector were many small grist mills—Florence Sherfey has identified more than thirty of these that were in operation in eastern Washington prior to 1880[29]—and a few large flour mills located in the larger urban areas. There were also a multitude of dairies and creameries; numerous foundries and blacksmith shops, that provided wrought iron products for home and farm, as well as thousands of horseshoes; dozens of saw mills in every well-settled area and an equal number of shake and shingle mills, plus the occasional sash and door factory; and, especially in Oregon, a number of woolen mills.[30] Virtually all of these manufacturing units, it should be noted, were simple processing operations, dependent for the most part on local supplies, waterpower, or coal or other fuel, and nearby markets. Not until transportation systems were markedly improved and businesses expanded to serve larger hinterlands was the distribution of these various industries to see dramatic change. And that, of course, happened long after 1880 in most industries.

Transportation services throughout the pioneer period were uniformly poor, seldom reliable, and always expensive. On land, especially in the wetter areas of western Washington, western Oregon, and coastal British Columbia with their heavy impermeable clay soils, travel was often almost impossible in the winter months, when trails and wagon roads became quagmires or worse. Plank roads and corduroy roads consisting of rough logs were constructed in some places, but it is hardly surprising, considering the appalling quality of most roads and the absence of any road in some parts, that so much pioneer travel and trade was by water or that most early settlements, especially those west of the Cascades, were located on navigable water.[31] Even the smallest stream was used if at all possible by flatbottom boat or canoe, while most of the larger rivers ran a variety of vessels and provided both passenger and freight services. Ferries crossed the Columbia, Snake, Willamette, and other rivers,[32] as well as many of the lakes of the region, as well as on Puget Sound[33] and the Lower Columbia, where they provided access to the many offshore islands. But with the coming of the railroads things were to change with startling rapidity.

PIONEER TOWN-BUILDING

Richard Wade has emphasized that towns were the true spearheads of the frontier.[34] Certainly this was the case in the Pacific Northwest, where the origin of most cities and towns can be traced to the mid- and late nineteenth century. Some, such as Vancouver and Boise, owed their existence to the Hudson's Bay Company, which developed them as trading posts, while Astoria was a creation of the Pacific Fur Company, but virtually all others were the product of pioneer settlers eager to plat and develop a site they hoped would become a viable town, if not a booming metropolis. However, to achieve success in these early years, a waterside location—whether at tidewater or on a navigable river—was mandatory. Surrounding the site there had to be some available resource—whether of forest or field, but preferably both—capable of quick commercial development. Acquiring status as a county seat or territorial capital was an additional boost to success, although frequent changes of county seats and even territorial capitals continued throughout the pioneer period. Oregon, for example began its territorial period with Oregon City as its capital, but a few years later the capital was transferred to Salem, where it has since remained.

In the 1850s and 1860s a multitude of *bonanza towns*, tied to some nearby and newly discovered mineral deposit, were founded overnight in places far distant from the, by now, well-settled western portion of the region.[35] As quickly as some of these appeared, just as quickly were they abandoned if the prospect proved less than promising, as it frequently did. For every mining town that survived, there could well be a dozen ghost towns.[36] For the survivors, trade in every-day necessities—including food and housing—and the development of nearby agriculture and timber supplies helped broaden the economic base and make more certain a long-term life.

Most early towns, as can be seen in the many panoramic or Bird's Eye View maps that began to appear in the later nineteenth century, were laid out on a regular grid pattern with one or two principal commercial streets and surrounding residential areas.[37] Zoning was virtually unknown, and what today might be regarded as especially choice view lots for housing could well be occupied by a lumber mill or other industrial establishment. Buildings were almost always wood-frame construction, many sporting ornate false fronts, although an occasional brick or stone building was erected to serve some public function as courthouse, hotel, or store. Streets were unpaved, sidewalks were few if any, and public utilities and essential services were generally nonexistent. Animals, such as milch cows, pigs, and chickens, were often kept by householders for personal use and consumption. Horses were also kept at home for personal use, although livery stables and saddlers' shops were among the earliest businesses established in many towns, along with taverns and hotels.

Two of the earliest settlements in Oregon were Oregon City, the first territorial capital and now part of Greater Portland, and Salem, the present state capital. Located just below the Falls of the Willamette on navigable water, Oregon City was chosen in 1829 by John McLoughlin as the site for a Hudson's Bay Company trading post. But it was not until 1842 that he actually named the town, after surveying the site and laying out a number of town lots. By that time, permanent settlers were arriving, and by 1847 Oregon City had become the largest settlement in the territory with some 600 to 700 inhabitants, two flour mills, two saw mills, and a number of stores—not to mention a flourishing regional commerce in a wide range of articles. Salem, which had begun in 1841 as the site of the new Methodist Mission and was then named

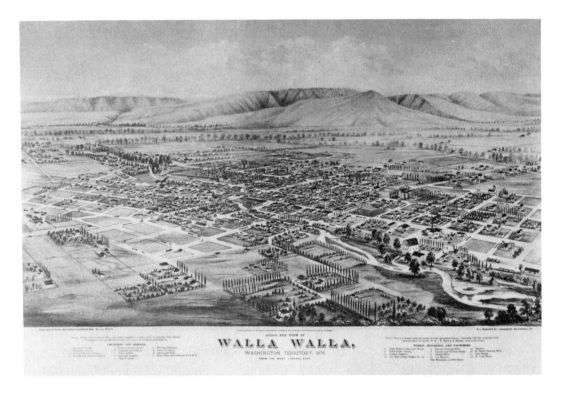

Figure 4. A birds-eye view map of Walla Walla, Washington.

Chemeketa, was chosen as the territorial capital in 1851. Although it was not to reach a population of 10,000 by 1863, as predicted by David Newsom in the previous decade,[38] Salem was by that date a thriving trading center and the site of the Pacific Northwest's oldest institution of higher learning: Willamette University.

Portland, Oregon's largest city today and one of the great cities of the American West, was also founded during the 1840s, but some miles downstream from the Falls of the Willamette on a site chosen by Asa Lovejoy and Francis Pettygrove in 1845. At first its growth was slow, but by 1850 Portland was recorded as being a town with "over 100 dwellings ... eighteen stores, six public boarding houses... (and) two large churches."[39] Thereafter its growth was no longer in doubt, as Portland quickly outdistanced all other Northwestern cities to become the principal port of the Pacific Northwest and the headquarters of many manufacturing, transportation, and commercial companies.[40]

Other Oregon towns established during this first decade of American settlement were Linn City in 1843, Multnomah City in 1844, Eugene in 1846, Milwaukee in 1847, Albany and Corvallis in 1848, and St. Helens in 1850. Astoria, originally established by the Pacific Fur Company decades before, was brought back to life in 1844 and by 1850 was a thriving port and urban community near the mouth of the Columbia.

Apart from Vancouver, which had been established as a fur trading post in 1824, most of Washington's early towns were founded a few years later than those in Oregon's Willamette Valley, even though American settlers had begun to arrive in the south Puget Sound region as early as 1845. Tumwater is credited with being the earliest American settlement north of the Columbia, but it was nearby Olympia, platted in 1850, that was to become the capital of the newly created Washington Territory a short three years later. Also founded in 1850 were Steilacoom and Port Townsend,[41] the latter platted by Francis Pettygrove, who by then had sold his interests in Portland real estate. Two years later, Seattle was founded on Elliott Bay,[42] and a year later Whatcom was established on Bellingham Bay.[43]

The official opening of the country east of the Cascade Mountains to permanent settlement in 1858 was preceded by the arrival of sizeable numbers of prospectors and miners a few years earlier. The resulting gold rushes led to mushroom growth, followed by almost immediate demise, of scores of mining camps and towns. A few did survive to become like Colville and Kettle Falls in northeastern Washington important market centers, as farming and lumbering took root in the wake of placer mining. Other towns, more conveniently located in terms of existing transportation routes but lying some distance from the actual mining districts, also became important market and mine supply centers. Foremost, in these early years, was Walla Walla, which became the main supply center for the Oro Fino mining district of central Idaho. Other such centers included Boise, Missoula, Bozeman, and La Grande. In the mining districts of Idaho, western Montana, and eastern Oregon a number of early mining towns survived to become major regional centers, among them Helena, Wallace, Idaho and Baker. Towns established in steep-sided and exceedingly narrow valleys frequently had grid patterns that were necessarily modified by the region's rough terrain. Burke, Idaho is perhaps the best example of such a town, restricted as it is to a single main street that also had to carry the railroad and with land steeply graded on both sides.

The lawless nature of these early bonanza towns with their irregular organization and unplanned growth are features that have been over-emphasized, not to say exaggerated, in scores of novels and other popular writings; more sober research has shown that most survivors quickly settled down after the first few months of frenetic activity, and the arrival of wives and families led shortly thereafter to the building of churches and schools.

THE RAILROAD ERA

Although a survey for a transcontinental railroad route across the northern part of the country, between the 47th and 49th parallels, had been undertaken in 1853 by Major Isaac I. Stevens, the newly appointed governor of Washington Territory, it was not until thirty years later that the Pacific Northwest was effectively joined to the rest of the country by a transcontinental railroad. In the preceding decade, however, interest in railroads reached fever pitch among businessmen of the region. In both Oregon and Washington—or perhaps more precisely, Portland and Seattle—plans were made to join their major cities to an existing railroad in California to the south or to another east of the Rocky Mountains, and also to build whatever short lines were needed to serve local areas and tap any potentially valuable natural resources found there.

The first railroad constructed was the Oregon Central, which had been granted by the Congress five million acres of land in western Oregon's Willamette, Rogue, and Umpqua valleys. Taken over in 1870 by Ben Holloday, of Overland Express fame, the first rails were laid between Portland and Salem that same year. Eugene was reached in 1871, and by 1872 the Oregon Central reached as far south as Roseburg. There for the next ten years the line ended, although a West Valley line paralleling the main line was constructed as far as Corvallis in 1872. In 1868, two years before the Oregon Central was started, Dr. Dorsey Baker, a pioneer physician and Walla Walla businessman, had organized the Walla Walla and Columbia Railroad Company, with the aim to make it easier and cheaper to ship wheat from the Palouse Hills to the coast.[44] It was not until 1875, however, that the so-called "rawhide railroad" was completed between Walla Walla and Wallula on the Columbia River.

The Northern Pacific Railroad, which had been empowered by the Congress some years before and granted millions of acres of the public domain to subsidize construction of a northern transcontinental line, also was making slow progress at this time. Tacoma, rather than Seattle, had been chosen as its ultimate western destination, track had been laid across parts of the Great Plains, and a route had been prepared for the line from Kalama on the Columbia River to Tacoma. In 1873, however, Jay Cooke, president and chief executive officer of the company, declared bankruptcy, bringing an abrupt halt to the transcontinental railroad project. Faring no better in this year of national depression was Ben Holloday, who had taken over the Oregon Central and by uniting it with his Californian interests had created the Oregon and California Railroad Company. At that point Henry Villard appeared on the scene representing the interests of a number of bondholders of the Oregon and California, and shortly afterwards Villard took control of Holloday's company, bought out his interest in Willamette River Steamers, and purchased the Oregon Steam Navigation Company. A few years later, Villard brought Baker's Walla Walla and Columbia Railroad into his combined river and railroad system, and so by 1882 the Palouse region was being served by railroad all the way from Walla Walla to Portland. At approximately the same time, negotiations involving various eastern capitalists and financiers led to Villard's acquisition of stock in the Northern Pacific Railroad. And a short time later, Villard was named a director and within a year was president of the company.

Once more the Northern Pacific was in business, and between 1881 and 1883 construction charged ahead full steam on both sides of the continental divide. In the process some of the most impressive railroad building ever undertaken anywhere in the world was successfully completed. This included the excavation of numerous extensive tunnels and cuttings and the crossing of almost as many canyon-like valleys by enormous wooden trestles that carried the rail lines. The traditional golden spike was driven by Villard at Independence Creek some miles west of Helena, Montana, on September 8, 1883, and a short time later trains were moving across the country from St. Paul, Minnesota, to Portland and on to Tacoma. It was Portland, however, rather than Tacoma, that reaped the initial benefits of the new transcontinental, as it was Portland that was at the receiving end of the corridor that funneled seaward the grain and other produce of the Columbia Basin: the Palouse Hills in particular. This led, quite naturally, to a demand from the Tacoma business community for a direct line across the Cascade Mountains from eastern Washington to Puget Sound.

Meanwhile, merchants and other business people in Seattle were far from idle in their promotion of railroads. Two incorporated at that time were the Seattle and Walla Railroad and the Seattle, Lake Shore, and Eastern Railroad, both of which were projected to cross the mountains and join up with the transcontinental lines there. In the end neither achieved its purpose, and in 1888 the Seattle, Lake Shore, and Eastern was bought out by the Northern Pacific. Completion of the Cascade Route the previous year by Northern Pacific meant that now both Tacoma and Seattle could be served by a transcontinental line. Tacoma's effort to establish itself as the principal west coast terminus of the railroad suffered yet another setback when, in 1893, James J. Hill's Great Northern Railway provided Seattle with another route across the country.

Built without the help of land grants or other subsidies from the federal government, and securing none from any of the states it passed through, the Great Northern was able to serve all the northern states from the Great Lakes westward, as well as parts of Canada just to the north, and to provide Seattle with the bulk of its export trade. Hill's Montana Central Railroad and a number of smaller lines—some of them narrow gauge lines—were built to connect the isolated valleys of the gold and silver-rich northern Rockies to his transcontinental. In doing so, the great copper deposits of Butte and Anaconda and the considerable lead and zinc deposits of northern Idaho were at last able to be exploited; and other natural resources, particularly those of the forests, effectively developed. In the process one of the West's best known vernacular regions—the Inland Empire—came into being with Spokane as its hub and major servicing center.[45]

Further south in Idaho, the Utah and Northern Railroad Company, which turned out to be one of the West's most financially rewarding railroads ever, began construction of its line north from Ogden, Utah, in 1878. Pocatello was reached the same year, and by 1884 the line had been extended another 300 miles or more north to meet the Northern Pacific at Garrison, Montana.

The Union Pacific Railroad, which in 1869 had joined with the Central Pacific Railroad to become the nation's first transcontinental, acquired a line to the Pacific Northwest when the Oregon Short Line from the Columbia across the Blue Mountains was joined with the Union Pacific Railroad in the Snake River valley. The Southern Pacific Railroad likewise moved into the Pacific Northwest when it acquired a number of the railroad companies of western Oregon, and was able to establish its Pacific Northwest terminus in Portland. Finally, the Milwaukee Road acquired various lines in northwest Washington, including the Bellingham and Northern Railroad, and was able eventually to reach tidewater in the Pacific Northwest in 1918.

The railroad history of the Northwest is, like that of so many other regions of the country, an extremely convoluted one, involving purchases and buy-outs, mergers and takeovers.[46] Easier to address is the influence of the railroads on the region's evolution and economic progress. Until the coming of the railroads, transportation was slow, uncertain, and costly; afterwards the costs of travel plummeted, and the movement of passengers and freight was greatly speeded up and made more reliable. This was due in large part to it being made to conform to more rigid timetables and various federally mandated safety regulations. Regions previously inaccessible, like the mountain valleys of Idaho and western Montana, and natural resources, like the copper of Butte and Anaconda and the silver-lead-zinc deposits of the Coeur d'Alenes of Idaho, were opened up. Moreover, a number of new settlements—railroad towns and junctions—were brought into being. And in the more settled areas of the region, the location of many industries was moved from waterside to railside while the already existing industries were provided with spur lines from railroad to factory and dock-side, which made easier the transshipment of a host of raw materials and processed or manufactured items.

There is no question that it was the arrival of the railroad that brought about a remarkable quickening of the economy of the Pacific Northwest, as well as exponential growth of the region's population in the immediately succeeding years. For example, between the 1880 Census and the 1890 Census—and particularly in the seven years following the opening of the Northern Pacific in 1883—the population of Tacoma and surrounding Pierce County increased almost fifteen times. Similar, though perhaps not so dramatic, increases occurred elsewhere in the region. It was the coming of the railroads that not only heralded, but made possible the transformation of the pioneer economy and the broadly based economy that followed and still prevails today.

THE EMERGENCE OF A MODERN COMMERCIAL ECONOMY

Changes of every sort—cultural, economic, social and political—came "fast and thick" in the Pacific Northwest in the last two decades of the nineteenth century. These changes helped to transform the region from a remote, and still somewhat inaccessible, colonial U.S. outpost to a region that at last could become an integral part of the nation. Here and there traces of the earlier pioneer past remained long after 1900, but from the 1880s on the Pacific Northwest was involved in political and economic activities and was beginning to involve itself in social movements that proved to be of more than regional significance.

The steel rails that now tied the region to the East and Midwest made it easier for increasing numbers of Americans and others, tourists and would-be settlers alike, to travel to what some have termed the "promised land."[47] The rails also hastened systematic exploitation and development of the region's many resources. By 1900 eastern and midwestern capital was rapidly supplanting or complementing San Francisco capital in many industries, and giant corporations were beginning to gain a major share in the lumbering, fish canning, and mining industries. This incursion of capital, quite naturally, was accompanied by a host of technological changes that helped initiate new agricultural ventures and establish new manufacturing industries across the region.

Agriculture

Not surprisingly, changes in this part of the primary sector came slowly in some regions but with amazing rapidity in others. These changes were so numerous that only the more significant of them will be discussed here, among them the beginnings of commercial farming, the expansion of irrigation, the transformation of the open range, the introduction of refrigeration, and the mechanization of farm operations. Others, such as the introduction of hybrid seeds, genetic engineering, fertilizers, herbicides and pesticides, will be considered in a later chapter.

(1) Commercial Farming: The transformation of the pioneer farm, with its mix of crops and animals, to one specializing in the production of a single crop or animal took place most rapidly in the vicinity of the more urbanized western part of the region. There the rapid growth of urban areas after 1883 led to greatly increased demand for regular supplies of a number of farm products, among them milk and milk products, fresh eggs, table birds, fruit, and vegetables. Reacting quickly to the changing conditions, many farmers and quite a few smallholders began to concentrate in such specialized activities as dairying, poultry farming, and truck farming: generally with marked success. The Green River Valley near Seattle, the Puyallup Valley near Tacoma, and the Willamette Valley in the vicinities of Portland, Salem, and Eugene were the first urban areas to be transformed; although the same phenomenon, on a somewhat smaller scale, was to be seen shortly thereafter in the vicinity of interior cities like Spokane, Boise, and Missoula. With the improvement of roads that began to occur at this time, as well as the building of electric inter-urban railroads, the hinterlands from which these products could be drawn rapidly expanded.

(2) The Expansion of Irrigation: Irrigation was not exactly a new practice in the Pacific Northwest of the 1880s. The Yakima Indians had long experimented with irrigation on a very modest scale, as had Marcus Whitman at the Waiilatpu Mission and Henry Spalding at the Lapwai Mission, again on a modest scale—just enough to provide water for the mission gardens. Most notably, the Mormons had initiated irrigation farming in the Salt Lake area and the Jordan Valley as early as 1847, and by the 1850s they had a few thousand acres irrigated. Expansion of irrigation, however, was seldom an easy and always an expensive project, as later pioneers in Idaho, Oregon, and Washington discovered. The Desert Land Act of 1877, which had been passed to encourage settlement of arid regions in the West, offered settlers 640 acres of land at $1.25 an acre, provided that they had the land irrigated within a three-year period. For virtually all of them, this was an impossible task, and as a means of encouraging settlement the act was a dismal failure.

However, in southern Idaho, where Mormon farmers had established settlements in Bear Lake, Brigham, Cassia, and Oneida counties, cooperative ventures proved much more effective. By 1890 there were some sixty-odd locally owned and locally operated irrigation canals, providing water to more than 4,300 farms, with a total acreage served of more than 217,000 acres. In Oregon and Washington, in contrast, cooperative ventures were far fewer, with most of the early projects being undertaken by individual farmers. Nonetheless, in 1890 Oregon had more than 3,100 farms involved, and the total area irrigated amounted to almost 178,000 acres. And Washington had just over 1,000 farms with a little under 50,000 acres irrigated.

The 1890s saw irrigation greatly expanded and federal government participation in irrigation projects greatly increased. In the first few years it was private investment that made possible a number of extensive projects, such as the Sunnyside Project of the Yakima Improvement Company. The Northern Pacific Railroad, eager to promote settlement in regions where it held large land grants, was an active participant in some of the projects undertaken. Other private companies, many of them financed from Salt Lake City or from New York and other eastern cities, were formed in Idaho and Oregon. The federal government's interest in irrigation projects was increased with the passage of the Carey Act in 1894; and although this helped the expansion of irrigation, especially in Idaho, it was the passage of the National Reclamation (Newlands) Act in 1902 that brought the federal government fully into the picture. This created a new agency, the Reclamation Service, that was willing and prepared to engage in the building of dams across many rivers and the construction of feeder canals from those dams. In the next few decades, projects were undertaken in the Yakima, Kittitas, Wenatchee, Okanogan, and Spokane Valleys of Washington, the Boise and Snake Valleys of Idaho, and the Umatilla and Klamath Basins in Oregon. The one large region not immediately helped by the Reclamation Service was the Big Bend country of Washington, but in the 1930s the inauguration of the Grand Coulee

Photo 1. Farming the Palouse, early 20th century.
Washington State Historical Society, Tacoma, Washington.

Project resulted in this area becoming by the 1960s one of the largest irrigation areas in the United States.

In extensive areas of both eastern Oregon and eastern Washington dry farming operations had resulted in the successful development of wheat and other small-grain farming, but with the expansion of irrigation other types of farming began to thrive including fruit farming and truck farming, which hitherto had been confined largely to western areas of both states.

(3) **Transformation of the Open Range**: Until the late 1880s and early 1890s vast areas of eastern Oregon and eastern Washington, as well as much of central and southern Idaho and portions of western Montana, were unfenced, unsettled, and only seasonally utilized public land, most of which was still open for homesteading or purchase. Other large, interspersed areas were part of the land grants of the Northern

Pacific Railroad. On these extensive open range lands large herds of cattle were grazed from the 1860s on and in later years, especially after 1880, by even larger flocks of sheep. Most of the animals grazed were owned by men who owned no land of their own, and who often had no fixed abode. Their cattle and sheep were regularly moved from winter pastures in the lower elevations of the Columbia Basin, particularly the Channeled Scablands, to summer pastures in the Cascades, the Okanogan Highlands, the Blue Mountains, the Rockies, and other upland areas. As settlement advanced across the Columbia Basin and similar areas in Idaho and Oregon, and as the nutritious bunch grass was used up and the land plowed under for the production of wheat and barley, the open range began to be fenced in and the opportunities for the long-practised transhumance to be drastically reduced.

A number of sheep herders like the McGregor Brothers of Hooper, Washington, purchased or leased large areas of winter range for their flocks from the Northern Pacific Railroad or from anyone else willing to sell or lease land. By the beginning of the twentieth century, the McGregor Brothers Company had purchased and fenced more than 30,000 acres of winter pasture in the Columbia Basin, and even larger areas of summer pasture were being leased in the mountains of northern Idaho. Other raisers of sheep who acquired large acreages between the 1893 Depression and the early years of the next century included the Harders of Kahlotus, Washington, the Drumhellers of Ephrata, Washington, Robert Jackson of Drayton and the Coffins of Yakima. Few cattlemen made the effort to purchase large acreages at this time, and by the turn of the century most had given up their herds for other farm activities.[48]

In other areas of the Pacific Northwest—eastern Oregon, southern and central Idaho, and western Montana—a similar transformation of the open range occurred.[49] Much of this land was either purchased and fenced in, or it was already included in national forest or other restricted federal or state areas. By World War I the free open range was a thing of the past in the Pacific Northwest, and wheat and sheep had become the principal products of much of the region.

(4) Refrigeration: Although refrigeration utilizing block ice came into use in the United States as early as 1851, it was not until the 1890s that serious attention was paid to this method of transportation in the Pacific Northwest. By then, the technology of refrigeration had been vastly improved, and with the opening up of markets across the country for lumber and wheat, it was inevitable that sooner or later refrigeration cars would be brought in to transport perishable items from the region. Milk products, meat, and fish were among the first items to be moved in refrigeration cars, but by the turn of the century various fruits and vegetables were also finding markets outside the area.

Most of the many regional market centers that sprang up to handle these farm products and prepare them for market quickly added bulky cold storage facilities and warehouses. These buildings were generally strategically located along a mainline railroad and clustered close to the downtown. In Chelan, Omak, The Dalles, Wenatchee, Yakima, and other centers, apple and other fruit packing and cold storage plants were the visible signs of an economy closely tied to a type of farming heavily dependent on the new technologies of refrigeration and irrigation.

(5) **The Mechanization of Farm Operations**: Of all the changes that have occurred on the farm in modern times, perhaps none seems as dramatic as the replacement of the horse and mule with machines. It should be emphasized, however, that although the first rather ungainly and expensive tractors began to appear on the farm in the first years of the present century, the passing of the horse and mule was long and protracted throughout the region. In the Palouse Hills, for example, the adoption of the tractor for plowing or harvesting was strung out over a number of decades due to the difficulties of negotiating these machines on the always changing and often very steep gradients.

From more than one million horses in the Pacific Northwest at the time of World War I, numbers declined fairly slowly to around 600,000 by the beginning of World War II. But thereafter the decline in numbers was precipitous, and after 1950 there were fewer than a quarter million horses left. The Clydesdales and Percheron horses of draft fame are almost all gone today, and the few tens of thousands that remain are bred largely for leisure riding and sport, no longer for work on the farm.

From the start of farm mechanization, the tractor was—as it has remained—the essential farm machine, but in later decades it was joined by a host of other machines able to perform many, if not all, of the tasks previously performed by the farmer and his farm laborers. But, like the horse and the mule, the latter have long been a fast-disappearing element of the farm scene, in which the work previously done by a score of men can today be completed just as well, if not better, by a mere handful; except, of course, for those few weeks of the year when additional hands may be needed for harvesting, shearing, or some other seasonal activity.

Lumbering

Changes in the lumber industry since the 1880s span a spectrum of activities from the felling of the trees to the production of pulp and paper and the manufacture of a host of wood products, including plywood and pressboard.

In the forests of the Northwest, the lumberjack's axe, which swung well into the 1880s, was supplemented by the crosscut saw and in more recent decades by ever more powerful chain saws.[50] The moving of the logs from forest to mill was initially undertaken by oxen or horses, aided by the use of skid roads; then came high-lead logging, donkey engines providing the power, and the lifting out of the logs to suitable sites from which they could be easily transported. By the 1890s this was already being done in many areas by steam locomotives and flat cars that ran on hastily constructed logging railroads, which could be removed and re-laid elsewhere as soon as the nearby timber was depleted. By the 1920s trucks were being used increasingly, as the road systems of many regions were improved and as specialized vehicles were developed to handle logs. In a few remote and inaccessible areas some use has been made in more recent times of helicopters for this purpose.

Photo 2. Jameson Logging Camp near Goshen, Washington in the 1880's.
Photo courtesy of Center for Pacific Northwest Sudies, Western Washington University.

Similar significant changes also occurred in the machinery and operations of saw, shake, and shingle mills, while pulp and paper manufacturing became established in the region as a major forest industry with a variety of by-products being produced for regional and overseas markets.

The most important change in lumbering, which gave rise in turn to many of the changes noted above, was the emergence after the 1880s of a number of giant companies using midwestern and eastern capital. The first of these was the St. Paul and Tacoma Lumber Company, and although the latter, like many other major companies at the time, began to buy up large acreages of forest land, it was not until the arrival of the Weyerhaeuser Company in 1900 that the modern period truly began. Following the purchase of 900,000 acres of forest land from the Northern Pacific Railroad in 1900 for approximately $6.00 an acre, the Weyerhaeuser Company entered the processing sector some two years later when it purchased the Bell-Nelson Lumber Company of Everett. Thereafter its progress was rapid, and within a few years it had surpassed all its competitors. As Robert E. Ficken points out: "The arrival of Weyerhaeuser in the Pacific Northwest transformed the region's timber industry.

Declining in importance, the old San Francisco-based companies sold out to midwestern investors or went out of business altogether, especially if they lacked sizeable timber holdings."[51] Of the older companies, only Pope and Talbot remained a major force in the industry, and it was joined in later years by such giants as Simpson, Georgia Pacific, Louisiana Pacific, ITT Rayonier, Crown-Zellerbach, and others. For more than seventy-five years a handful of companies has dominated all facets of the forest products industries of the region.

The Fisheries & Fishing Processing

Although overshadowed from the start by both the forest and agricultural industries, fisheries and fish processing have played a significant role in the economies of Oregon and Washington since the 1880s.

Early commercial activities in fisheries were only slightly different from those of the coastal Indians, with canoes being used to troll for Chinook salmon in the spring and silvers in the fall. Small sailing trollers replaced the canoes in the 1880s, and by the first decade of this century, gas and diesel-powered vessels were being used. Dip nets and reef nets were the most common equipment in the early years, but the gill net, which had been introduced in 1851, eventually overtook the other methods and was widely used in Lower Columbia and Puget Sound waters. The method of purse seining, which was introduced somewhat later, required larger vessels and was used with greatest success in Alaskan waters. One other method, which became especially common in the Puget Sound, was the fish trap. Such traps occupied hundreds of acres of offshore waters in many parts of the region, but when rapid depletion of salmon stocks in the 1920s threatened the industry, the move was made to have them outlawed and they were declared illegal in all Washington waters in 1934.

Technological advances in the final years of the nineteenth and the first few years of the twentieth centuries helped revolutionize the cannery industry. Most significant of these new machines was one that skinned, eviscerated, and prepared salmon for canning. Better cans and more efficient ovens also were introduced in the first decade of the century. The new machines greatly reduced the need for large numbers of workers, many of whom were Chinese.

The small-scale individualism and competition that was typical in the early days of the industry began to be replaced at the turn of the century by the increasing dominance of large firms. This occurred not only in fish processing but also in the catching of the fish, especially where companies were able to acquire licenses for large numbers of traps. Firms such as Pacific American Fisheries, which for years boasted of having the largest salmon cannery in the world in South Bellingham (Fairhaven), and Alaska Packers, with headquarters at Semiahmoo near Blaine, Washington, were two of the giants. As Pacific Northwest fisheries became severely depleted and as more rigid state rules began to take effect, these and other companies greatly expanded their operations in Alaskan waters. By this time also refrigeration was becoming increasingly important in the industry, and a number of firms turned

Photo 3. Unloading salmon at a dock in Puget Sound. Early 20th century.
Photo courtesy of Center for Pacific Northwest Studies, Western Washington University.

their attention to frozen products and from salmon alone to a number of other varieties of fish.

The Mining Industry

Like each of the other resource industries, mining benefitted from a plethora of technological innovations that transformed the methods of mining, the preparation of the ores, and the smelting and refining process. It is not necessary here to detail those innovations, except to note that while all of them aimed at greater efficiency of operation—usually successfully—they also greatly increased the capital needed to open and operate a mine or to run a smelter. At the same time, stricter government regulations, both state and federal, necessitated other changes that increased costs, even though improving safety.

Photo 4. Salmon canning operation, Pacific American Fisheries, South Bellingham, Washington. Early 1900's.
Photo courtesy of Center for Pacific Northwest Studies, Western Washington University.

Because of the huge costs entailed in the mining and smelting of such non-ferrous metals as copper, lead, and zinc, which by 1900 had become the major minerals mined in the Pacific Northwest, a small number of giant companies came to dominate the mines and the smelters, most notably Anaconda and Guggenheim. Whole towns, like Butte and Anaconda, became dependent largely on the wages paid out by a single company while state politics were increasingly beset with the influence-peddling operations that accompanied the power of the mining companies. Not that these operators always had it their way. Strikes, which became more and more common in the industry in the last decade of the nineteenth century—Coeur d'Alenes for example suffered the so-called Mining War of 1892[52]—were severely and quickly repressed initially by company leaders who were often supported by state officials. But over the long haul, great concessions were gained by the mineworkers unions. Meanwhile in the lumber industry, strikes were beginning to occur with ever greater frequency, and in the second decade of the century they were to lead to the Centralia and Everett

Massacres which eventually led to the widespread condemnation of the principal union, the International Workers of the World (the IWW or "Wobblies").

Manufacturing

Throughout the pioneer period most manufacturing could be said to have been no more than simple processing of one or other of the region's natural resources, although here and there a boat or some items of household furniture might be made for sale. Processing industries continued to dominate the secondary sector of the economy of all four Pacific Northwest states, as well as British Columbia, long after the beginning of the commercial age, and indeed they can be said to do so today in some parts of the region where lumbering or food processing are the only manufacturing industries.

During the 1890s, however, larger regional markets led to a broadening of the manufacturing base and the production of various consumer products, including a variety of clothing and household goods. Most of the firms were small and undercapitalized, but a few such as the Carnation Company, which pioneered the manufacture of canned evaporated milk and other milk products, survived to become major firms in later decades.

Of all the manufacturing companies started in the Pacific Northwest in the early decades of this century, none is more famous and more significant to the region's economic well-being than the Boeing Company. Founded in 1916 in the now famous Red Barn on the banks of the Duwamish River, Boeing came into its own in World War II. A transportation equipment industry which increased in importance during World War I was shipbuilding, although it was World War II that established the Pacific Northwest as a major shipbuilding and ship repairing region.

Impressive as the early developments in manufacturing may have been, it was not until the Second World War, as Gerald Nash has convincingly argued, that manufacturing, other than simple processing, became a major aspect of the region's economy.[53] It was at that time that shipbuilding, aircraft production, and aluminum smelting helped propel the region to national and even international prominence in manufacturing. And the key factor in this transformation of the industrial sector was the bringing on line of the large amounts of electricity being produced at the Bonneville and Grand Coulee Dams on the Columbia River. Since then a vast array of industries has boosted manufacturing output and helped broaden a still heavily weighted resource-based economy.

REFERENCES

1. Taylor, Herbert C., Jr., "The Utilization of Archeological and Ethnohistorical Data in Estimating Aboriginal Populations." *Bulletin of the Texas Archeological Society* 32 (1962): 121–40.

2. A good overview of this matter is provided in Angelo Anastasio's "The Southern Plateau: An Ecological Analysis of Intergroup Relations." *Northwest Anthropological Research Notes* 6 (1972): 109–229, but see particularly pp. 119–141.
3. Ross, Alexander. *Adventures of the First Settlers on the Oregon.* Milo M. Quaife, editor. New York: The Citadel Press, 1969. (Reprint of the 1849 edition).
4. Among others the missionary Samuel Parker and botanists Thomas Nuttall and John Kirk Townsend.
5. See particularly Carl O. Sauer, "Man's Dominance by Use of Fire." *Geoscience and Man* X (1975): 1–13. Reprinted in Carl O. Sauer. *Selected Essays 1963–1975.* Berkeley: Turtle Island Foundation, 1981, pp. 129–156.
6. The standard work on the Spanish explorers in the Pacific Northwest is Warren L. Cook. *Flood Tide of Empire: Spain and the Pacific Northwest, 1543–1819.* New Haven: Yale University Press, 1973.
7. The standard work on James Cook is J.C. Beaglehole. *The Life of Captain James Cook.* Stanford: Stanford University Press, 1974. See also Robin Fisher & Hugh Johnston, eds. *Captain James Cook and His Times.* Seattle: University of Washington Press, 1979, particularly the article by Christon I. Archer, "The Spanish Reaction to Cook's Third Voyage," pp. 99–119.
8. See especially Edmond S. Meany. *Vancouver's Discovery of Puget Sound.* Portland: Binfords & Mort, 1957.
9. An excellent brief account of this period is provided in Barry M. Gough. *Distant Dominion: Britain and the Northwest Coast of North America, 1579–1809.* Vancouver: University of British Columbia Press, 1980.
10. Books by two geographers, one American the other British, cover important aspects of overland exploration: John Logan Allen. *Passage Through the Garden: Lewis and Clark and the Image of the American Northwest.* Urbana: University of Illinois Press, 1975 and E.W. Gilbert. *The Exploration of Western America 1800–1850: An Historical Geography.* New York: Cooper Square Publishers, 1966.
11. Lewis, Meriwether & William Clark. *The History of the Lewis and Clark Expedition.* Edited by Elliott Coues. New York: Dover Publications, volume II, p. 702.
12. The most useful history of Western exploration in the nineteenth century is William H. Goetzmann. *Exploration and Empire: The Explorer and the Scientist in the Winning of the American West.* New York: Alfred A. Knopf, 1967.
13. The most recent and most authoritative work on Astoria and the Pacific Fur Company is James P. Ronda. *Astoria and Empire.* Lincoln: University of Nebraska Press, 1990.
14. There are literally dozens of works dealing with the Hudson's Bay Company, and as many more that deal with its chief factors and traders. The official history of the company is E.E. Rich. *Hudson's Bay Company: 1670–1870.* 2 volumes. London: The Hudson's Bay Record Society, 1968. More recently Pater C. Newman has published another two-volume work on the company: *Company of Adventurers.* Markham, Ontario: Viking/Penguin, 1985 & 1987. A major work by a Canadian geographer is Richard I. Ruggles. A *Country So Interesting: The Hudson's Bay Company and Two Centuries of Mapping, 1670–1870.* Montreal & Kingston: McGill-Queen's University Press, 1991.
15. The most useful treatment of the company's agricultural activities is the work of another Canadian geographer: James R. Gibson. *Farming the Frontier: The Agricultural Opening of the Oregon Country, 1786–1846.* Seattle: University of Washington Press, 1985.

16. The best treatment of this is in John S. Galbraith. *The Hudson's Bay Company as an Imperial Factor, 1821–1869*. Toronto: University of Toronto Press, 1957, but see also Frederick Merk. *The Monroe Doctrine and American Expansionism, 1843–1849*. New York: Alfred A. Knopf, 1966.
17. A brief, but somewhat pedestrian coverage of the missionary era is given in Cecil P. Dryden. *Give All to Oregon!: Missionary Pioneers of the Far West*. New York: Hastings House, 1968.
18. The most scholarly treatment is that of Robert J. Loewenberg. *Equality on the Oregon Frontier: Jason Lee and the Methodist Mission, 1834–43*. Seattle: University of Washington Press, 1976.
19. Among the many writings on the American Board missionaries by Clifford M. Drury are *Marcus and Narcissa Whitman and the Opening of Old Oregon*. 2 volumes. Seattle: Pacific Northwest National Parks and Forests Association, 1986; *Nine Years With the Spokane Indians: The Diary, 1838–1848, of Elkanah Walker*. Glendale, CA.: The Arthur H. Clark Company, 1976; and *The Diaries and Letters of Henry H. Spalding and Asa Bowen Smith Relating to the Nez Perce Mission, 1838–1842*. Glendale, CA: The Arthur H. Clark Company, 1958.
20. A well-documented volume on the "Black Robes" is Maria Ilma Rauffer. *Black Robes and Indians on the Last Frontier*. Milwaukee: The Bruce Publishing Company, 1966. A useful chronology of the Catholic presence in the Pacific Northwest is provided by Wilred P. Schoenberg. *A Chronicle of Catholic History of the Pacific Northwest, 1743–1960*: Spokane: Gonzaga Preparatory School, 1962.
21. The only authoritative history of the overland crossings is that of John D. Unruh. *The Plains Across: The Overland Emigrants and the Trans-Mississippi West. 1840–60*. Urbana: University of Illinois Press, 1979.
22. Most useful here is the atlas and commentary prepared by William A. Bowen. *The Willamette Valley: Migration and Settlement on the Oregon Frontier*. Seattle: University of Washington Press, 1978. This deals with events in the Willamette Valley to 1850.
23. Maps of the donation lands of Oregon and Washington are to be found in: Samuel N. Dicken & Emily F. Dicken. *The Making of Oregon: A Study in Historical Geography*. Portland: Oregon Historical Society, 1979, p. 98. and James W. Scott & Roland L. DeLorme. *Historical Atlas, of Washington*. Norman: University of Oklahoma Press, 1988, Map 30.
24. Important works covering this topic are the work respectively of a geographer and a historian, both of them natives of the Columbia Basin: Donald W. Meinig. *The Great Columbia Plain: A Historical Geography, 1805–1910*. Seattle: University of Washington Press, 1968, and Alexander Campbell McGregor. *Counting Sheep: From Open Range to Agribusiness on the Columbia Plateau*. Seattle: University of Washington Press, 1982.
25. The best coverage of the Pacific Coast lumber industry in the nineteenth century is Thomas R. Cox. *Mills and Markets: A History of the Pacific Coast Lumber Industry to 1900*. Seattle: University of Washington Press, 1974. Works on many of the major companies have also been written. Among the best are Edwin T. Coman & Helen M. Gibbs. *Time, Tide and Timber: A Century of Pope & Talbot*. Stanford: Stanford University Press, 1949, and Murray Morgan. *The Mill on the Boot: The Story of the St. Paul & Tacoma Lumber Company*. Seattle: University of Washington Press, 1982.
26. The classic work on western mining is William J. Trimble. *The Mining Advance Into the Inland Empire*. Fairfield, WA.: Ye Galleon Press, 1986. (First published by the University of Wisconsin in 1909). The best modern history is Rodman W. Paul. *Mining Frontiers of the Far West, 1848–1880*. New York: Holt, Rinehart & Winston, 1963.

27. The well known lexicographer Williard R. Espy has written a fascinating memoir of life on the Long Beach Peninsula of Washington in the later nineteenth century. This provides some information on the early oyster industry of the region. Williard R. Espy. *Oysterville: Roads to Grandpa's Village*. New York: Clarkson N. Potter Publishers, 1977.
28. There is no good modern book-length work on the Pacific Northwest fisheries. A useful article is Vernon Carstensen, "The Fisherman's Frontier on the Pacific Coast: The Rise of the Salmon-Canning Industry" in *The Frontier Challenge: Responses in the Trans-Mississippi West*. John G. Clark, editor. Lawrence: University of Kansas Press, 1971, pp. 57–79. The best monograph on any aspect of the industry is Gordon B. Dodds. *The Salmon King of Oregon: R.D. Hume and the Pacific Fisheries*. Chapel Hill: University of North Carolina Press, 1959.
29. See Florence E. Sherfey. *Eastern Washington's Vanished Gristmills and the Men Who Ran Them*. Fairfield, WA.: Ye Galleon Press, 1978.
30. The first volume of Alfred L. Lomax's two volume history of Oregon's woolen mills covers the pioneer period: *Pioneer Woolen Mills in Oregon: History of Wool and the Woolen Textile Industry in Oregon, 1811–1875*. Portland: Binfords & Mort, 1941.
31. The classic treatment of these topics is Oscar O. Winthur. *The Old Oregon Country: A History of Frontier Trade, Transportation and Travel*. Stanford: Stanford University Press, 1941.
32. Particularly useful is Randall V. Mills. *Stern-Wheelers Up Columbia*. Paolo Alta: Pacific Books, 1947. See also Donald Abbott, editor. *Steamboat Days on the Rivers*. Portland: Oregon Historical Society, 1969, and Howard McKinley Corning. *Willamette Landings*. Portland: Oregon Historical Society, 1947.
33. Gordon Newell. *Ships of the Inland Sea*. Portland: Binfords & Mort, 1960, provides much useful material on the "mosquito" fleets of the Puget Sound region. Also of use is Patrick S. Grant. "The Evolution of Ferry Service in the San Juan Islands" Unpublished M.S. Thesis in Geography, Western Washington University, 1981.
34. Richard C. Wade. *The Urban Frontier: The Rise of Western Cities, 1790–1830*. Cambridge: Harvard University Press, 1959.
35. See John W. Reps. "Bonanza Town Planning on the Western Mining Frontier" in *Pattern and Process: Research in Historical Geography*. Ralph E. Ehrenberg editor. Washington, D.C.: Howard University Press, 1975, pp. 271–89.
36. Although there are scores of volumes available on ghost towns of the West, most are of little historical or geographical value. One of the few good ones is Muriel S. Wolle. *The Bonanza Trail: Ghost Towns and Mining Camps of the West*. Bloomington: Indiana University Press, 1958.
37. Two of the best works available are both by John W. Reps: *Panoramas of Promise: Pacific Northwest Cities and Towns on Nineteenth-Century Lithographs*. Pullman: Washington State University Press, 1984, and *The Forgotten Frontier: Urban Planning in the American West Before 1890*. Columbia: University of Missouri Press, 1981.
38. David Newsom. *David Newsom: The Western Observer, 1805–1882*. Portland: Oregon Historical Society, 1972, p. 52.
39. Charles H. Carey. *A General History of Oregon Prior to 1861*. Volume 2, p. 655, as quoted in Reps. *The Forgotten Frontier*. p. 42.
40. A brief, highly readable history of Portland in this period is Eugene E. Snyder. *Early Portland: Stump-Town Triumphant*. Portland: Binfords & Mort, 1970.
41. Port Townsend has been well treated by two geographers: Herbert L. Combs. "The Historical Geography of Port Townsend, Washington" Unpublished M.A. Thesis, Uni-

versity of Washington, 1950, and John G. Newman. "Port Townsend, Washington: A Humanistic Interpretation of Its Historical Geography" Unpublished M.S. Thesis, Western Washington University, 1989. There also are many other volumes by historians and journalists on the city. A well-illustrated work is Peter Simpson & James Hermanson. *Port Townsend: Years That Are Gone.* Port Townsend: Quimper Press, 1979.

42. An excellent academic work by a geographer on early Seattle is Thomas W. Pohl. "Seattle, 1851–1861: A Frontier Community" Unpublished Ph.D. dissertation, University of Washington, 1970. Among published works on Seattle two of the best are Murray Morgan. *Skid Road: An Informal Portrait of Seattle.* New York: Viking Press, 1951, and Roger Sale. *Seattle Past to Present.* Seattle: University of Washington Press, 1976.

43. See James W. Scott & Daniel E. Turbeville. *Early Industries of Bellingham Bay and Whatcom County.* Bellingham: Fourth Corner Registry, 1980, for a brief introduction to the early Bellingham Bay towns.

44. D.W. Meinig has much to say on this topic in his *Great Columbia Plain,* but see also his article "Wheat Sacks Out to Sea" *Pacific Northwest Quarterly.*

45. The work that deals most directly with this is John Fahey. *Inland Empire: D.C. Corbin and Spokane.* Seattle: University of Washington Press, 1965, but see also D.W. Meinig. *The Great Columbia Plain.*

46. In addition to the many histories of virtually every one of the major, and many of the minor, railroads, see Robert E. Riegel. *The Story of the Western Railroads: From 1852 Through the Reign of the Giants.* Lincoln: University of Nebraska Press, 1964 (first published in 1926) for a useful overview, though now somewhat outdated, of the complex history of the western railroads.

47. There is a good discussion of this in G. Thomas Edwards & Carlos A. Schwantes. *Experiences in a Promised Land: Essays in Pacific Northwest History.* Seattle: University of Washington Press, 1986, pp. xii-xiv.

48. See Alexander Campbell McGregor. *Counting Sheep* for the most complete treatment of this topic.

49. The range lands of Oregon are well covered in Phil F. Brogan. *East of the Cascades.* Portland: Binfords & Mort, 1964.

50. A well-written and informative popular work about the American lumberjack is Stewart Holbrook. *Holy Old Mackinaw.* Sausalito, CA.: Comstock editions, 1980.

51. An excellent article is Robert E. Ficken. "Weyerhaeuser and the Pacific Northwest Timber Industry, 1899–1903" in *Experiences in a Promised Land.* G. Thomas Edwards & Carlos A. Schwantes, editors. Seattle: University of Washington Press, 1986, pp. 139-152. The quotation is from p. 151.

52. See Robert Wayne Smith. *The Coeur d'Alene Mining War of 1892: A Case Study of an Industrial Dispute.* Corvallis: Oregon State University Press, 1961.

53. Gerald D. Nash. *The American West Transformed: The Impact of the Second World War.* Bloomington: Indiana University Press, 1985.

The Physical Environment

LANDFORMS Chapter 2

Eugene P. Kiver and Dale F. Stradling
Eastern Washington University

INTRODUCTION

The Pacific Northwest includes Washington, Oregon, Idaho, and most of the mountainous area of western Montana. The diversity of geologic structure and process, as well as significant differences in the timing of formative events, has produced a varied and complex topography in the region that in turn controls or influences climate, soils, vegetation, and land use.

The regional definition used here utilizes political boundaries including the Canadian border; the east edge of Flathead, Lewis and Clark, Broadwater, and Gallatin counties in Montana; the Idaho-Wyoming border; and the southern borders of Idaho and Oregon. These political boundaries cut indiscriminately across the region's natural boundaries. This vast region has an area of nearly 2,000,000 km^2 (774,000 mi^2) and is drained by the Columbia River and its tributaries except for a small area east of the continental divide in southwestern Montana where drainages flow into the Missouri River system and those streams in the far west that discharge directly into the Pacific Ocean.

The Pacific Northwest is divided into distinct landscape units called geomorphic provinces that form the major headings for this chapter (Figure 1). The ideal criteria for province designation are continuity, similarity of landforms, and distinct topographic and structural boundaries. Each province is further subdivided into sections to better emphasize more geographically specific structural and geomorphic characteristics. The Northern Rocky Mountain Province and its geomorphic features are described first, followed by the Columbia Intermontane, Basin and Range, a small part of the Middle Rocky Mountains in southeastern Idaho, Cascade Mountains, and the Pacific Border provinces.

NORTHERN ROCKY MOUNTAINS PROVINCE

Geomorphic Sections

While the subdivision of the province into the five sections (Figure 2) used here has not been formally proposed before, both Thornbury (1965) and Hunt (1974)

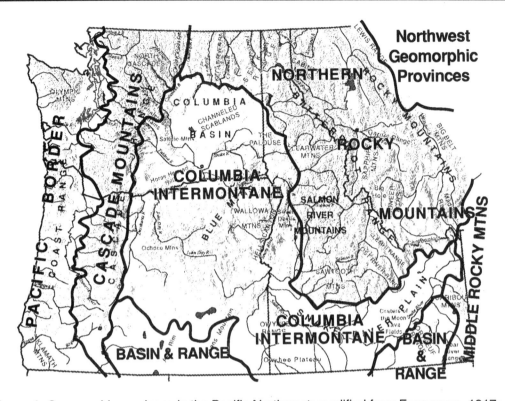

Figure 1. Geomorphic provinces in the Pacific Northwest; modified from Fenneman, 1917; landform map by Erwin Raisz. Key to numbers: 1—Pacific Border; 2—Cascade Mountains; 3—Columbia Intermontane; 4—Northern Rocky Mountains; 5—Basin and Range; and 6—Middle Rocky Mountains.
Base Raisz map used with permission by Raisz Landform Maps, Melrose, Mass. Copyright by Erwin Raisz, 1957.

describe four general structural-topographic regions. Northern sections are separated from southern sections (Figure 2) by the Lewis and Clark Line or Lineament. The line is a broad zone of geologic and geophysical features associated with an alignment and parallelism of topographic elements that extends from the south edge of the Northern Rocky Mountains in Washington to beyond Helena and Three Forks in Montana (Figure 2). It is a complex of strike-slip, or San Andreas type, and other faults and discontinuities that are major structural features in this part of the continent. A significant part of this natural corridor was used by engineers and designers of the Interstate 90 right-of-way in western Montana and northern Idaho.

The prominent north-south lineament called the Rocky Mountain Trench further subdivides the northern sections into the Disturbed Belt and the northwest sections. This trench is a fault-bounded lowland more than 7,000 km (4,350 mi) long that extends from Alaska to the south end of the Bitterroot Valley (see Figure 3 in Oldow and others, 1989). Portions of eight different rivers occupy the trench including the

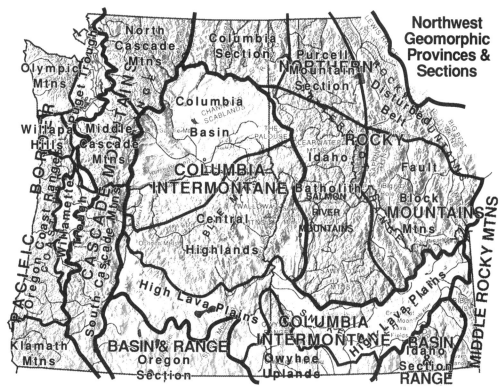

Figure 2. Geomorphic Province subdivisions in the Pacific Northwest.

Stillwater, Flathead, and Bitterroot rivers in Montana. The continuation of the trench south of the Lewis and Clark Line along the Bitterroot Valley further subdivides the Idaho Batholith Section from the Fault-Block Mountains Section.

The northwest part of the Northern Rocky Mountains shown in Figure 2 is divided here by the Purcell Trench into the Purcell Mountains to the east and the Columbia Mountains to the west. The Trench is a major structural feature that branches from the Rocky Mountain Trench in Canada and extends 1,400 km (870 mi) southward to the north end of Lake Coeur d'Alene where it apparently terminates against the Lewis and Clark Line.

Maximum elevations are mostly below 2,135 m (7,000′) in the north but soar to 3,859 m (12,662′) in the southeast part of the province at Mt. Borah, the highest point in Idaho. Intermediate elevations close to 2,745 m (9,000′) in the north part of the province occur only in Glacier National Park and the Bitterroot Range near Missoula.

The Idaho Batholith Section (Figure 2) is dominated by granite, granodiorite, and related igneous rocks that produce an overall dendritic drainage pattern. This section, as well as the Purcell Mountains Section, are almost entirely in mountain slopes producing what Alt and Hyndman (1986) call "claustrophobic topography." The Salmon River ("the river of no return") is the only river that completely crosses the section.

The Fault-Block Mountain Section (Figure 2) contains a wide variety of rocks and about 27 individually named ranges, most of which trend northwest. The western ranges are blocks that slid eastward from the top of the Idaho batholith leaving topographic gaps marked by the Bitterroot, Big Hole, and other valleys. The eastern ranges are basement uplifts and Basin and Range type mountains distinguished by large normal faults.

The Disturbed Belt (Figure 2) is characterized by relatively thin thrust sheets. The abrupt topographic break that marks the base of the Front Range in Glacier National Park and south corresponds to the east edge of large thrust faults. North-south oriented normal faults later blocked out the Flathead, Mission, and Swan ranges.

The Purcell Mountain Section contains thick blocks of late Precambrian sedimentary rocks (Belt rocks) that formerly covered the Kaniksu batholith west of the Purcell Trench. Uplifting of the batholith allowed thick blocks of overlying rock to detach and move eastward along thrust faults.

Stretching of the crust west of the Purcell Trench, emplacement of gneiss domes, and collision of a large microcontinent with North America created many of the folds and faults in the Columbia Mountains Section. As in the Disturbed Belt, north-south oriented normal faults developed later helping to outline the linear ranges of northeastern Washington and northern Idaho.

Geology

Figure 3 briefly summarizes the geologic and geomorphic evolution of the Northern Rocky Mountains and also provides a generalized geologic time scale. Although geomorphology is considered by some to be concerned only with relatively recent landscape changes, all geologic events beginning with the early Precambrian have influenced the present topographic expression of the Rocky Mountains.

Continents contain a record in the rocks that indicates that they grow by the addition of materials around their perimeters. Internal processes in the earth produce surface movements that drive these accretionary events and also account for splitting off and migration of crustal masses. These processes and their effects fall under the study known as plate tectonics.

One of the nuclear areas of the ancient North American continent is represented by the 2.7 Ga (billions of years ago) metamorphic rocks found in the Fault-Block Mountain Section. These are the oldest rocks in the Pacific Northwest and were part of an ancient highland that provided one of the sources of sand and mud sediment that accumulated in a late Precambrian (about 1.8–0.8 Ga) basin to the north. These younger rocks are known as the Belt Supergroup and derive their name from the Little Belt Mountains where equivalent rocks were first described. The Belt basin split apart about 750 Ma (millions of years ago). The missing western part may have accreted to Antarctica.

GEOLOGIC TIME UNITS			YEARS BEFORE PRESENT	GEOLOGIC MATERIALS	SIGNIFICANT EVENTS
CENOZOIC	Quaternary	Holocene	.01	Loess	Warmer and drier
		Pleistocene		Glacial, stream, lake deposits	Cooler and wetter; alpine and continental glaciers; glacial lakes form; stream terraces; drainage changes, erosion
	Tertiary	Piocene	2.5 Ma	Flaxville Conglomerate	Dry climate, valley fill (10–2.5 Ma); Intermountain Seismic Belt, fault-block mountains (~5 Ma)
		Miocene		Sixmile Gravels	Warm, moist, near tropical climate
		Oligocene			
		Eocene		Renova and Tiger formations (~40 Ma)	North-south normal faults block out northern ranges; dry climate, Laramide Orogeny ends
		Paleocene		Challis and Sanpoil volcanics, plutons (50–40 Ma)	North Cascade microcontinent docks (~50 Ma) volcanism slabs and blocks slide eastward folding thrust faults

Figure 3. Significant geologic and geomorphic events in the Northern and Middle Rocky Mountains.

GEOLOGIC TIME UNITS	YEARS BEFORE PRESENT	GEOLOGIC MATERIALS	SIGNIFICANT EVENTS
MESOZOIC	65 Ma	Idaho, Boulder, Kaniksu, Colville and other batholiths (90–70 Ma)	Laramide Orogeny begins, microcontinent collides (100 Ma)
PALEOZOIC	245 Ma	Marine strata	Eastern Oregon and Washington at west edge of continent
PRECAMBRIAN	600 Ma	Belt Supergroup	Continent splits, sediments deposited on trailing edge of continent
		Metamorphic rocks ~2.7 Ga	Forms ancient nucleus of North America

Figure 3. (continued)

The new western edge of the North American plate was then located in what is now eastern Washington and Oregon. A maximum of a few thousand meters of Paleozoic sediment accumulated on top of the Belt rocks as the North American plate continued its slow, eastward movement.

In early Mesozoic time, westward movement of the North American plate produced a subduction zone and collisions with island-like land masses. Subduction zone activity created enormous volumes of granitic magma and ultimately crustal stresses that produced the episode of mountain building known as the Laramide Orogeny. The Okanogan microcontinent accreted to the west edge of North America about 100 Ma and the North Cascades microcontinent docked about 50 Ma.

Numerous intrusions of granitic magma were injected into the western Cordillera from late Mesozoic through mid-Cenozoic time, creating the huge Idaho batholith as well as smaller batholiths elsewhere. Sliding of the cover rocks from the uplifted batholiths and compression of the crust produced folding and numerous thrust faults. Later crustal extension produced normal faults and gneiss domes. Oldow and others (1989) suggest that a decoupling occurs between the deeper crust and the overlying rock materials allowing very complex structural deformation to occur in the detached block. Thus, the major framework of the modern topography was established by events that occurred 100–50 Ma ago.

Volcanism about 50 Ma produced some localized volcanic mountains, another mass of granitic intrusions, and a dike swarm that extends northeast from Boise to central Montana. Filling of the Laramide age valleys and basins with sediments about 40 Ma was interrupted about 20 Ma as wetter, more tropical conditions prevailed. Widespread development of laterite soils and the nature of the fossil fauna and flora also indicate that wetter and warmer conditions prevailed during the middle Miocene. Valley filling resumed about 10 Ma as climates once again became more arid.

The cooler, wetter climates of the Quaternary triggered downcutting by streams and episodes of glaciation. Numerous alpine glaciers were generated in the higher mountain ranges when continental glaciers from Canada spilled southward into the north-trending structural valleys in the northern sections.

Some Special Features

Generation of molten rock and the availability of favorable host rocks and structures dramatically increased the probability of creating mineral deposits. The Northern Rocky Mountains contain numerous ore deposits, some of which are world famous. Discovery of gold in the Clearwater drainage in 1860 and the Bannack and Argenta areas near Dillon in 1862 caused large numbers of people to enter the region. As demand for copper increased the mines at Butte became the world's largest producer and "richest hill on earth." The Coeur d'Alene district began lode mining in 1884 and has produced more silver than any other mining district in the world.

A broad zone of active faulting, known as the Intermountain Seismic Belt extends from Utah north along the Idaho-Wyoming border and apparently ends near Kalispell,

Montana. Fault-block mountains, typical of the Basin and Range Province, are abundant in this zone. Earthquakes were reported soon after the first European settlers arrived and continue unabated. Hundreds of shocks have been recorded in the Flathead Lake, Helena, Madison, and Yellowstone areas, and occasional damaging earthquakes have occurred since 1869.

The August 17, 1959, Hebgen Lake earthquake is Montana's largest recorded with a surface-wave magnitude of 7.5 on the Richter scale. A number of new geomorphic features formed including 29 km (18 mi) of fault scarps and a large landslide that buried 26 campers and dammed the Madison River canyon.

The Intermountain Seismic Belt also extends westward to the Stanley Basin in central Idaho. Idaho's largest recorded earthquake occurred on October 28, 1983, near Borah Peak. The shock had a surface wave magnitude of 7.3 on the Richter scale and produced some 36 km (22 mi) of fault scarps. Two deaths, the destruction of several buildings, and a 50 ton boulder that rolled into Challis were also a result of the earthquake.

Cordilleran ice entering from Canada followed the ready-made fault valleys generating distinct ice lobes in each major valley. Ice covered much of the mountainous topography near the Canadian border but was confined to the trenches and valleys in the south, except for the Okanogan lobe that spilled onto the Columbia Plateau.

Knowledge of ice cap and alpine glacier chronology south of the Cordilleran ice sheet is sparse, but deposits studied so far seem to reasonably correlate with the dated moraine sequence at Yellowstone. As in other areas of the Cordillera, the late Wisconsin glaciation 43–12 Ka (thousands of years ago) was less extensive than the penultimate (Pre-Wisconsin) advance about 150–140 Ka.

The Purcell Trench lobe blocked the Clark Fork valley in northern Idaho producing an immense ice-dammed lake named Lake Missoula. Lake silts extend considerable distances up tributary valleys, and wave-cut benches (Photo 1) on valley-side slopes record a lake basin containing over 2,000 km^3 (600 mi^3) of water. Failure of the ice dam generated catastrophic floods of unprecedented proportions that swept across the adjacent Columbia Intermontane Province.

Drainage anomalies include the transverse drainage of the Jefferson River between Whitehall and Three Forks. Formerly regarded as a superimposed river, recent work (1987) favors an antecedent river explanation. Other drainage anomalies occur in the Purcell Trench and the Kootenai, Pend Oreille, Colville, Columbia, and Spokane River valleys. Explanations include impoundment and diversion by Miocene lava flows and drainage reversals due to glaciation.

Photo 1. Wave-cut strandlines above Missoula, Montana indicate that glacial lake Missoula was about 305 m (1,000′) deep here.

COLUMBIA INTERMONTANE PROVINCE

Topography

Fenneman (1917) called the large, topographically diverse area between the Basin and Range and the Cascade, Northern Rocky, and Middle Rocky mountains the Columbia Plateau. The more detailed classification by Freeman and others (1945) that is used here names the area the Columbia Intermontane Province and divides the province into the Columbia Basin, Central Highlands, High Lava Plains, and Owyhee Upland subprovinces and twelve sections (Figure 4). More recently, it has been referred to as the Columbia Intermontane region. A unifying characteristic of the province is the presence, and in most areas dominance, of basalt lava flows. Except for the more transitional boundary with the Basin and Range Province in southern Oregon, the province margins are topographically distinct and are located where lava flows lap against the surrounding mountains.

The master streams are the Snake and Columbia rivers which mostly collect runoff from the surrounding highlands where precipitation is higher. The rain shadow of the

Figure 4. Divisions of the Columbia Intermontane Province proposed by Freeman and others (1945).

Cascade Mountains produces a semiarid climate here except in the Central Highlands where significant orographic precipitation occurs. The Harney-High Desert Section in Oregon has mostly internal drainage. Small, mostly intermittent streams are common elsewhere in the Columbia Intermontane. Local areas of sand dunes occur near Moses Lake in Washington, Bruneau in Idaho, and elsewhere. Semiarid steppe conditions combined with the relative recency of volcanic and other geologic events contribute to an unusually clear landform record.

Elevations are highest in the Central Highlands (Eagle Cap Peak in the Wallowa Mountains, 2,900 m; 9,505′) and the Owyhee Upland (Hayden Peak, 2,560 m; 8,403′).

Except for occasional higher peaks, ridges, and buttes, elevations range from 150 to 760 m (500′-2,500′) in the Columbia Basin and from 1,070 to 1,370 m (3,500–4,500′) in the High Lava Plains.

Geology

The early Tertiary and older rock record in the Columbia Intermontane Province is concealed by young rocks except in the Central Highlands and in the steptoes in the eastern Columbia Basin. The older geologic history is similar to that interpreted for the nearby Northern Rocky Mountains. The western edge of North America during Paleozoic time extended through eastern Washington and western Idaho. A collision zone formed here during the Mesozoic and large microplates were added by accretion to the continent edge.

The Central Highlands subprovince contains accreted rocks, including the oldest rocks in Oregon (Devonian) in the Strawberry-Aldrich mountain area. Late Mesozoic batholiths formed in the Central Highlands and a relatively complete record of Tertiary age volcanic and terrestrial sedimentation, occurs in the John Day area of northern Oregon. The sediments contain vertebrate and plant fossils that indicate that mild, subtropical conditions prevailed during middle Tertiary time. A generalized summary of important Cenozoic events in the Pacific Northwest is shown in Figure 5.

Crustal extension in the northeastern Central Highlands and the southeastern Columbia Basin produced one of the world's largest flood basalt regions (Columbia River Basalt Group) in a relatively short time during the Miocene. Fissures in southeastern Washington and northeastern Oregon produced hundreds of individual flows, some of which exceed 1,000 cubic kilometers. These remarkably voluminous flows buried much of the pre-existing landscape, diverted rivers from their channels, and created the vast lava plains of the Columbia Basin Subprovince. Although lava flows here range in age from about 17–6 Ma, over 90 percent of these occurred from about 16–14 Ma. The present location of the Columbia and Spokane rivers in the northern Columbia Basin very nearly follows the northern edge of this vast "lava sea" where Miocene lavas "pushed" the rivers against the Northern Rocky Mountains and North Cascades. This area is now known as the Big Bend. Many of the dam sites on the Spokane and Columbia rivers, including Grand Coulee Dam, are located where the rivers are superimposed from the basalt surface onto the granitic rocks below.

The extremely thick (over 3,050 m; 10,000′) lava section in southern Washington subsided thousands of meters producing a slope reversal. Where enormous lava floods formerly flowed downslope to the north and west during the Miocene, the surface now slopes gently to the south.

The basalt, particularly in central Washington and northern Oregon, was warped and folded during the late Miocene and Pliocene into a number of anticlines, synclines, and basin structures known as the Yakima Folds. As a result of this deformation, the Columbia River in central Washington acquired a more easterly course. In

GEOLOGIC TIME UNITS			Ma BEFORE PRESENT	NORTHERN AND MIDDLE ROCKY MOUNTAINS	COLUMBIA INTERMONTANE
CENOZOIC ERA	Quaternary Period	Holocene	.01	Warmer and drier—minor loess deposition, downcutting streams	Warmer and drier—minor loess deposition, sand dunes activated locally, local volcanism in Snake River Plain
		Pleistocene		Stream terraces, local drainage changes Cordilleran glaciers occupy northern area, alpine glaciers at high elevations, glacial Lake Missoula and Columbia form	Large landslides on Yakima Folds. Channeled Scabland eroded, major loess deposition, pluvial lakes in Oregon, alpine glaciers in Central Highlands, volcanism at Newberry, Craters of the Moon, and elsewhere in the High Lava Plains
	Tertiary Period	Pliocene	2.5	Fault-block mountains begin (~5 Ma). Terrestrial sediments containing vertebrate fauna fill valleys; climate varies from humid subtropical (Miocene) to drier and cooler (Pliocene)	Yakima Folds cause antecedent stream development
		Miocene	5		Voluminous Columbia River Basalt flows and structural changes produce major drainage diversions
		Oligocene	25		
		Eocene	38	North-south normal faults delineate northern ranges (~40 Ma) Laramide Orogeny ends; North Cascade microcontinent docks ~50 Ma Volcanism Plutons Slabs and blocks slide eastward Thrust faults Folding	Terrestrial sedimentation in John Day area leaves long record of faunal, floral, and volcanic history
		Paleocene	55		
			65		

Figure 5. Generalized geomorphic history of the Pacific Northwest.

BASIN AND RANGE	CASCADE MOUNTAINS	PACIFIC BORDER
Semiarid climates, localized volcanism at Klamath Lake, Diamond Craters, Jordan Craters, and Saddle Butte	Numerous recent and historic volcanic eruptions. Numerous alpine glaciers in the North Cascades and high peaks elsewhere. Mt. Mazama ash blankets much of Pacific Northwest 6800 years ago	Coastal landforms including bars, spits, stacks, and sea cliffs; sea level rise produces some submergent coastal features; large earthquakes every few 100 years; extensive coastal dunes in Oregon; measurable uplift and local subsidence continues
Alpine glaciation in higher ranges, large pluvial lakes in basins, Lake Bonneville flood ~15,000 yrs ago. Volcanism and basin and range faulting continue, lava disrupts drainage	Cordilleran glaciers in north, large mountain icecaps elsewhere, abundant alpine glaciers. Mostly basalt eruptions with localized, large andesite and dacite stratovolcanos in last few 100,000 years (Mt. Baker, Rainier, Hood, Mazama, etc.). Uplift continues	Puget Sound glacial lobe, local alpine glaciers, wave-cut terraces produced during major sea level and tectonic changes, landslide lakes form. Coast Range uplift continues, tilts eastward. Volcanic eruptions in the Portland area
Both Basin and Range faulting and silicic ("rhyolitic") volcanism (associated with caldera formation) begin ~17 Ma and migrate eastward.

Steens Mountain basalt contemporaneous with Columbia River Basalt.

Albion Mts. gneiss dome forms in SE Idaho (40–20 Ma) | Shield and cinder cone platform with large stratocones (mostly eroded now). Volcanism shifts from western Cascades to graben areas in eastern or High Cascades (~7 Ma). Sea retreats to west as uplift, westward tilting, and folding continues. Columbia River Basalt flows across area near present Columbia River Gorge. Volcanism begins in western Cascades (~40 Ma); sediments deposited into trench formed behind clockwise-rotating Coast Range block. North Cascade microcontinent docks about 50 Ma (adds rocks as old as Precambrian to North America), produce NW trending faults in Middle Cascades that control location of many present valleys | Coast Ranges uplifted, distinct topographic boundaries between troughs and mountains established, sea retreats westward

Columbia River Basalt locally flows into area.

Resistant dikes and sills (38–29 Ma) later localize high peaks in Oregon Coast Range

Coast Range block separates from OWL and rotates clockwise. Volcanic and marine deposits into subduction trench, oldest rocks about 64 Ma |

Figure 5. (continued)

some cases segments of the Columbia and some of its tributaries maintained their courses in spite of the rising anticlinal ridges although adjustment of rivers to developing structural lows or saddles occurred prior to the cutting of the numerous watergaps in the western Columbia Basin Subprovince. Notable examples include Sentinal Gap through the Saddle Mountains and Wallula Gap through the Horse Heaven Hills.

The High Lava Plains subprovince also contains abundant basalt flows as well as other volcanic features, but most are of much younger age. The occurrence of relatively recent Miocene through Holocene (Figure 6) volcanism, combined with semiarid conditions, results in a relatively undissected topography except along major rivers. Silicic volcanism migrated from the Harney Basin-Steens Mountain area (Miocene eruptions) northwestward to Newberry volcano where many volcanic rocks are younger than one million years old. A clockwise rotation occurred as western Oregon was dragged northward when the Pacific and North American plates converged obliquely; major fault systems such as the Brothers Fault Zone in the High Lava Plains resulted. Numerous cinder cones, lava flows, and other volcanic features are associated with this major structure including the spectacular 6,000 year old lava flows at Lava Butte and the 2,000 year old Big Obsidian flow in the caldera atop of Newberry shield volcano in Newberry Volcanic National Monument.

Interior drainage is restricted to the Harney-High Desert Section of Oregon. Harney Basin is located on the Brothers Fault Zone and is the largest closed depression in the province (2,050 km^2; 795 mi^2). Malheur, Harney, and Mud lakes are small remnants of the giant Lake Malheur that existed here during the Pleistocene. Subsurface drainage through the Snake River Plain aquifer as well as the dry climate contributes to the lack of dissection there. Disappearing streams along the north edge of the Snake River Plain reappear as enormous springs, with some groups discharging more than 34 m^3/s (1,200 ft^3/s) of water along the north wall of the Snake River Canyon.

Lava dams in the Malheur-Boise Section (Figure 4) caused thick sections of stream and lake sediments to accumulate and form a rolling, low-relief topography. Cinder cones, volcanic buttes, and playa lake basins occur in the Oregon part of the section.

The Snake River Plain contains a series of volcanic centers that first erupted during the Miocene (about 17 Ma) in southwest Idaho and migrated northeast through Pleistocene and Holocene times. The cycle is initiated by rhyolitic eruptions and culminates with basaltic eruptions that leave a relatively flat plain dotted with shield and cinder cones. Migration of volcanic activity to the northeast and development of Basin and Range faults appear to be linked. An appealing explanation for these relations is that the North American plate is moving westward over a geographically fixed zone of heat that emanates from deep within the earth known as a hotspot. However, explanations involving breaking or rifting of the continent along the axis of the Snake River Plain, crustal extension closely associated with the formation of Basin and Range faults to the south, or a combination of factors is also possible.

Photo 2. The Snake River flows along the Oregon-Idaho border through the nearly 2,440 m (8,000′) deep Hells Canyon, the deepest canyon in North America.

Some Special Features

Many intriguing geomorphic features and problems occur in the Columbia Intermontane Province. A major geomorphic element involves the drainage evolution of the Columbia and Snake rivers and their tributaries. Diversion of the Columbia River by lava flows and the formation of watergaps by antecedent rivers in the Columbia Basin Subprovince have already been discussed. Several studies summarize these and other drainage changes in more detail including the establishment of the Snake River through Hells Canyon (Photo 2), the deepest canyon in North America. Livingston (1928) believed the Snake River was antecedent, and Wheeler and Cook (1954) believed that headward erosion and crustal warping produced ponding and that the resulting spillover caused the river to abandon its former, more southerly course.

Pleistocene lava flows caused numerous diversions along the Snake River Plain and shallow magma contacting groundwater produced explosion craters at Hole-in-the-Ground and Fort Rock (Photo 3) in the Harney-High Desert Section. Holocene volcanism in the Columbia Intermontane Province of Oregon occurred at Newberry Volcano, Lava Butte, Diamond Craters, and Hole-in-the-Ground. In Idaho Holocene

Photo 3. Fort Rock is a tuff ring formed explosively during the early Pleistocene when rising magma contacted groundwater. Wave-cut benches along its flanks and a breach in the rim were produced when the Fort Rock basin was later flooded by a shallow Pleistocene lake.

eruptions, some as recently as 2,000 years ago, occurred in the eastern Snake River Plain along the Great Rift at Craters of the Moon and Kings Bowl and at Hells Half Acre lava field near Pocatello. The Kings Bowl Rift opened up about 2,130 years ago to a depth of at least 244 m (800′).

Today numerous landslides occur on the steeply dipping north slopes of the Yakima Fold anticlines (Photo 4). Poorly consolidated sedimentary beds located between and beneath the basalt flows and the more effective moisture of the late Pleistocene contributed to the development of the landslides. Glacial lacustrine sediments in the Columbia River valley and glacial-flood steepened slopes are particularly vulnerable. Many slopes failed during the wetter climatic regime of the Pleistocene.

Pleistocene loess up to 75 m (250′) thick in the Palouse Hills Section displays a steep, rolling topography reminiscent of very large swells at sea. Locally the topography may have a dune or wind-controlled origin, or it may be produced by dissection of a flat, loess plain. Following each of the many outburst floods or jokulhlaups that

Photo 4. Large landslide on the Toppenish Ridge anticline in the Yakima Folds Section occurred during the Holocene. The slide diverted Toppenish Creek to the north.

inundated the nearby Channeled Scabland, deposition was episodic with increased sedimentation.

Alpine glaciation was locally important in high areas in the Central Highlands where cirques, horns, and other landforms were produced. The Wallowa Creek glaciers deposited pre-Wisconsin and impressive Wisconsin moraines around Wallowa Lake, as well as a few hundred to over 6,800 year old moraines in higher cirques.

Cordilleran ice was funneled through the north-south trending Okanogan, Columbia, and Chamokane valleys in the Northern Rocky Mountains and spilled onto the north edge of the Columbia Basin Subprovince. Excellent examples of drumlins, eskers, kettle holes, and other ice sheet landforms occur in the Waterville Plateau Section.

Indirectly related to the presence of glaciers in the northeastern part of the Columbia Basin Subprovince is one of the most unique and bizarre landscapes in the world. Numerous catastrophic releases of ice-dammed glacial Lake Missoula sent gigantic floods across eastern Washington producing the complex of channels, dryfalls (Photo 5), bars, giant current dunes, and other landforms characteristic of the Channeled Scabland. The relationship of flood sediments to ash layers erupted from

Photo 5. A section of the Dryfalls complex at the head of the lower Grand Coulee was last swept by Lake Missoula floodwaters during the very late Wisconsin glaciation. The rim is over 4.8 km (3 miles) wide and about 107 m (350') high. The plunge pool lakes are as much as 27 m (90') deep.

Glacier Peak and Mount St. Helens volcanoes indicates that the last floods occurred between 13,000 and 11,200 years ago.

Another flood of epic proportions occurred about 15,000 years ago when cooler climatic conditions permitted water from Pleistocene age Lake Bonneville in the Great Salt Lake basin of northern Utah to overtop a drainage divide and spill into the Snake River valley. The upper 108 m (350') of the spillway eroded rapidly unleashing a catastrophic flood that produced scabland features that are recognizable as far downstream as Lewiston, Idaho.

A relatively small but abundant landform in the northern Columbia Basin Subprovince is the enigmatic silt or Mima mounds. Their origin has been variously attributed to Pleistocene gophers, glacial meltwater, periglacial processes and many other explanations. A recent study suggest that shock waves from earthquakes can organize thin layers of loose silt overlying a hard substratum into mound forms. If true, the presence of mounds in what is now considered to be an area of low seismic risk could have important implications.

BASIN AND RANGE—MIDDLE ROCKY MOUNTAINS PROVINCES

Topography

Southern Oregon and Idaho include parts of the Great Basin section of the Basin and Range Province, Columbia Intermontane Province, and the extreme southeastern corner of Idaho has a small extension of the Middle Rocky Mountain Province (Fig. 1). As defined in this volume, only that part of the Middle Rocky Mountain Province and the Basin and Range Province within the confines of the Oregon and Idaho borders are considered part of the Pacific Northwest.

The boundaries between the Middle Rocky Mountains, Basin and Range, and Columbia Intermontane provinces in southeastern Idaho are relatively sharp and distinct. Province boundaries in southeastern Oregon are more transitional and therefore more arbitrary.

Individual ranges of the Basin and Range in Oregon such as Walker Rim, Winter Rim, Abert Rim, Warner Peak, Hart Mountain-Poker Jim Ridge, and Steens Mountain are widely separated compared to the closely spaced ranges in southernmost Idaho. The northern topographic boundary of the Basin and Range in Oregon clearly swings around the north end of the ranges; however, the boundary between the adjacent basins and the low relief topography of the Columbia Intermontane Province is arbitrarily drawn across the north ends of the intervening closed basins without regard to drainage divides. Although the topographic boundary is transititional and imprecise, a major structural boundary exists where ranges in the Basin and Range terminate against the High Lava Plains Section of the Columbia Intermontane Province. Where the Basin and Range Province swings south into Nevada around the Owyhee Upland, the Owyhee River drainage divide is used as the boundary.

Except for the Klamath River basin in the extreme western part of the province and relatively short north-flowing streams entering the Snake River in Idaho, the drainage is internal and is part of the Great Basin which was first recognized and named by John C. Fremont in 1844. Summer, Goose, Alvord, and Lake Abert are significant playa lakes in Oregon. South-flowing streams south of the Snake River Plain enter similar playas in Utah and Nevada. Basin and valley elevations range from 1,262 m (4,139') at Klamath Lake in the west to about 1,524 m (5,000') elsewhere and mountain peaks range from 2,134 m (7,000') to as high as 3,150 m (10,335') at Cache Mountain in the Albion Range of Idaho. Steens Mountain crest (2,948 m; 9,670') is the highest point in the Basin and Range of Oregon.

The Snake River flows from Jackson Hole across the ranges near the Idaho-Wyoming border through an impressive canyon. Segments of numerous drainages in the main part of the Middle Rocky Mountain Province to the east, as well as the Snake River canyon along the Wyoming-Idaho border, were superimposed onto a mountainous topography that was buried by Tertiary-age sediment.

Geology

The edge of the North American continent was located in western Idaho and eastern Oregon during late Precambrian and Paleozoic times and shifted westward during the Mesozoic Era as plate movement added new land masses to the west edge of the continent. Late Mesozoic orogenic activity produced detachment faults that enabled thin slices of Paleozoic and Mesozoic sedimentary rocks to move tens of kilometers eastward and stack up in the Middle Rocky Mountain overthrust belt of southeastern Idaho and western Wyoming.

Erosion helped produce a low-relief topography before Pliocene time and Tertiary lava flows also helped level much of the land surface that was destined to become part of the Basin and Range Province. Crustal stretching began about 17 Ma in eastern Oregon and initially produced large silicic calderas along the Oregon-Nevada border and the voluminous Steens Mountain and Columbia River Basalt farther north. The over 12,500 km^3 (3,000 mi^3) of Steens Mountain Basalt in southeastern Oregon is unusual in that it erupted in perhaps only 50,000 years. Silicic volcanic activity on the north edge of the Basin and Range Province in Idaho was characterized by eastward-migrating explosive volcanism followed by an episode of quieter, voluminous effusions of basaltic lava. Volcanism also migrated northwest from the Steens Mountain area as indicated by progressively younger silicic volcanic rocks towards the Cascade Mountains. Numerous faults accompanied the crustal extension blocking out the present ranges in the southern Idaho and Oregon Basin and Range.

Quaternary basalts in southeastern Idaho disrupted numerous drainages including part of the Raft River which was diverted eastward. Holocene lava flows occurred in the Klamath Lake basin on the west edge of the province and at Jordan Craters where eruptions occurred as recently as 4,000 years ago. Partly based on the distribution of fossil mollusks and fish, it has been suggested that the Snake River formerly flowed through southeast Oregon to northern California. Drainage disruption by block faulting and volcanism diverted the river northward through Hells Canyon in the Columbia Intermontane Province about 2 Ma.

The late Cenozoic extensional faults characteristic of the Basin and Range are superimposed on older structures and rock units of the Rocky Mountains and lava plains of southeastern Oregon. The east-west extension or splitting of the continent began about 17 Ma in central Nevada and migrated eastward at a 3.5–6.5 cm/yr (1.38–2.56 in/yr) rate. Simultaneously, the northwest migration of silicic volcanism in southern Oregon progressed at the slower rate of 1.3–3.8 cm/yr (0.5–1.5 in/yr). Stretching of the crust in the Basin and Range may amount to 100 percent during the late Cenozoic. The currently active zone of extension is in the Intermountain Seismic Belt where the province is expanding at the expense of the adjacent Rocky Mountains.

The recency of geologic activity and the slow rates of denudation in this arid to semiarid climatic regime dictates that "interpretation of late Cenozoic tectonics requires an understanding of the morphological development of the region." An obvious corollary is the reverse; understanding the morphology of the region requires an understanding of its tectonic development.

Some Special Features

Relatively untapped geologic resources of the Basin and Range Province include the large reserves of phosphate present in the late Paleozoic age Phosphoria Formation of southeastern Idaho and areas of high geothermal heat in both Oregon and Idaho. Thinning and fracturing of the crust produced numerous thermal springs. Drill holes have encountered water at 93 °C. (200 °F.) and some, such as Hunters Hot Spring and Crump Geyser in Oregon, erupt or formerly erupted as human-produced geysers.

An elusive target so far is in the overthrust belt where Idaho has its best chance of locating a producing oil or gas well. The same age rocks and structural features in adjacent Wyoming are excellent petroleum producers. In spite of numerous exploratory wells, Idaho still lacks a producing oil or gas well.

Cooler Pleistocene intervals produced large lakes in the closed basins of the province. Shoreline remnants indicate that Pleistocene Lake Modoc covered over 2,580 km^2 (1,000 mi^2) and that Summer and Abert Lakes were connected and water in Pleistocene Lake Chewaucan was as much as 107 m (350′) deep and covered 1,190 km^2 (461 mi^2). Northern embayments of the huge Pleistocene Lake Bonneville in the Salt Lake Basin extended into southeastern Idaho. The rising lake waters finally spilled through Red Rock pass south of Pocatello about 15,000 years ago. The upper 108 m (354′) of soft rock in the spillway was rapidly eroded permitting a catastrophic torrent to rip through the Marsh Creek valley to the Portneuf valley and into the Snake River valley.

Pleistocene alpine and cirque glaciers occurred throughout the higher, and more northern ranges of the Basin and Range Province. The Kiger Canyon glacial valley at Steens Mountain in Oregon is particularly impressive because of its classic U-shaped cross sectional profile. Because of its scenic and scientific values the area has been proposed as a national park.

CASCADE MOUNTAINS PROVINCE

Topography

The province is divided into North, Middle, and South sections along distinct topographic boundaries (Figure 2). We place the boundary between the Middle and South sections along the Columbia River Gorge rather than near North Klamath Lake as Fenneman (1931) proposed. The Columbia River boundary is a distinct topographic division but may not be as geologically significant as the topographically indistinct Klamath Lake boundary. The prominent northwest-southeast oriented Olympic-Wallowa Lineament is marked by a coincidence of topographic and structural features and forms a useful boundary between the North and Middle sections. The Oregon Cascade Mountains can be further subdivided into the older Western Cascade and the younger eastern High Cascade sections along a major topographic and structural boundary.

Mountain peak elevations, exclusive of the major Pleistocene stratocones, are frequently between 2,440–2,740 m (8,000–9,000') in the North Cascade Section and less than 2,130 m (7,000') in the Middle and South Cascade sections. The Pleistocene age volcanic peaks are spaced about 80 km (50 mi) from each other and many tower as much as 1,800 m (6,000') above the surrounding mountains. Spacing reduces significantly in central Oregon near the Three Sister volcanic complex. Significant volcanic peaks include Glacier Peak and Mt. Baker in the North Cascade Section; Mt. Rainier, Mt. Adams, and Mount St. Helens in the Middle Section; and Mt. Hood, Mt. Jefferson, Mt. Washington, Three Sisters, Mt. Thielsen, Mt. Mazama, and Mt. McLoughlin in the South Cascade Section. Mt. Rainier (4,393 m: 14,410') is the highest peak in Washington and Mt. Hood (3,428 m; 11,245') is the highest in Oregon.

Rivers on the west side of the Cascade drainage divide flow into the Columbia River, except north of Mt. Rainier where they flow into Puget Sound and in southern Oregon where the Umpqua and Rogue rivers drain directly into the Pacific Ocean. East side rivers also flow into the Columbia River except south of Mt. Thielsen where they enter the internal drainage system of the Basin and Range Province. A significant drainage feature occurs along the middle-south Cascade boundary where the Columbia River flows directly through the Cascade Mountains in a spectacular gorge along the Washington-Oregon border.

Geology

Except for the North Cascade Section and a small area in the Middle Cascade Section, all of the rocks and geologic events in the Cascade Mountains are Eocene (about 50 Ma) or younger in age. The accretion of the North Cascades subcontinent about 50 Ma resulted in a westward shift of the collision or subduction zone between North America and the oceanic plate.

The North Cascade block contains rocks as old as Precambrian that were metamorphosed during the Precambrian, Paleozoic, and Mesozoic eras prior to its collision with North America. The colliding block was moving northeasterly and produced north and northwest-oriented folds and faults about 50–40 Ma. For example, faults associated with the Chiwaukum and Methow grabens and the Lake Chelan valley contribute to the prominent northwest-oriented topographic grain in the North and Middle Cascade sections.

Intense volcanism began in the western Cascade Mountains during the late Eocene and Oligocene (about 36 Ma). By Miocene time (about 17 Ma) the mountains were still relatively low and Columbia River Basalt flowed westward in a broad zone along the present Columbia River valley. By about 7 Ma volcanism was ending in the

Western Cascade Section and shifting eastward. Basalt and basaltic-andesite rocks formed a platform of overlapping shield and cinder cones upon which the imposing andesite and dacite stratovolcanoes formed during the last few 100,000 years. Although built on subsiding graben structures, the regional uplift and constructive volcanic activity far exceed the rates of erosion and graben subsidence. Numerous historic events are recorded on many of the Cascade Mountain volcanoes including the famous 1980 eruption of Mount St. Helens.

Some Special Features

The 120 km (75 mi) long Columbia River Gorge through the Cascade Mountains is North America's finest example of a transverse valley produced by an antecedent river. The Cascade crest elevations decrease markedly between Mt. Hood and Mt. Adams providing a natural corridor for water flow. A topographic low and the ancestral Columbia River were located here about 16 Ma as indicated by the distribution of distinctive gravel deposits and Columbia River Basalt flows of this same age. The lava was funnelled westward across this low area into western Oregon and Washington where it eventually reached the Pacific Ocean. The ancestral Columbia was "pushed around" by erupting volcanoes in the Columbia Gorge area during the last five Ma but was able to maintain its path across the uplifting Cascade Mountains. At one time the Columbia River flowed across the present site of the Mt. Hood stratovolcano. Floodwaters from Glacial Lake Missoula crested some 305 m (1,000') above the bottom of the gorge helping to create hanging valleys, spectacular waterfalls, and unstable walls that helped spawn the Bonneville and other large landslides. The gorge was declared a National Scenic Area in 1986 to limit environmentally and scenically destructive development. The experiment so far has had only limited success.

An excellent representation of geomorphic features associated with volcanism is readily accessible on many lands managed by the National Park Service and U.S. Forest Service. Stratovolcano morphology can be examined at Mt. Rainier (Photo 6) and Crater Lake National parks and at Mount St. Helens Volcanic Monument. Shield cones, recent lava flows, and other volcanic features are abundant in the Middle and South Cascade sections. Good access along the McKenzie Pass highway in central Oregon (Photo 7) combined with the recency of volcanic features and the sparse subalpine zone vegetation provide an excellent opportunity to observe textbook examples of volcanic landforms. A 19 km (12 mi) long blocky basalt flow from Belknap shield volcano blocked the McKenzie River 1,600 years ago. Eruption of cinder and lava from nearby Collier Cone about 400 years ago produced some of Oregon's newest pieces of real estate! Washington's youngest rock was added in 1986 on the plug dome inside the crater of Mount St. Helens.

Following the deadly 1980 eruption of Mount St. Helens, interest in the Cascade volcanoes and their associated hazards has greatly increased. Historic events along

Photo 6. Mount Rainier (4,392 m; 14,410′) is the highest and most extensively glaciated peak in the Cascade Mountains. Holocene activity includes a massive mudflow generated by failure of the summit cone about 5,700 years ago and subsquent eruptions that constructed the upper 366 m (1,200′) of the mountain. Remnants of the former cone visible here on the mountain flanks project upward and indicate that the former summit may have been as much as 610 m (2,000′) higher.

with geomorphic features indicate that most of these volcanoes have had Holocene age eruptions and are likely to erupt again. One of the world's largest known mudflows raced down the northeast flank of Mt. Rainier about 5,700 years ago into Puget Sound. A similar event today would endanger over 100,000 people.

The older, Tertiary age western Cascade volcanic rocks retain no expression of their original volcanic form. Pliocene-Pleistocene edifices such as Mt. Thielsen ("the lightning rod of the Cascades") and Mt. Washington are considered extinct partly because their flanks are deeply furrowed and greatly modified by glaciers and streams. As long as the oceanic plate continues to dive beneath North America in the subduction zone along the Washington and Oregon coasts, the Cascade landscape will continue to be dramatically modified and influence human activity.

High heatflow occurs under much of the High (East) Cascade Range. A belt of hot springs, perhaps controlled by a boundary fault, is located along the West and

Photo 7. Lava flows, shield cones, cinder cones, and other volcanic landforms as young as 400 years occur along the McKenzie Pass area. The Belknap Flow and tree-covered steptoes shown here are about 1,600 years old.

East Cascade border in Oregon. Hot spring temperatures increase towards central Oregon where waters at Belknap and Breitenbush hot springs are 82–88 °C (180–190 °F). Geothermal exploration continues with encouraging signs but no major development has occurred so far.

A mountain icecap, only broken in the Columbia River Gorge area, blanketed the high Cascades in Washington and as far south as Mt. McLoughlin in Oregon during intervals of Pleistocene glaciation. Cordilleran ice overwhelmed the margins of the North Cascade Section. The Cowlitz glacier which was spawned from the Pleistocene icefields on Mt. Rainier extended as far as 105 km (65 mi) from the peak and glaciers farther south near Mt. Mazama advanced 27 km (17 mi) from the summit. Datable materials in glacial deposits have not yet been discovered but the stratigraphic relationship of glacial and datable volcanic materials, as well as use of weathering rind measurements and other relative dating methods, has enabled workers to establish a tentative chronology. Glacial advances occurred over 616 ka but most drift is of Illinoian (370–130 ka) or late Wisconsin (30–10 ka) age. As in other high areas of the Cordillera, an early Holocene, early Neoglacial (3–2 ka), and late Neoglacial (last few hundred years) activity is recognized.

As expected, the present glacier cover on the high peaks is proportional to elevation and latitude. Ice volumes decrease rapidly from Mt. Rainier (6.7 km^3; 1.6 mi^3) to Mt. Hood (0.3 km^3; .08 mi^3) and southward to the Three Sisters (0.17 km^3; .04 mi^3) in central Oregon. Over 50% of the approximately 750 glaciers in the North Cascade Section were expanding from 1944–1976 but by 1977 all were in retreat due to a 15% decrease in winter precipitation and a 1.0 °C (1.5 °F) rise in average summer temperature. Recently it has been predicted that if global warming continues, 690 of the 750 North Cascade glaciers will disappear during the next century with an accompanying decrease in surface runoff and water supplies.

PACIFIC BORDER PROVINCE

Topography

The area between the Cascade Mountains and the Pacific Ocean is divisible into the Puget Trough, Willamette Trough, Olympic Mountains, Willapa Hills, Oregon Coast Range, and Klamath Mountain sections (Fig. 2). The Puget Trough Section is relatively flat and is mostly less than 150 m (490′) above sea level. The myriad of channels and islands in Puget Sound represents a glaciated topography drowned by the approximately 125 m (410′) rise of sea level since the late Wisconsin glacial maximum. The main channels permit ocean-going vessels to enter one of the finest natural harbors on the west coast.

The Puget Trough narrows and ends just south of Olympia but related structural features continue south and control the location of parts of the Chehalis, Cowlitz, and Columbia rivers. The Cowlitz River flows south and enters the Willamette Trough Section near Longview, Washington.

The Willamette Trough Section is about 120 m (400′) in elevation at its south end and is near sea level at Portland at the north end. The Willamette River drains the relatively flat Oregon part where about 70% of the states population resides. The Willamette River is a relatively low energy, underfit stream that merely occupies a lowland created by structural processes rather than a broad valley created by its own erosive power.

The Coast Range is divisible into the Olympic Mountains, Willapa Hills, Oregon Coast Range, and Klamath Mountains sections. From north to south, the Chehalis, Columbia, and Coquille-upper Umpqua rivers are designated here as the boundaries. Rugged, high relief topography characterizes both the north and south ends of the Coast Range and a wide variety of coastal landforms including headlands, sea stacks, wave-cut terraces, spits, and sand dunes produce a diverse landscape.

The Olympic Mountain Section forms the nucleus of Olympic National Park. Its northerly location, with elevations up to 2,425 m (7,954′), promoted extensive Pleistocene glaciation and currently supports about 60 modern cirque and valley glaciers. Like spokes on a wheel, rivers drain radially from the central uplift into the large bodies of water around the peninsula (Fig. 1). Maritime air masses produce over 350 cm (140″) of yearly precipitation and create a unique temperate rain forest condition on the west slope.

Elevations in the Willapa Hills Section are mostly less than 610 m (2,000′) with Boistfork Peak in the southeast reaching 948 m (3,110′). West-flowing drainages are longer reflecting steeper stream gradients to the ocean and heavier precipitation on western slopes.

The Oregon Coast Range Section also has longer streams on the western slopes. Elevations are higher than the Willapa Hills with Mary's Peak just west of Corvallis reaching an elevation of 1,249 m (4,097′).

Mountain peaks in the Klamath Mountains Section rise from 762 m (2,500′) near the coast to 2,296 m (7,530′) at Mt. Ashland on the east side. In addition to higher elevations, the coast range geology drastically changes in the Klamath Mountains. The Rogue River is antecedent across the late Cenozoic Klamath Mountains uplift and flows through a spectacular gorge noted by boaters for its challenging rapids.

Geology

The oldest rocks in western Oregon are found in the Klamath Mountain Section of the Coast Range. A linear belt of Paleozoic-Triassic marine rocks is in thrust contact with a belt of Jurassic age rocks. These rock masses were added to the western edge of North America along the northwest-trending Mesozoic plate collision zone. Basaltic oceanic crust was caught in the plate tectonic vise and granitic plutons similar in age and origin to the Sierra Nevada batholith occur throughout the range.

The Oregon Coast Range, Willapa Hills, and Olympic Mountain sections contain Eocene to Miocene volcanic and coastal plain rocks. Collision with a volcanic island chain and clockwise rotation of the entire Coast Range to its present north-south orientation occurred during the last 50 million years. The distribution of Cenozoic sedimentary rocks records a continually shrinking marine embayment in western Oregon and Washington. By late Miocene-Pliocene time the coast was close to its present position.

Large areas of the coast range contain detached segments of oceanic plate overlain by relatively thin marine sediments and underlain by thick masses of sedimentary materials that were jammed into a former trench between colliding plates. The thick sedimentary trench filling eventually uplifted the coast range and is well exposed in the core of the Olympic Mountains. The basaltic oceanic crust wraps around the Olympic Mountains like a giant horseshoe open to the sea.

The massive Columbia River Basalt flowed into western Washington and Oregon near the present Columbia River valley region. Flow remnants occur as far south as Newport at Seal Rocks along the Oregon coast and perhaps as far south as Eugene in the Willamette Trough. Late Pliocene and Pleistocene eruptions of the Boring lavas produced about 100 small cinder cones and other volcanic vents in the Portland area.

Late Cenozoic uplift and eastward tilting of the Coast Range greatly emphasized the structural and topographic divisions between the low-lying troughs, the Coast Range, and the Cascade Mountains. Stream incision and cutting of marine terraces occurred in response to Pleistocene sea level changes and tectonic adjustments that continue today. Pleistocene marine terraces are uplifted and complexly warped. The oldest recognizable wave-cut terrace in Oregon is about 490 m (1,660') above sea level and is about 1.0 million years old. Younger, more distinct terraces are seaward of the older terraces.

Some Special Features

The abundance of sedimentary rocks of shallow marine origin ensures that some organic materials, and therefore some oil and gas, is present. Oil and gas seeps, and the presence of "smell muds" in the Olympic Mountains Section and hundreds of dry holes in the Coast Range area, many with oil and gas shows, have frustrated oil drillers as far back as 1901. The discovery of the Mist gas field in 1979 along the Newhalem River in Oregon has provided new inspiration.

Coal beds, mostly Eocene in age, formed in former coastal swamps and have been extensively mined in the Coos Bay and Eden Ridge areas in Oregon and the Chehalis area in Washington. Shore processes produce "black sands" which are concentrates of sand-size mineral particles with high densities. Local mining of these concentrates along streams and on elevated terraces, especially in Oregon before World War Two, produced gold, platinum, chromite, and other minerals.

Continuing plate movement adds crustal stresses that produce uplifting as high as 2.5 cm/yr (1'/yr) in the Cape Blanco area of Oregon and 2.5 cm/15 yrs (1'/15 yrs) in the Neah Bay area of Washington. Other areas such as Seattle and Vancouver are sinking by as much as 2.5 cm/yr (1'/yr). More rapid release of stresses can produce damaging earthquakes such as the 1949 (M=7.1) and 1965 (M=6.5) earthquakes in the Puget Trough. Atwater describes disturbing geologic evidence that great earthquakes, with magnitudes perhaps as large as 8 or 9, occurred at least three times during the past 2,000 years in western Washington and Oregon. The last major great earthquake occurred about 300 years ago, prior to settlement by Caucasians. Thus, although great earthquakes have not occurred historically, the area will likely experience a devastating earthquake in the future.

Mima Prairie in the Puget Trough is the type of locality of the enigmatic silt hummocks known as Mima mounds. These features are about 1 to 3 m high and 3 to 15 m in diameter and occur in large concentrations where a shallow silt or fine sediment deposit overlies gravel or other hard substratum. Numerous hypotheses including extinct Pleistocene gophers, wind, periglacial, or combination processes have been proposed. The concentration of mounds in many areas of known seismic risk such as the Puget Sound and the ease at which these features can be experimentally produced by mechanical impacts led Berg (1990) to propose a seismic origin for these intriguing landforms.

Extensive coastal dune belts, especially along the Oregon Coast from Florence to Coos Bay, owe their sand source to sea-cliff erosion and the Columbia and other coastal rivers. Longshore currents seasonally bring sediment southward where offshore winds blow the sand inland. A chain of freshwater lakes including the 7,812 hectare (3,164 acre) Siltcoos Lake south of Florence is contained in the dune belt. A number of state parks and the Oregon Dunes National Recreation Area make excellent sites to examine these dunes although many areas are becoming stabilized because of a lessened sediment supply due to hydropower dams on the Columbia River and importation of the tenacious European beach grass.

A wide variety of coastal landforms occur along the west edge of the Coast Range. Emergent coast features such as stacks, headlands, and marine terraces are mixed with submergent coastal features including embayments, spits, baymouth bars, coastal lakes, and other features (Photo 8). Although tectonic uplift continues, submergent features developed since the end of last glaciation due to the slow but persistent rise in sea level. Sea cave development produced the Depoe Bay spouting horn and collapsed cave features such as the Devils Punchbowl and natural bridges. Sea cliff formation encourages numerous landslides along Puget Sound and coastal areas. Commercial development on spits and other ephemeral features continues to produce significant property and environmental losses.

Cordilleran glaciation in the Puget Trough Section left an excellent stratigraphic record that provides the best-dated glacial sequence in western North America. Pre Wisconsin drift is between 2.48–0.7 Ma and the Salmon Springs drift has some finite dates of about 840 ka. Radiocarbon dates indicate that early and middle Wisconsin drift was deposited between 70–35 ka and late Wisconsin glaciation occurred from 30–10 ka. Alpine glaciation was extensive in the Olympic Mountains and restricted to the high peaks elsewhere in the Coast Range.

About 1,000 m (3,000') of glacial debris fills the Puget Trough and numerous lake basins, moraines, glaciolacustrine sediments, and other features were produced by the invasion of continental ice. Water ponded in the south end of the Puget Trough drained south through the now underfit Chehalis River valley.

The Pleistocene course of the Willamette River was significantly influenced by lava flows, small volcanoes, alluviation, and recent folding and faulting. The catastrophic Lake Missoula floods sweeping out of the Columbia River gorge deposited coarse material in the Portland area, icebergs carrying house-size boulders up to 145 metric tons (160 tons) as far south as Salem, and the rhythmically bedded Willamette Silt in the Willamette Trough.

Photo 8. Coastal landforms including sea stacks, cliffs, arches, and wave-cut benches occur along the Olympic coast. Similar features are found along parts of the Oregon coast.

REFERENCES

Allen, John E., 1979. The magnificent gateway; a layman's guide to the geology of the Columbia River gorge. Timber Press, Portland, Ore., 144 p.

———, 1988. Volcanic hazards from High Cascade peaks. Oreg. Geol., v. 50, p. 56–63.

Allison, Ira S., 1982. Geology of pluvial Lake Chewaucan, Lake County, Oregon. Oregon State Univ., Studies in Geology, 11, 78 p.

Allmendinger, R.W., 1982. Sequence of late Cenozoic deformation in the Blackfoot Mountains, southeastern Idaho. Idaho Bur. Mines and Geol. Bull. 26, pp. 505–516.

Alt, David, and Hyndman, D. W., 1978. Roadside geology of Oregon. Mountain Press, Missoula, Montana, 272 p.

———, 1984. Roadside geology of Washington. Mountain Press, Missoula, Montana, 282 p.

———, 1986. Roadside geology of Montana. Mountain Press, Missoula, Montana, 427 p.

———, 1989. Roadside geology of Idaho. Mountain Press, Missoula, Montana, 393 p.

Armstrong, R.L., Harakal, J.E., Brown, E.H., Bernardi, M.L., and Rady, P.M., 1983. Late Paleozoic high-pressure metamorphic rocks in northwestern Washington and southwestern British Columbia: The Vedder complex. Geol. Soc. Amer. Bull., v. 94, pp. 451–458.

Atwater, Brian F., 1992. Geologic evidence for earthquakes during the past 2000 years along the Copalis River, southern coastal Washington. Jour. Geophys. Res., v. 97, pp. 1901–1919.

Baldwin, Ewart M., and Perttu, R., 1980. Paleogene stratigraphy and structure along the Klamath borderland, Oregon. Geologic Field Trips in western Oregon and southwestern Washington. Oreg. Dept. Geol. and Min. Ind. Bull. 101, pp. 9–38.

Barksdale, Julian D., 1975. Geology of the Methow Valley, Okahogan County, Washington. Washington Division of Geology and Earth Resources, Bulletin 68.

Bennett, E.H., Siems, P.L., and Constantopoulos, J.T., 1989. The geology and history of the Coeur d'Alene mining district, Idaho. *In*, Chamberlain, Valerie E., Breckenridge, Roy M., and Bonnichsen, Bill, eds., Guidebook to the geology of northern and western Idaho and surrounding area. Idaho Geol. Surv. Bull. 28, pp. 137–156.

Berg, Andrew, 1990. Formation of Mima mounds: A seismic hypothesis. Geol., v. 18, pp. 281–284.

Billingsley, P.R., and Locke, A., 1939. Structure of ore deposits in the continental framework. American Institute of Mining and Metallurgical Engineers, 51 p.

Bretz, J Harlan, 1923. The channeled scablands of the Columbia Plateau. Jour. Geol., v. 77, pp. 505–543.

———, 1969. The Lake Missoula flood and the channeled scabland. Jour. Geol., v. 77, pp. 505–543.

Bussaca, Alan, J., 1991. Loess deposits and soils of the Palouse and vicinity. *In*, Morrison, Roger B., ed., Quaternary nonglacial geology; conterminous U.S. Geol. Soc. of Amer., The Geology of North America, v. K-2, pp. 216–228.

Christiansen, Robert L., 1989. The Yellowstone Plateau-Island Park region. *In*, Ruebelmann, Kerry L., ed., Snake River Plain-Yellowstone volcanic province. Amer. Geophys. Union, Field trip guidebook T305, pp. 14–37.

Chitwood, L.A., Jensen, R.A., and Groh, E.A., 1977. The age of Lava Butte. The Ore-Bin, v. 39, pp. 157–164.

Christiansen, Robert L., and McKee, Edwin H., 1978. Late Cenozoic volcanic and tectonic evolution of the Great Basin and Columbia intermontane regions. *In*: Smith, Robert, and Eaton, Gordon P., eds., Cenozoic tectonic and regional geophysics of the western Cordillera. Geol. Soc. Amer., Memoir 152, pp. 283–311.

Connors, John A., 1976. Quaternary history of northern Idaho and adjacent areas. Ph.D. thesis, University of Idaho, 504 p.

Covington, H.R., Whithead, R.L., and Weaver, J.N., 1985. Ancestral canyons of the Snake River; Geology and hydrology of canyon-fill deposits in the Thousand Springs area, south-central Snake River Plain, Idaho. Geol. Soc. of Amer., Rocky Mountain Section, 38th Annual Meeting Guidebook, 30 p.

Crandell, Dwight R., 1965. The glacial history of western Washington and Oregon. *In*, Wright, H.E. jr., and Frey, D.G., eds., The Quaternary of the United States. Princeton University Press, Princeton, N.J., pp. 341–353.

———, 1967. Glaciation at Wallowa Lake, Oregon. U.S. Geol. Survey, Prof. Paper 575-C, pp. C145–C153.

Crone, A.J., 1987. Surface faulting associated with the 1983 Borah Peak earthquake at Doublespring Pass Road, east-central Idaho. *In*, Beus, S.S., Geol. Soc. Amer., Decade of North American Geology, v. 6, pp. 95–98.

Dalquest, W.W., and Scheffer, V.B., 1942. The origin of the Mima mounds of western Washington. Jour. of Geol., v. 50, pp. 68–84.

Driedger, Carolyn L., and Kennard, Paul M., 1986. Ice volumes on Cascade volcanoes: Mount Rainier, Mount Hood, Three Sisters, and Mount Shasta. U.S. Geol. Surv. Prof. Paper 1365, 28 p.

Easterbrook, Don J., 1986. Stratigraphy and chronology of Quaternary deposits of the Puget Lowland and Olympic Mountains of Washington and the Cascade Mountains of Washington and Oregon. *In*, Sibrava, V., Bowen, D.Q., and Richmond, G.M., Quaternary glaciations in the northern hemisphere. Pergamon Press, New York, pp. 145–159.

Fenneman, Neville M., 1917. Physiographic divisions of the United States. Association of American Geographers, Annals, v. 6, pp. 19–98.

Freeman, O.W., Forrester, J.D., and Lupher, R.L., 1945. Physiographic divisions of the Columbia Intermontane Province. Association of American Geographers, Annals, v. 35, pp. 53–75.

Graf, William L., ed., 1987. Geomorphic systems of North America. Geol. Soc. Amer. Centennial Special Volume 2, 643 p.

Greeley, Ronald, 1987. The Snake River Plain. *In*, Graf, W.L., ed., Geomorphic systems of North America. Geol. Soc. Amer., Centennial Special Volume 2, pp. 444–464.

Gresens, Randall L., 1982. Early Cenozoic geology of central Washington state: 1. Summary of sedvimentary, igneous, and tectonic events. Northwest Science, v. 56, no.3, pp. 218–229.

Harris, Stephen L., 1988. Fire mountains of the west: The Cascade and Mono Lake volcanoes. Mountain Press Pub. Co., Missoula, Mont., 379 p.

Hunt, Charles B., 1974. Natural regions of the United States and Canada. W.H. Freeman, San Francisco, 725 p.

Janda, Richard J., 1969. Age and correlation of marine terraces near Cape Blanco, Oregon. Geol. Soc. America Abstracts for 1969, Pt.3, pp. 29–30.

Kays, M.A., 1992. Geologic guide for the northern Klamath Mountains—Part 1, Cow Creek to Red Mountain. Oreg. Geol., v. 54, no.2, pp. 27–33.

Keller, Sarah A.C., ed., 1981. Mount St. Helens, One Year Later. Symposium Proceedings, East. Wash. Univ. Press, Cheney.

———, 1986. Mount St. Helens, Five Years Later. Symposium Proceedings, East. Wash. Univ. Press, Cheney.

Kiver, Eugene P., 1974. Holocene glaciation in the Wallowa Mountains, Oregon. *In*, Mahaney, William C., Quaternary environments, Proceedings of a symposium. Geographic Monograph no.5, York University, Canada, pp. 169–195.

———, 1982. The Cascade volcanoes—comparison of geologic and historic records. *In*, Keller, Sarah A.C., ed., Proceedings of the Mount St. Helens one year later symposium. Eastern Washington University Press, pp. 3–12.

———, and Stradling, Dale F., 1986. Lake Roosevelt shoreline study—glacial geology, terraces, landslides, and lineaments. U.S. Bureau of Reclamation, Grand Coulee Project Office, 96 p.

———, Stradling, Dale F., and Moody, Ula, L., 1989. Glacial and multiple flood history of the northern borderlands. *In*, Joseph, Nancy L., ed., Geologic guidebook for Washington and adjacent areas. Washington Division of Geology and Earth Resources Information Circular 86, pp. 321–335.

Komar, Paul D., 1992. Ocean processes and hazards along the Oregon coast. Oreg. Geol., v. 54, no.1, pp. 3–19.

Livingston, D.C., 1928. Certain topographic features of northeastern Oregon and their relation to faulting. Jour. Geol., v. 36, pp. 694–708.

Luedke, Robert G., and Smizth Robert L., 1991. Quaternary volcanism in the western conterminous United States. *In*, Morrison, R.B., ed., Quaternary nonglacial geology; Conterminous U.S. Geol. Soc. Amer., The Geology of North America, v. K-2, pp. 75–92.

Mabey, Don R., 1982. Geophysics and tectonics of the Snake River Plain, Idaho. *In*, Bonnichsen, Bill, and Breckenridge, Roy M., Cenozoic geology of Idaho. Idaho Bureau of Mines and Geology Bull. 26, pp. 139–153.

Malde, Harold E., 1991. Quaternary geology and structural history of the Snake River Plain, Idaho and Oregon. *In*, Morrison, Roger B., ed., Quaternary nonglacial geology; Conterminous U.S. Geol. Soc. Amer., Geology of North America, v. K-2, pp. 251–281.

McDowell, Patricia F., 1987. Geomorphic processes in the Pacific coast and mountain system of Oregon and Washington. *In*, Graf, W.L., ed., Geomorphic systems of North America. Geol. Soc. Amer., Centennial Special Volume 2, pp. 539–549.

McKee, Bates, 1972. Cascadia, the geologic evolution of the Pacific Northwest. McGraw-Hill, New York, 394 p.

Misch, Peter, 1980. Involvement of ancient continental basement and of alpine peridotite in complex imbricate zone beneath blueschist-facies Shuksan thrust plate, northwest Cascades, Washington. Geol. Soc. Amer., Abstracts with Programs, v. 12, p. 142.

Morrison, Roger B., 1991. Quaternary stratigraphic, hydrologic, and climatic history of the Great Basin, with emphasis on Lakes Lahontan, Bonneville, and Tecopa. *In*, Morrison, R.B., ed., Quaternary nonglacial geology; Conterminous U.S. Geol. Soc. Amer., v. K-2, pp. 283–320.

Mullineaux, Donal R., Wilcox, Ray E., Ebaugh, Walter F., Fryxell, Roald, and Rubin, Meyer, 1978. Age of the last major Scabland flood of the Columbia Plateau in eastern Washington. Quat. Res., v. 10, pp. 171–180.

Newcomb, R.C., 1952. Origin of the Mima mounds, Thurston County region, Washington. Jour. Geol., v. 60, pp. 461–472.

Nickman, Rudy, 1979. The palynology of Williams Lake fen, Spokane County, Washington. M.S. thesis, Eastern Washington Univ., 71 p.

Oldow, J.S., Bally, A.W., Ave Lallemant, H.G., and Leeman, W.P., 1989. Phanerozoic evolution of the North American Cordillera; United States and Canada. *In*, Bally, A.W., and Palmer, A.R., eds., The Geology of North America—an overview. Geol. Soc. Amer., The Geology of North America, v. A.

Orr, Elizabeth L., Orr, William N., and Baldwin, Ewart M., 1992. Geology of Oregon. Kendall-Hunt, Dubuque, Iowa, 254 p.

Palmer, Stephen P., and Lingley Jr., William S., 1992. Petroleum geology of the Washington continental margin. Wash. Div. Geol. and Earth Res. Newsletter, v. 17, no.1, p. 22–30.

Pardee, J. T. 1942. Unusual currents in glacial Lake Missoula, Montana. Geol. Soc. Amer. Bull., v. 53, pp. 1569–1600.

Pelto, Mauri S., 1991. North Cascade glaciers; Their recent behavior. Dept. of Env. Sci., Nichols College, Maine, 17 p.

Pierce, Kenneth L., Obradovich, J.D., and Friedman, I., 1976. Obsidian hydration correlation and dating of Bull Lake and Pinedale glaciations near West Yellowstone, Montana. Geol. Soc. Amer. Bull., v. 87, pp. 703–710.

Prinz, Martin, 1970. Idaho rift system, Snake River Plain, Idaho. Geol. Soc. Amer. Bull., v. 81, pp. 941–947.

Qamar, A.I., and Stickney, M.C., 1983. Montana earthquakes 1869–1979; historical seismicity and earthquake hazard. Montana Bureau of Mines and Geology, Memoir 51, 79 p.

Raisz, Erwin, 1945. The Olympic-Wallowa lineament. Am. Jour. Sci., v. 243–A, Daly Volume, pp. 479–485.

Rau, Weldon W., 1980. Washington coastal geology between the Hoh and Quillayute rivers. Wash. Div. of Geol. and Earth Res. Bull. 72, 57 p.

Richmond, Gerald M., 1986. Tentative correlation of deposits of the Cordilleran ice-sheet in the Northern Rocky Mountains. *In*, Sibrava, V., Bowen, D.Q., and Richmond, G.M., Quaternary glaciations in the northern hemisphere. Pergammon Press, New York, pp. 129–144.

Ringe, Donald, 1970. Sub-loess basalt topography in the Palouse Hills, southeastern Washington. Geol. Soc. of Amer. Bull., v. 81, pp. 3049–3060.

Schmidt, R.G., 1986. Geology, earthquake hazards, and land use in Helena area, Montana—a review. U.S. Geol. Surv. Prof. Paper 1316, 64 p.

Schmidt, C., Aram, R., and Hawley, D., 1987. The Jefferson River canyon area, southwestern Montana. *In*, Beus, S.S., ed., Geol. Soc. Amer. Centennial Field Guide, v. 2, pp. 63–68.

Scott, William E., McCoy, W.D., Shroba, R.R., and Rubin, M., 1983. Reinterpretation of the exposed record of the last two cycles of Lake Bonneville, western United States. Quat. Res., v. 20, pp. 261–285.

———, Gardner, Cynthia A., and Johnston, David A., 1990. Field trip guide to the central Oregon High Cascades. Part 1: Mount Bachelor-South Sister area. Oregon Geol., v. 52, no.5, pp. 99–114.

Smith, Robert B., and Sbar, M.L., 1974. Contemporary tectonics and seismicity of the western United States with emphasis on the Intermountain seismic belt. Geol. Soc. Amer. Bull., v. 85, pp. 1205–1218.

Stradling, Dale F., 1980. The geomorphic history of the Columbia River system. *In*, Stradling, D.F. —ed., The Columbia River: an inexhaustible resource? Northwest Institute for Advanced Studies, East. Wash. Univ., pp. 1–24.

———, and Kiver, Eugene P., 1986. The significance of volcanic ash as a stratigraphic marker for the late Pleistocene in northeastern Washington. *In*, Keller, Sarah A.C., ed.,1986. Mount St. Helens; Five years later. Eastern Washington Univ. Press, pp. 120–126.

Stump, Edmund, 1992. The Ross orogen of the Transantarctic Mountains in light of the Laurentia-Gondwana split. Geol. Soc. Amer., GSA Today, v. 2, no.2, pp. 25–31.

Taylor, Edward M., 1981. Central High Cascade roadside geology, Bend, Sisters, McKenzie Pass, and Santiam Pass, Oregon. U.S. Geol. Survey, Circular 838, pp. 55–58.

Thornbury, William D., 1965. Regional geomorphology of the United States. John Wiley and Sons, New York, 609 p.

Tolan, Terry L., Reidel, Stephen P., Beeson, Marvin H., Anderson, James L., Fecht, Karl R., and Swanson, Donald L., 1989. Revisions to the estimates of the areal extent and volume of the Columbia River Basalt Group. *In*, Reidel, S.P., and Hooper, P.R., eds., Volcanism and tectonism in the Columbia River flood-basalt province. Geol. Soc. Amer., Special Paper 239, pp. 1–20.

U.S. Geological Survey, 1964. The Hebgen Lake, Montana earthquake of August 17, 1959. U.S. Geol. Surv. Prof. Paper 435, 242 p.

Vink, G.E., Morgan, W.J., and Vogt, P.R., 1985. The Earth's hot spots. Scient. Amer., v. 252, no.4, pp. 50–57.

Waitt, Richard B., and Swanson, Donald B., 1987. Geomorphic evolution of the Columbia Plain and river. *In*, Graf, W.L., ed., Geomorphic systems of North America. Geol. Soc. Amer., Centennial Special Volume 2, pp. 403–416.

Washburn, A.L., 1988. Mima mounds; an evaluation of proposed origins with special reference to the Puget Lowlands. Wash. Div. of Geol. and Earth Res., Rep. of Inv. 29, 53 p.

Wells, Merle W., 1961. History of mining in Idaho. *In*, Idaho's Mineral Industry. Idaho Bur. of Mines and Geol. Bull. 18, p. 9–17.

Wheeler, Harry E., and Cook, Earl F., 1954. Structural and stratigraphic significance of the Snake River capture, Idaho-Oregon. Jour. Geol., v. 62, pp. 526–536.

Zoback, M.L., and Thompson, G.A., 1978. Basin and range rifting in northern Nevada—clues from a mid-Miocene rift and its subsequent offsets: Geol., v. 6, pp. 111–116.

CLIMATE

Chapter 3

Claude W. Curran
Southern Oregon State College

INTRODUCTION

Pacific Northwest climates are more varied and complex than in most regional areas in the United States. Contrasts in climate within the region are controlled in varying degrees by a variety of factors. All of the controls in combination, or separately, produce climatic variations from north to south, and especially from west to east. Significant factors controlling climate are latitude, proximity to the Pacific Ocean and ocean currents, landform barriers and altitude, and the presence of semi-permanent cells of high and low pressure.

There are few regions in the United States that display as varied climate patterns as the Pacific Northwest. In the nearly 800 miles from the shores of Pacific Ocean in southwestern Oregon to southeastern Idaho, climate can range from mild oceanic in the west to continental in the east. A transect from west to east along the Canadian border reveals a similar variation of climates; while north-south transects illustrate much less variation from the Canadian border to the southern limits of the Pacific Northwest along the 42nd parallel. How can these patterns be accounted for? Why is there more west to east variation than there is north to south?

In order to understand the complexity of climate in the region it is necessary to consider the climatic controls that help to delineate the Pacific Northwest as a unique geographic region.

The Controls

Latitudinal position (latitude) is the primary control of most climates, and the Pacific Northwest is no exception. The 45th parallel divides the region into two slightly unequal halves: of no particular importance in itself, but the fact that the Pacific Northwest lies astride the 45th parallel half-way between the equator and the north pole is important. The length of daylight period and subsequent receipt of solar insolation and its distribution are critical. Daylight periods are approximately 15 hours in length during the high sun period (June 21, the summer solstice) and approximately 9 hours in length during the low sun period (December 21, the winter solstice).

The primary effect of latitude is to control temperature regimes by producing seasonal variation in the amount of solar energy received. It becomes apparent, however, that the amount of solar radiation available for reception at the earths' surface does not vary greatly between the Canadian border and the southern margin of the Pacific Northwest. Other climatic controls create differences within the region depending upon such factors as altitude, proximity to the Pacific Ocean, and daily weather phenomena such as cloud or fog cover.

Temperature

Average annual temperatures display rather uniform levels when weather records are examined from stations located along an longitudinal line extending from north to south. For example, in the extreme southwestern corner of the Pacific Northwest at Brookings, Oregon, the average annual temperature is 53 °F. Bellingham, Washington, which is about 550 miles north of Brookings, has an average annual temperature of 49 °F. Average annual temperature is a crude measurement because the high monthly average tends to cancel out the low monthly average. A better picture of the actual difference is achieved when the range between those two months is also considered. In the case of Brookings the warmest monthly average, in September, is 59 °F, and for Bellingham, in July, is warmest reaching 61 °F. The coolest month for both is January with 47 °F in Brookings and 37 °F in Bellingham. The annual range in temperature, the difference between the warmest and coolest months in temperature for each location, is 12 °F in Brookings and 24 °F in Bellingham. This variation in range can be explained in part by latitudinal differences and other controlling factors including ocean currents, fog, and cloud cover.

At more continental locations, average annual temperatures are reduced from coastal areas but not by as much as might be expected. Twin Falls, Idaho, is approximately 500 miles east of Brookings and only a few miles further north, yet average annual temperature is 50 °F, only 3 °F less than Brookings. The annual range in temperature, however, is 46 °F ranging from 73 °F in July to 27 °F in January. In the Idaho Panhandle, Kellogg also has a 46 °F range in temperature: July at 66 °F is several degrees cooler than Twin Falls. January is the coldest month in Kellogg averaging 20 °F which is 7 °F cooler than Twin Falls. When compared to Bellingham, Kellogg's interior location accounts for almost twice the range: Bellingham 24 °F and Kellogg 46 °F. This is a clear demonstration of the effect of continentality for Kellogg and proximity to the Puget Sound/Pacific Ocean at Bellingham.

The effects of latitude as a constant on climate is less pronounced in the Pacific Northwest than at similar locations in the continental interior. It is apparent, however, that other maritime or continental influences produce a more pronounced temperature variation than that which occurs when simply taking into account the latitudinal expanse of the Pacific Northwest.

Oceanic Influences

The effects of the Pacific Ocean on the Pacific Northwest are, of course, very important. The north Pacific current bifurcates in the eastern Pacific several hundred miles offshore at about 40 ° latitude. One branch of the current nearly parallels the coast as it circulates northward out into the Gulf of Alaska; thus the Alaska current transports warm water along the western border of the Pacific Northwest, British Columbia, and Alaska. The southerly flowing California current transports water southward along the coasts of California and Baja California transporting cool ocean waters to the subtropics.

The ocean currents transport massive amounts of energy from the western Pacific to the west coast of North America. Ocean currents, at least in part, then distribute that energy both north and south. Obviously this is one of the reasons that Brookings and Bellingham display the closeness in similarity of temperature regimes.

As the ocean currents approach the coast, an upwelling of cold water occurs, which reduces the sea-surface temperature along shore. Air flowing across the water is cooled, characteristically producing fog which persists along the entire coast throughout most of the year. Coastal locations tend to display uniform temperature characteristics because of the cooler water and especially because of the fog, which reduces the receipt of solar energy. Fall and spring seasons at coast locations experience less fog persistence than during summer and winter. Lessening of fog at these times is generally due to similarity in regional air pressures between the land mass and the Pacific and because air temperatures over both land and water are equalized. Transient low pressure systems and associated storms also dissipate fog during their duration of influence, which is ordinarily from October through April or May. And, finally, occasional periods of a few days or more are fog free during both winter and summer when offshore airflow occurs. Usually such a flow is produced by stagnant intense high pressure systems extending over the northern Great Basin into the northern Rockies. Air flowing off the south and southwestern margins of the high either drives the fog westward out to sea or directly evaporates it into the drier air.

An excellent example of this phenomenon is the so-called "Brookings effect" at Brookings in southwest Oregon. Air from the interior is channeled down the Chetco River Canyon which meets the Pacific at Brookings-Harbor. The warm, dry air clears the coastal atmosphere of fog and pollutants, creating conditions where the beauty of the Pacific and coastal mountains may be fully appreciated due to the clarity of the atmosphere. It is not uncommon during these conditions in summer to have the mercury soar into the 100 °F range! That is a marked contrast from the fog-controlled daily temperature that ordinarily would be in the high 50 °s or 60 °s (F). The effect usually does not produce such a dramatic temperature change in winter, but clear atmospheric conditions prevail. On occasion a winter day under such conditions may experience a temperature into the mid-to-high 70 °s.

Landforms

Major mountain ranges in the Pacific Northwest have a pronounced effect on the weather patterns and climatic regimes found in the region. Perhaps the most significant factor is the north-south trend of the mountains. (See Chapter 2, Landforms) Coastal mountains, despite their relative low altitudes, still are a significant barrier over which moisture-laden winds from the Pacific must rise. Generally the mountains are below 4,000 feet in elevation except the Olympic Peninsula where elevations range from 4,000 to 8,000 feet above sea level, culminating in a mountainous knot.

Further inland the Cascade Range extends the full length of the Pacific Northwest and beyond. Most passes across the Cascades are above 4,000 feet elevation with the general summit level being above 5,000 feet. Major volcanic cones such as Mt. Baker, Mt. Rainier, Mt. Adams, Mt. Hood, and The Three Sisters range in elevation from 10,000 feet to a little over 14,000 feet above sea level. Finally, the northern Rockies trend southeastward from northeastern Washington to southern Idaho. Interim mountain ranges such as the Blue and Wallowa mountains and fault block mountains including Hart and Steens all produce more localized conditions.

The weather patterns and climatic attributes of northwest mountain ranges are similar to those associated with significant ranges found around the world. There are few places, however, where three mountain ranges parallel one another aligned generally in a north-south orientation along the west coast of a continent in the middle latitudes. This is a very important fact, primarily because of a latitudinal position where the prevailing winds are in a westerly direction. Atmospheric disturbances are blocked by the three ranges resulting in pronounced sub-regional patterns of temperature and moisture.

The Cascade Range is an effective barrier blocking both the flow of marine air to the interior as well as continental air from the interior toward the Pacific Borderlands. Wintertime continental air masses situated between the Cascade Range and the northern Rockies are seldom as bitterly cold as regions east of the Rockies situated in the northern Great Plains. Without the Rocky Mountains arctic air from northern Canada could easily invade much farther west than it does now, despite the prevailing westerlies. The flow of cold air toward the Pacific Coast would occur only occasionally, however, the result would be unusually low temperatures. Summers are less effected because there is far less temperature contrast over the land mass than that occurring during the winter season.

The only significant physiographic break in the coastal ranges and the Cascade Range is the Columbia Gorge. The Gorge is a conduit for the flow of air either east or west up or down the Gorge and produces substantial localized effects. During the winter season, cold air of continental origin from the interior flows westward down the Columbia bringing unusually cold weather to metropolitan Portland, Vancouver, and other areas located in proximity to the Gorge. As a result, Portland, Oregon, and Vancouver, Washington, and surrounding areas experience occasional weather characterized by cold easterly wind, freezing rain or snow, and greatly reduced daily temperatures.

Conversely, warm, moist marine air flows eastward up the Columbia Gorge creating a narrow zone along the river almost to the Idaho border that is warmer than surrounding environs. The main effect of the warm air is to lengthen the frost-free period in agricultural areas astride the Columbia River.

Altitude

Altitudinal variation is responsible for creating a complex pattern of temperatures. Coastal borderlands, western interior valleys, and much of central and eastern Washington are situated within a thousand feet of mean sea level. Coastal mountains, the Cascade Range, the Rockies, and the high Columbia Intermountain range soar between 2,000 to 5,000 feet above sea level, except for the high ranges. Thus the effect of attitudinal reduction of temperatures affects most of the Pacific Northwest.

Altitude not only reduces the length of the frost free period, but it may also indirectly create conditions conducive to greater radiational loss of energy. Fog and low clouds are a common cool season feature of much of the region west of the Cascade Mountain Range, thereby reducing the surface receipt of solar radiation. Most areas above 2,000 feet mean sea level, especially east of the Cascade Range, receive a rather high annual percentage of possible solar radiation.

Semi-permanent High and Low Pressure Cells and Wind

Semi-permanent pressure cells are a very important control on weather patterns and resultant climates. A high pressure cell located over the eastern Pacific dominates summer weather and a low pressure cell situated in the mid-Pacific and Gulf of Alaska greatly influences winter weather.

A cell of high pressure known as the Hawaiian High is centered at about 40 °W latitude approximately 1,500 to 2,000 miles offshore. (Note: this is the mean location of the pressure cell but actual location can vary some from year to year.) The atmospheric pressure at the center of the Hawaiian High is about 15 millibars above the standard sea-level pressure. Areas dominated by high pressure characteristically are dry with a tendency toward clear skies. During the summer season, the high pressure cell shifts northward of its winter position and intensifies at the same time. During winter the high migrates southward centered on 30 ° to 35 °N latitude and at the same time experiences diminished intensity. During the winter period, it has a greater effect on Southern California and Baja California than it does further north.

Another important feature of the Hawaiian High is its shape: Isobars encircling the high reveal an elliptically shaped dome with the smaller end aligned in a westerly direction pointing toward Asia. The broader, blunter eastern end tends to parallel the Pacific shore from the Alaskan Panhandle to the Central California coast. The significance of this is that air flowing off the high either descends inland without flowing in direct contact with ocean waters before flowing over the land mass, or else it flows parallel to the coastline, again insuring its low absolute humidity. In addition, air that descends in altitude is heated by compression, creating winds capable of desiccating surfaces and organisms.

Finally, the high pressure cell also is an effective barrier to transient middle latitude cyclonic storms. To be sure, fewer storms develop in the summer, except in the high latitudes, nonetheless seldom can such a storm penetrate into a sea dominated by the high pressure cell.

All of the features of the Hawaiian High pressure cell in combination create a summer weather pattern dominated by clear skies and warm to hot sunny days, except in fog shrouded coastal areas or clouded mountain peaks.

During the winter season, when the high pressure has migrated southeastward to its position off Baja California and decreased in intensity and areal extent, the low pressure cell in the Gulf of Alaska develops to its greatest intensity. This low pressure center known as the Aleutian Low is shifted northwestward in summer over Siberia and the Arctic Ocean and is at its weakest state. By early winter, the Aleutian Low migrates southeastward and intensifies. By December or January it dominates the entire north Pacific Ocean, from about 35° latitude to 60° to 65°N.

The center of the Aleutian Low ordinarily is about 15 millibars below average sea level pressure; as a result the north Pacific Ocean is a region characterized by cyclogenesis, the birthplace of middle latitudinal cyclonic storms. Storms develop in the Aleutian Low as the result of conflict between polar and subtropical air masses. Conflicting air masses pass across broad expanses of the Pacific Ocean where air comes into direct contact with the ocean surface. As these westerly winds sweep in an easterly direction toward the North American continent, vast amounts of energy and moisture are incorporated into the air masses and transported eastward. Both energy and moisture are released, at least in part, by storms that invade the Pacific Northwest. The stormy season is at its maximum during mid-winter, especially in December and January.

The seasonal variations of the semi-permanent high and low pressure systems explain the march of summer into winter and then winter into summer. For example, Bellingham receives about one inch of precipitation in July, the driest month, and nearly five inches in December, the wettest month. In Medford, July and August each receive about one-quarter inch of precipitation and about three and one-half inches in December, which is the wettest month. Middle-latitude cyclonic storms progressively work southward into lower latitudes as summer fades into fall and fall into winter. The first storms of fall are not strong enough to penetrate the receding Hawaiian High with as much vigor in the south as in the north. Later in the fall season, as storms are strengthening in the Aleutian Low and the Hawaiian High is weakening, storms penetrate farther and farther south. In October Bellingham receives three-and-one-half inches of precipitation while Medford receives just under two inches. During late winter and early spring, the Hawaiian High intensifies and migrates northward, weakening or blocking storms out of the Pacific. Concurrently, the Aleutian Low begins to weaken with the approach of summer and storm tracks move northward out of the Pacific Northwest into northern British Columbia and Alaska. As a result, Bellingham's winter, where there are eight months with two or more inches of precipitation, is "longer" than Medford's with only four months of this type of weather. Locations between Bellingham and Medford would experience a logical progression between the two extremes, except where localized conditions might create pronounced changes.

Jet Streams

Middle latitude or polar jet streams are an extremely important factor in the development of cool season weather throughout the Pacific Northwest. Upper level westerly winds are concentrated in a relatively narrow, meandering belt of high velocity winds on an intercontinental scale. Meteorologists refer to the most significant jet stream in the middle latitudes as the circumpolar jet stream. This convoluted river of air is located at the 500 millibar level in the ambient atmosphere, which ordinarily translates to approximately 18,000 feet in altitude above mean sea level.

Winds in this narrow stream of air by definition exceed 60 miles per hour as a minimum flow, and it is not uncommon to find wind speeds in excess of 120 miles per hour during the heart of the winter season. Wind speeds increase with increased altitude to a velocity of approximately 170 miles per hour. It appears that as cyclonic and anticyclonic storms are generated, disturbances in the atmosphere that the circumpolar jet tends to accentuate, creating giant waves that flow in a circuitous pattern. These waves transport vast amounts of energy and moisture poleward from the subtropics and cold, dry air southward from higher latitudes.

The circumpolar jet is largely responsible for boosting middle latitude cyclonic storms on their way from east to west around the globe in the middle latitudes. The Pacific Northwest is in the heart of this generalized flow of air from west to east and, as a result, typically receives stormy periods during mid-winter that may persist from a few hours to several days. During January, if the jet stream flows in a giant wave into western Canada then shoots offshore into the eastern Pacific, a cold snap will descend upon the entire west coast until the jet stream turns inland to continue its easterly course across western North America toward the interior. Conversely, if a giant wave in the jet is situated a few hundred miles off the coast of the Pacific Northwest, storms born in the Gulf of Alaska are ushered southward over the Northwest. These storms may produce plentiful snow at intermediate to higher altitudes in all mountains except lower coastal mountains and hills parallel to the Pacific Ocean. Occasionally the jet has shifts far south of its usual path across the northwest into central or southern California.

Subtropical air from the southern eastern Pacific may become entrained in this system resulting in copious rainfall at lower elevations and heavy snow in the Cascade Range and especially the Sierra Nevada Mountains of California. A knowledge of the vagaries of movement of the jet stream during the winter season is the key to understanding and predicting the weather of the Pacific Northwest.

As the winter-season begins to wane after the vernal equinox, the circumpolar jet commences to lose its strength, resulting in reduced intensity of precipitation in most years. This explains why December and January are the wettest months in the Pacific Northwest: this period coincides with the greatest intensity of the circumpolar jet. By late spring and early summer the jet begins to split into more than one stream of air that tends to further weaken its ability to transport moisture and energy from east to west. By summer, the circumpolar jet becomes discontinuous with several branches, further stripping it of its intensity.

Semipermanent high-pressure cells in the eastern Pacific act as a barricade, virtually eliminating any significant source of moisture during the period ordinarily extending from June into early September, depending upon latitude and local orogenic conditions. After the autumnal equinox, the weaker branches of "the jet" disappear, and there tends to be only one split rather than several. Later in the fall period and early winter as the jet continues to develop, more and more atmospheric disturbances are boosted westward through the Pacific Northwest and across the rest of the continent before moving out across the North Atlantic ocean.

Typically cold snaps on the west coast are contrasted with unseasonably warm temperatures over the eastern half of the United States and over southeastern Canada as far north as the Gulf of St. Lawrence. Conversely, unseasonably warm winter temperatures along the Pacific coast may result in a polar outburst in the same regions of the eastern United States and southeastern Canada. In any event, it is safe to say that the prediction of winter weather any place to the west of the observer's location is likely to provide the key to understanding the potentials for storminess at that location. Along the coast of western North America the weather during any given period in winter is closely linked to the events happening in the Gulf of Alaska. Pursuing this phenomenon a little further to the west indicates that Pacific coast weather is directly tied to the evolution of storm systems as they move across Pacific waters from the Asian landmass.

These imperfectly understood rivers of air in the upper atmosphere, jet streams, are of the utmost importance in the production of seasonal storm patterns, temperature regimes, and moisture patterns. More of the mystery about the jets that affect our regions under study will unfold as satellites with electronic sensing devices more accurately map and measure the key factors that generate the circumpolar jet stream. Perhaps in the not too distant future our ability to understand and predict weather phenomena over the Pacific Northwest will be enhanced as more information is garnered, enabling us to understand the complex dynamics in the atmosphere. Our ability to predict both short-term and long-term weather phenomena will enhance our ability to understand the complex and unique weather and climate and other physical attributes of this region.

REGIONAL PATTERNS

Temperatures

Mean annual temperatures in the Pacific Northwest range from 50 °F to 55 °F along the coast downward to about 35 °F in the mountainous regions of south central Idaho. Proximity to the ocean, interior location, altitude, the affect of mountain barriers, and local topography all affect the mean annual temperature. Most of the entire region east of the Cascade Range experiences mean annual temperatures of 40 °F to 45 °F. Coastal regions, western interior valleys in Oregon and Washington, and the lower elevations of the Columbia Basin are characterized by mean annual temperatures of 50 °F to 55 °F.

Seasonal mean temperatures display a significant departure from the annual mean. In July, a narrow coastal strip is dominated by fog in the mid 50's; while inland as far as the Cascade Range, the mean temperature is 60 °F to 65 °F. Most of the Columbia intermountain region east of the Cascade mountains is 65 °F to 70 °F with a few small areas of 70 °F to 75 °F. In each of the regions there tends to be sheltered enclosed valleys at lower altitudes that depart from the regional norm: examples include the Rogue Valley, Lower Snake Plain, and along the Columbia River.

January mean temperatures west of the Cascade Range are above freezing, ranging from 40 °F along the coast to 35 °F along the western foot of the Cascade Range. East of the mountains most of the region is below freezing except for the narrow zone extending up the Columbia and lower Snake Rivers to the vicinity of Walla Walla, Washington. The coldest mean temperatures are found in the high Cascades and the mountains of east-central Idaho.

Frost free periods are of shortest duration in mountainous areas and at the eastern and northern interior margins of the Pacific Northwest. The longest frost free periods are along the coast and range from 250 to 300 days, depending upon local conditions. Proximity to the ocean and persistent fog and low clouds mitigate harsh conditions. Toward the interior factors such as altitude, mountain slopes, air drainage and predominant clear skies with attendant radiation loss result in a much shorter frost free period.

The primary controlling factor is the Cascade Range, which effectively blocks warm, marine air from passing to the Interior. The Puget-Willamette Trough has frost free periods somewhat reduced from those along the coast, however, the growing season is sufficiently long to permit the cultivation of a diversity of agricultural crops. The average season is 200 to 240 days, depending upon localized conditions. East of the Cascade Range the frost free season is generally 80 to 120 days, except for a narrow zone along the Snake River in southeastern Washington and along the Columbia River northward almost to the Canadian border where the season ranges upward to approximately 150 days.

Daily temperature means and extremes also warrant some discussion. Coastal areas are moderate year-round with a 20 °F to 25 °F between the monthly maxima for January and July. Astoria, for example, has a daily maximum of 44 °F in January while it is 68 °F in July. The extreme high temperature was 58 °F in January and 89 °F in July. The lowest January minima was 15 °F. The respective monthly means are 40 °F in January and 61 °F in July, resulting in a 21 °F annual temperature range. At western valley locations the temperature range is 30 °F to 40 °F with more severe extremes. Medford and Olympia have January maxima of 45 °F and daily minima of 30 °F respectively. The lowest temperature in Olympia was 4 °F while it was -3 °F at Medford. There is greater departure in Medford in July temperatures from those of Olympia. Olympia's July mean is 64 °F while Medford's is 72 °F. The July daily maximum for Medford was considerably higher than Olympia: 115 °F compared to 100 °F. Olympia's mean annual range is 26 °F whereas Medford's is 35 °F.

Even greater disparities occur east of the Cascade Range. Spokane has a January mean of 25 °F but with a daily minimum that has dropped as low as -13 °F. July in Spokane is a 70 °F mean with a daily maximum of 102 °F. The mean annual temperature range for Spokane is 45 °F, the same for Boise, but there is a 50 °F range for Pocatello (January 22 °F and July 72 °F).

These data clearly illustrate the transition from moderate coastal regimes to more severe conditions as continentality becomes a more significant factor (altitude is also important; for example, Astoria and Olympia are at sea level, Medford 1,300 feet MSL Boise 2,850 feet MSL, Spokane 2,350 feet MSL and Pocatello is at nearly 4,500 feet MSL).

Climatic Data—Select Stations

Station	Month	Daily Maximum*	Daily Minimum*	Mean Monthly*	Extreme High Sta.	Extreme Low Sta.
Olympia	Jan	45	31	38	60	4
	July	79	48	64	100	35
Astoria	Jan	44	36	40	58	15
	July	68	55	61	89	42
Medford	Jan	45	30	37	68	-03
	July	89	55	72	115	40
Spokane	Jan	31	19	25	49	-13
	July	86	55	70	102	39
Boise	Jan	36	22	29	63	-17
	July	91	59	75	11	41
Pocatello	Jan	31	13	22	55	-19
	July	90	55	72	100	37

Figure 1. Climatological Data. Climates of the states, NOAA, Asheville, North Carolina, 1993.

Moisture

Moisture regions also reflect a strong influence from landform orientation; plus the fact that the entire region is in the heart of the westerlies, which are characterized by copious cool season precipitation and persistent cloud cover along with coastal borderlands, coastal mountain ranges, and the Cascade Range. Locations interior to the Cascades receive less cloud cover and precipitation but a greater percentage of possible solar radiation.

The greatest mean annual precipitation occurs along coastal areas and in the northern Cascade mountain ranges of Washington. Coastal areas receiving the most precipitation are in northern Washington where 100 inches or more is common. Northward from Willipa Bay to the Olympic Peninsula heavy winter rains received from middle latitude cyclonic storms ensure ample rainfall. It is not uncommon to receive five inches of rain or more from a single storm. Generally speaking, coastal areas from southwest Oregon to the central Washington coast receive from 70 to 100 inches average annual precipitation with the vast majority coming from rain and on rare occasion, from light snowfall. Local variations do occur, for example, Brookings, Oregon, receives approximately 80 inches mean annual precipitation while Bandon, a few miles to the north, records only about 55 inches average annual precipitation. On the Washington coast, Willipa Harbor receives 86 inches to more than 130 inches on the Quinalt River, which is located on the windward side of the Olympic mountains.

Immediately inland, coastal mountains may receive much higher mean annual precipitation because of an orographic effect. Air that is forced to rise over a mountain barrier is cooled yielding greater precipitation. Some areas receive substantially more than 100 inches average annual precipitation. If weather recording stations were established in as intricate a network in mountainous areas as they are on coastal lowlands and valleys, it is not difficult to conceive of at least a few locations that would receive in excess of 150 inches average annual precipitation. For example, Aberdeen, Washington, has a mean average precipitation of 85 inches. In the coastal mountains to the east the average annual precipitation has been reported to be slightly in excess of 130 inches.

The Puget-Willamette lowlands receive markedly less precipitation than either coastal lowlands or coastal mountains. For example, Bellingham and Seattle, Washington, each receive nearly 35 inches mean annual precipitation. Portland, partially due to its location at the mouth of the Columbia Gorge, receives a little over 40 inches and Eugene about 37 inches. Farther to the south in the Umpqua Valley, Roseburg receives slightly more than 30 inches. Finally, Medford, in the Rogue River Valley, receives 20 inches mean annual precipitation.

In comparison between coastal locations and those of valleys west of the Cascade Range, it is clear the range in mean annual precipitation is markedly less. Coastal variance is as much as 60 or 80 inches between north and south but only about 20 inches in western valleys. The Rogue River Valley, which is transitional between

Marine West Coast and Mediterranean climactic types, makes the difference appear greater than it is for most of the western valleys. Note the difference in mean annual precipitation for Roseburg, Oregon, and Bellingham, Washington, is only about 5 inches.

Mountainous regions are a nightmare for climatologists, and the Cascade Range and other mountainous areas in the Pacific Northwest are no exception. The Cascade Range displays elevational differences from mountain passes at 4,500 feet to massive volcanoes from 10,000 feet to more than 14,000 feet above mean sea level. Variation in altitude, directional orientation and slope account for radical differences in a very short distance. A classic example is the very high mean annual precipitation (150 inches or so) on the windward side of the Olympic mountains in northwestern Washington as compared to Sequim, Washington, on the Strait of Juan de Fuca, which has average annual precipitation of less than 10 inches! At various locations throughout the Cascade Range from north to south, the average annual precipitation varies from 55 inches to over 110 inches.

Unlike most of the area west of the Cascade Range, except for the Olympic mountains, snow is an important source of precipitation in the Cascades. Two examples should suffice. The first is Crater Lake National Park, Oregon, where the mean annual snowfall is 528 inches at the headquarters area, which is 6,500 feet MSL. However, in the 1932–33 season 879 inches of snow fell. Snow has accumulated on the ground to a depth of 242 inches. The second is Mt. Rainier National Park where, in 1955–56 more than 1,000 inches of snow was recorded at the 5,500 feet MSL at Paradise. It is safe to generalize that most of the high Cascade Range in Washington and Oregon receives from 300 to 500 inches mean annual snowfall. Suffice it to say that maximum precipitation in these latitudes occurs between about 6,000 and 7,000 feet mean sea level.

The lee side rain shadow east of the Cascade Range in both eastern Washington and eastern Oregon as well as Idaho is an illustration of classic geographic regionalization based on physical factors. The vast majority of the entire Pacific Northwest region is situated east of the Cascade Range, which serves as a barrier to both moisture and surface advection of energy by prevailing winds. Immediately east of the crest of the Cascade precipitation amounts start to decline. Weather stations at the eastern base of the mountains in Oregon, such as Bend, Keno, and Chiloquin, have mean annual precipitation from 12 inches to almost 9.0 inches. In Washington, Ephrata records only about 8.5 inches and there is about 7.5 inches at the Yakima airport. The higher, broader Cascade range in Washington creates the maximum rain shadow effect in the interior of the region.

Average annual precipitation for the remainder of the vast area east of the Cascade is very similar from east to west and north to south except for mountainous areas such as the Salmon Mountains and Seven Devils Mountains and that portion of the Rocky Mountains in Idaho. In Oregon the Wallowa, Elkhorn, and Blue-Ochoco mountains display the most significant departure from the remainder of the region. Idaho's panhandle and central mountains receive 25 to 35 inches of mean annual precipitation. Deep snow accumulates in winter throughout the mountainous regions, however, it has generally an annual snowfall of 100 to 150 inches which is one-half to one-third the amount recorded in the Cascade Range.

Lower elevations throughout the region east of the Cascade mountains receive from as few as 7.5 inches mean annual precipitation to 18 inches. Spokane, for example, receives about 18 inches while Prosser, which is located on the Yakima River near the Tri-Cities, receives only a little more than 7.5 inches. Other locations with 10 to 12 inches include Boise, Burns, Hermiston, Ritzville, Pocatello, Caldwell, and Payette. Klamath Falls and Lakeview, each receive about 14 inches, which is higher than most of the remainder of the region due primarily to lake effects. Warmer, moister air in the vicinity of Klamath Lake and other water bodies in the Klamath Basin and Goose Lake near Lakeview enhance the opportunity for slightly increased mean annual precipitation.

Seasonality of precipitation is worthy of consideration. In the area from the Cascade westward to the coast, winter is the season of precipitation maximum. Winter rains begin in earnest in October and continue through March, creating at least a six-month rainy season. At some locations, as many as ten months consecutively receive at least two inches of precipitation per month. Aberdeen receives 85 inches mean annual precipitation. Only July and August receive less than two inches mean monthly precipitation. Brookings has nine consecutive months with at least two inches mean monthly precipitation. The wettest month ordinarily is December, but both November and January are close seconds. For example, Newport, Oregon receives 15 inches mean monthly precipitation in December, a little over 11 inches in November, and 9.5 inches in January. Aberdeen's December mean is 14.5 inches but has 11 inches in November and 12.75 inches in January.

Western valleys have fewer months with two inches or more of mean monthly precipitation: six months in Seattle, seven in Centralia, seven in Eugene, seven in Roseburg, and only four in Medford. In all of these locations, December is the wettest month. This discussion serves to establish that there is one maximum period of precipitation from the Cascade westward: the dead of winter. Conversely, the month receiving the least precipitation is July, in the heart of summer! The amount of July precipitation is greater in the north: Seattle 0.6 inch, Eugene 0.26 inch, and Medford about 0.15 inch. Aberdeen receives 1.5 inches in July while Brookings records a little over one-half inch. It is apparent that most places west of the Cascade Range have dry summers except for the northwest Washington coast.

The precipitation regime east of the Cascade Range displays two seasonal maxima. There is one in mid-winter, most often it is December, but occasionally January receives a little more than December. The second maxima occurs in late spring or early summer. Pendelton illustrates this phenomena rather well. The wettest month, December, is 1.7 inches mean monthly precipitation. Each month declines until May, which receives 0.95 inches. June spikes up to 1.25 inches, and July is a meager quarter inch. The mid-winter maxima is attributable to precipitation from middle latitude cyclonic storms while the spring-early summer maxima is attributable to a combination of cyclonic activity augmented by convectional precipitation. Indeed, farther east in Idaho, Idaho Falls in December receives 0.8 inch and 0.9 inch in January. The wettest month, however, is June with 1.15 inches. An even more striking example is Grangeville, Idaho, where December mean precipitation is 1.8 inches and May is 3.3 inches followed closely by June with just over 3 inches.

The significance of this discussion will be more fully realized when water resources are considered. Streams and rivers of each region reflect precipitation regimes in runoff patterns, as do agriculture, forestry, and recreation. For example, the Palouse of Washington and Idaho receive ample late spring and summer precipitation to support one of the more productive dry land wheat farming areas in the entire United States. Rafting, kayaking, and other water sports flourish on rivers west of the Cascades earlier in the summer than they do east of the Cascade Range.

Fog is another feature of the moisture regime that warrants a further understanding. Despite the fact that meteorologists distinguish between several types, a discussion of two basic types will be sufficient to understand the influence of fog on the weather and climate of the Pacific Northwest. The two types are advection fog and radiation fog. Advection fog is formed when comparatively warm and moist air flows horizontally across a cooler surface. Radiation fog is formed in a much different situation. Generally it occurs under calm wind conditions and is the result of comparatively warm and moist air becoming chilled from the bottom by radiational loss of contained or latent energy.

Here's how it works. During long, cold winter nights the earth loses terrestrial radiation outward through the atmosphere. As a result, the earth's surface becomes chilled and a distinct temperature disparity occurs between the surface and the air immediately above it. Energy is then radiated from the air to the ground, resulting in cooling of the air mass. If adequate, cooling occurs to drop the air temperature near the surface below the dew-point, temperature condensation in the form of minute water droplets will form. The droplets are too light to be drawn to earth by gravity, therefore they are in suspension and simply "float" around in the lower atmosphere. Radiation fog is seldom more than a few tens of feet to a few hundred feet in depth. During daylight hours, when air temperature is likely to increase, fog droplets are evaporated back into the surface layer of air. Occurrence of wind will also dissipate radiation fog for two reasons: (1) often the invading moving air is warmer than the resident air, and (2) turbulence in moving air creates kinetic energy and subsequently warms the air and increases the dew point temporarily.

Advection Fog

The entire Pacific Coast of Oregon and Washington and the Puget Sound area are effected by advection. Smaller areas in proximity to water bodies and rivers may also experience some advection fog during the cool season of the year. The formation of fog along coastal areas is linked to the upwelling of colder waters off the southern Oregon and northern California coasts. Cool water offshore finds its way to the surface cooling the air that flows across it. The key to advection fog is a consistent source of air. During summer, air flowing from under the Hawaiian High blows consistently from out of the northwest at 10 to 30 miles per hour. Coastal regions have night and morning fog throughout the summer based on northwesterly air flow. During winter both northwesterly air flow and southwesterly air are a source for the formation of advection fog. Unless an atmospheric disturbance moves on shore, there is persistent fog during many winter days. Under storm conditions, invading air mixes with resident air under ensuing low pressure and enables the fog to "lift." Summer conditions allow dissipation during the late morning or early afternoon due to radiational heating of the air, which causes evaporation of the water droplets.

Coastal fog occurs in a narrow zone along the Pacific Borderlands. During the summer season, daytime temperature maxima in fog shrouded areas is greatly retarded by both the fog and the chilled air. For example, the maximum temperature in Coos Bay on a fog shrouded summer day when sun does not break through could be 60 °F to 65 °F. Upper reaches of coastal valleys where fog dissipates could reach into the 80s °F; whereas western interior valleys, such as the Umpqua Valley and Roseburg, could well have a maximum on that same day of 100 °F or more. During brief periods in the fall and spring, coast areas may be devoid of advection fog primarily because of equality of temperatures and air pressures both over the sea and land. Cloudless skies, clear atmosphere, and mild temperatures are the trademark of these conditions. A variation of advection fog of importance in the Pacific Northwest is upslope fog. Upslope fog occurs when surface air that is relatively moist is forced to rise over an orographic barrier. As the air rises, it is cooled and condensation occurs producing fog droplets. A dense but thin blanket of fog is not an uncommon winter phenomena in the mountains of the Pacific Northwest, although its existence is rather ephemeral. A variation on this occurrence is pogonip (a Paiute term) or ice fog. The air is so cold that water vapor sublimates to form ice particles rather than water droplets. Pogonip is most likely to occur in high mountain basins or valleys, especially in the Great Basin and in the interior mountains of Oregon and Idaho.

Radiation Fog

Regional areas with persistent or frequently recurring radiation fog are western interior valleys, the Columbia Basin, and mountain valleys below approximately 4,000 feet mean sea level. Generally radiation fog occurs under clear, cold, long winter nights where high pressure dominates. Fog tends to form during early evening hours becoming "thicker" during the night. During the atmospheric conditions conducive to radiation fog, "pea-soup" fog may be common during nighttime and early morning hours. Pea-soup fog may persist throughout the daytime, especially during mid-December to mid-January, which tends to be the coldest "month" in many of the areas likely to receive radiation fog. There just simply is not enough insolation to heat surfaces and air so that the fog droplets would be evaporated. These thick fogs are sometimes referred to locally as "river fog" or "tule fog." Radiation fog is most common during the winter season and would be quite rare during the warm season.

Frost

Frost is the final form of moisture addressed here. Frost, like fog, is a form of condensation rather than a type of precipitation. The primary effect of frost is as a natural hazard rather than a moisture source. The hazard created by frost tends to effect agriculture, horticulture, and conifer seedlings that have been planted to regenerate burned or harvested forests. Certainly frost is linked to the growing season concept, also referred to as the frost free period, previously discussed. Methods of frost mitigation vary depending upon the vegetation threatened, however, a primary means is to warm the air, preventing it from cooling to the critical threshold temperature. Application of water, the use of orchard heaters, wind machines, and other ingenious methods and devices are often helpful in preventing or at least diminishing the full effect of frost.

Pressure and Wind

The overall pressure and wind patterns of the west coast and eastern Pacific Ocean were discussed earlier. Generalized pressures and winds explain most weather conditions occurring in the Pacific Northwest, although there are regional and seasonal variations which are more localized than the controlling Hawaiian High in summer and Aleutian Low in winter.

During winter, high pressure dominates most of the Pacific Northwest. Primarily high pressure is an effect of cooling over the land mass in response to long winter nights and more rapid loss of radiation from the land mass as compared to the Pacific Ocean. Low pressure dominates over the sea. Much of the region east of the Cascade Range is characterized by clear skies in December and January except for transient storms that rapidly move across the entire region. Pressures are reversed in summer when high pressure develops over the eastern Pacific and low pressure is common over the land mass.

First let us consider the winter season when the Aleutian Low is an extremely important factor in spawning middle latitude cyclonic storms that bring generalized storminess to the entire Pacific Northwest. The process by which these storms develop, cyclogensis, is at its maximum in mid-winter in the Gulf of Alaska. Storms sweep eastward out of the Gulf where high mountains of panhandle Alaska and coastal British Columbia, coupled with intense high pressure inland, drive storms southeastward along the Pacific Coast. Eventually the storms penetrate inland across Washington, Oregon, and northern California, or they are driven as far south as coastal or southern California before moving inland. The most well developed of these storms may produce heavy precipitation lasting 12 to 36 hours, and occasionally twice that length of time or more.

Coastal locations commonly experience wind speeds in excess of 50 miles per hour, and of 100 miles per hour or more when fronts associated with the strongest depressions cross the coast. Coastal areas often receive the most precipitation from these storms with 24 hour rainfall ranging from three inches to five inches or more. Copious precipitation also occurs inland with heavy snows at higher elevations in the coastal mountains and generalized snowfall in the Cascade Range, central mountains, and in the mountains of Idaho. After a storm has passed, high pressure once again dominates the region.

Typical midwinter conditions under widespread high pressure include clear skies or fog, cool to cold temperatures, and blockage of weak storms in their attempt to move eastward across the region. Despite the blocking effect by continental high pressure, transient storms that are well developed are able to invade the interior; as a result most of the region east of the Cascade Range during winter is characterized by brief stormy periods followed by clear skies, except at lower elevations where fog persists under those conditions.

The high sun (summer time) season is also dominated by high pressure on the west but trending toward low pressure on the eastern margins of the Pacific Northwest, including most of Idaho and the eastern third of Washington. The Hawaiian High dominates the eastern Pacific and coastal areas at least as far east as the Cascade Range. This region of high pressure is not thermally induced but occurs as part of dynamic earth processes, whereas low pressure to the east is of thermal origin.

First, let us consider the simplest of the two zones of summer pressure. At interior locations with abundant sunshine, air is heating causing it to expand and rise with resultant lower pressure. Further evidence of this is the bimodal distribution of precipitation east of the Cascade Range, where there are both winter and late spring or early summer maxima. This was discussed under the moisture regime: Spokane, Grangeville, and Idaho Falls all display early warm season precipitation of note.

Now for the more difficult area: high pressure during July. By that time of the year the Hawaiian High is dominating the entire Pacific Coast from Baja California northward, including all of the Gulf of Alaska. The high evidences an asymmetrical profile aligned with the large rounded end over the region and the pointed end in the Central pacific north of Hawaii. The center of the ridge of high pressure ordinarily is at about 35° to 40° north latitude approximately 2,500 miles offshore (Figure 1). Air that flows off the high does so in a clockwise manner in the northern hemisphere.

As a consequence of that and the blunted eastern end of the high, air flows around the high parallel to the coast of the Pacific Northwest. At the same time, it is descending in altitude and as a consequence is being heated through compression. Therefore, air that flows in over the region in summer does not flow across the oceans' surface so it is not humidified; and as it is heated by compression, its ability to evaporate available moisture is greatly increased. Both of these conditions mitigate against precipitation and instead result in loss of moisture through intense evaporation with resultant droughty conditions. This situation is well illustrated by summer rainfall west of the Cascades at places like Eugene and Medford with approximately one-quarter inch of precipitation in July and a little over one-half inch in Seattle.

Storms

Special consideration of storms is warranted because they are the vehicle by which moisture and energy are transported into the region. The two major systems are middle latitude cyclonic storms and convectional systems. Middle latitude cyclonic storms have been previously discussed to a certain extent, but to elaborate further it must be emphasized that these disturbances are primarily responsible for cool season precipitation. The movement of a full-fledged middle latitude cyclonic storm across the region is regarded by many as a predictable phenomenon inherently beautiful yet fraught with potential danger. The first signs of such a storm are wind shifts to the southeast with high, thin cirrus clouds. Within a few hours wind velocities increase and middle and lower clouds rapidly move from west to east with the first precipitation. As the storm center with its associated fronts approaches, the action is at its greatest with gusty southeast winds, heavy rain at lower elevations, and heavy snow in the colder reaches. After passage of the warm front, winds shift to the southwest and precipitation diminishes or subsides. Soon after the cold front rips through bringing heavy rain or snow for a short duration followed by a wind shift to northwesterly with a rapid decline in temperature.

Even though the storm has passed across coastal areas, the Cascade Range and points eastward, there often is a large pool of cold, unstable air that continues to move inland. This air results in shower activity that may persist for two or three days following the passage of the fronts. Showers are an important source of moisture, especially in mountainous areas where a combination of altitude and cold air produce snowfall—often a significant enhancement of the snow pack created by the main storm.

Convectional precipitation is the second type to be considered and it occurs primarily during the summer period when low pressure is dominant east of the Cascade Range. Air masses not dominated by the Hawaiian High contain ample moisture to foster the development of convectional cells throughout the region, especially over the mountains. As is typical with thunderstorms, clouds form and rise to significant vertical heights, which results in rapid cooling and moderate to heavy precipitation: either rain or hail, or both. On occasion, especially in mid to late summer, moisture laden air masses generated by the Gulf of California and the Arizona low, make their way into the region east of the Cascade. Often there is localized heavy precipitation associated with these conditions; at other times, they are the source of lightning storms which pose potential fire threat from lightning strikes in forests, grasslands, and some agricultural endeavors. Another variation on these conditions exists west of the Cascade, usually sometime between mid-summer and early fall. On occasion an upper-level depression develops off the coast of California. Typically these "upper level devils" wander north and south and then finally dissipate. At other times they move inland creating thunderstorms over coastal mountains as well as the Cascade Range, particularly in late summer and fall in western Oregon and to a lesser extent in western Washington, where intense summer heating and drought have evolved into an artificial fire hazard situation. The severe wildfires in southern Oregon in 1978 and 1988 were a direct result of plentiful lightning with no significant accompanying rainfall to dampen ignition of forests and woodlands by lightning.

Fronts

During some seasons of the year, air masses not associated with depressions invade the region. This is most common in winter when Arctic fronts originating in the Canadian north push southward. Usually the Pacific Northwest is on the western margin of such occurrences, especially the borderlands west of the Cascade Range. On occasion the alignment of the jet stream boosts westward the high pressure and cold air behind the front with unusually cold, dry conditions over the entire region. Outbursts can result in near-zero readings west of the Cascade and bitter cold in the interior. Fortunately, these outbursts occur only once or twice in a decade. The converse of the cold front in winter is the heat wave of summer. Air that originated over the California deserts occasionally ushers in unusually high temperatures, where the mercury soars above 100 °. For example, under these conditions Seattle has recorded 100 °F, Portland 107 °F, and Medford 115 °F.

Canyon Convergence

Any time air converges the potential for condensation and precipitation or both occurs. A very localized effect, yet one of great significance in the Pacific Northwest, is canyon convergence. As air flows or is pushed up a canyon, it becomes more constricted as well as forced to rise. As a result, air cooled below dew point condenses producing clouds and often precipitation. At the same time, wind velocities increase due to the constriction of the canyon. As a consequence, especially in the coast ranges and the Cascade Range, the multitude of canyons occupied by westward flowing streams create conditions conducive to increased cloudiness and precipitation. During winter, at higher elevations, greatly increased snowfall may result. The effect on the natural landscape may be pronounced as reflected in landform development, soils, and biota, especially native vegetation. Human utilization of the landscape also is impacted since most surface transportation routes are constructed in canyons leading to passes over the mountains. Highway and railway maintenance are increased, plus there are the additional costs of snow removal and sanding of icy higher elevation routes and passes. Despite its low elevation, the Columbia River Gorge provides an excellent example of canyon convergence with fog, clouds, precipitation, and high mean wind velocities.

Air Stagnation

Some atmospheric conditions coupled with local or regional typographic features result in air stagnation. Typically air stagnations occur under atmospheric high pressure. The longer high pressure persists, the greater the opportunity for air quality to deteriorate. Dominant high pressure occurring both in summer and winter have been discussed earlier in this chapter. Generally western interior valleys and mountain valleys and basins experience the worst conditions. In western Oregon, the Umpqua Basin and Rogue Basin experience severe air quality problems both in summer and winter. Large communities east of the Cascade experience poor air quality in winter. Bend, Tri-Cities, Spokane, Boise, and communities along the Snake River have varying periods of poor air quality. Transient storms destroy inversion layers responsible for pollution, hence air quality in winter is restored from time to time due to a basic change in the air pressure regime. Air quality in winter in many communities is adversely affected by particulates and aerosols emitted by wood burning stoves and fireplaces. As a result, many communities have mandated measures to alleviate the most severe conditions. Summer, however, can be dominated for many weeks without change in the well-developed high pressure, which results in an obscured atmosphere and severe air pollution.

Air Drainage

Cooled, dense air has a tendency to flow downslope in response to gravity. Many places in the Pacific Northwest situated in mountainous areas experience air drainage. Once again mountain valleys and basins would be most effected. During summer, air drainage is welcomed because it results in significant cooling and respite from daytime heat. In winter, just the opposite occurs: colder, denser air pools in topographic depressions resulting in lower minimum temperatures. Horticulturists along Lake Okanagan in north central Washington take advantage of air drainage to protect their fruit crops from cold, frosty conditions in late spring. Orchards are planted on hillslopes above the lake and are protected by the cold, heavy air that flows near the surface and preserves warmer conditions in the branches that support the blooms or immature fruit.

Chinook

Chinook winds occur when relatively cold or cool air flows from high altitude down a long mountain slope and is radically heated by compression. As a result, a day that starts out at sub-zero temperatures may experience a drastic rise in the thermometer to above freezing. One of the greatest changes on record occurred along the Rocky Mountains in mid-winter in Browning, Montana, where there was a 100 °F differential between the daily minimum and maximum. Chinooks are generally associated with the lee side of mountain ranges and are a welcome relief during winter. Cold, stagnant air, fog, and air pollution are "scoured" out of valleys with the onset of a well developed Chinook. During summer, Chinooks are less welcome because warm or hot days may be rendered even hotter.

Summary and Conclusions

Climatic features and weather phenomena of the Pacific Northwest display both small scale features, regional characteristics, and extraregional patterns that are fundamental to the understanding of the weather and climate of the Pacific Northwest. Always consider the occurrence and magnitude of the controls on climate and the weather elements responsible for creating variation throughout this geographically diverse region. The combination of the Pacific Ocean generating a marine climate coupled with orogenic features which modify and often intensity weather patterns producing highland climates, and the interior locations where pronounced rain shadow and continentality create more severe conditions leads one to conclude that the Pacific Northwest has one of the most diverse and distinctive climates of any region in the conterminous United States.

REFERENCES

Baldwin, James L. *Weather Atlas of the United States*, Environmental Data Service, 1989.

Climactic Atlas of the United States. United States Department of Commerce, Environmental Sciences Services Administration, Washington, D.C. 1992.

Climatological Data. Climates of the States. National Oceanic and Atmospheric Administration, Asheville, North Carolina. 1993.

Decker, Fred. *The Weather of Oregon*. Oregon State University Press, Corvallis, Oregon, 1960.

Freeman, Otis. *The Pacific Northwest*. Wiley and Sons, New York, 1954.

Mather, John R. *Climatology: Fundamentals and Applications*, MacGraw-Hill, New York, 1974.

Netboy, Anthony. *The Pacific Northwest*. Doubleday, Garden City, New York, 1963.

Rue, Walter. *Weather of the Pacific Coast, Washington, Oregon and British Columbia*. Writing Works Inc., Mercer Island, Washington, 1978.

Reiter, Elmar R. *Jet-stream Meteorology*. University of Chicago Press, Chicago, 1963.

———. *Jet Streams: How Do they Affect Our Weather?* Doubleday, Garden City, New York, 1967.

VEGETATION OF THE PACIFIC NORTHWEST   Chapter 4

Keith S. Hadley
Western Oregon State College

INTRODUCTION

To many people, mention of the Pacific Northwest elicits an image of unbroken forest extending from the Pacific Ocean to the snow crowned summits of the high Cascades and beyond. Such an image, however, belies the incredible diversity of vegetation that clothes the region. Depending on location, one might just as easily envision a landscape of expansive steppe, or a landscape of forest intermixed with wetlands, chaparral, or alpine tundra. Further bestowed upon this diverse landscape is the change wrought by the human activities during the past one and a half centuries. Consequently, the Pacific Northwest today is no longer the primeval landscape encountered by Lewis and Clark but a landscape forged by the early pioneers and their successors.

The goal of this chapter is to explore the diverse nature of this living landscape, and to seek an understanding of the processes that helped shape it. It begins with an introduction to the basic concepts of vegetation and the dynamic processes of vegetation change. These contemporary processes are then placed into an evolutionary perspective as we review the changing features of the regional vegetation since the end of the Cretaceous period approximately 65 million years ago. The chapter culminates with a review of the contemporary vegetation types of the Pacific Northwest and their ecology.

VEGETATION

Vegetation may be defined as the mosaic of plant communities that covers a landscape. As a living mosaic, vegetation is responsive to several factors including climate, soils, topography, biotic interactions, and time. While external or exogenous factors such as climate, soils, and topography are important in determining the character of vegetation over large areas, "internal" or endogenous processes such as plant succession are equally important at the community level. Natural and human disturbance such as fire and logging are also important factors determining the composition and physical characteristics of plant communities.

Succession and Disturbance

Succession, in the traditional sense, refers to the more or less orderly and predictable replacement of species within a plant community over time. Commonly cited examples of succession are pond succession and succession on recently exposed areas following deglaciation. In either case, pioneer species invade a site, alter the local environment, and are replaced by seral or intermediate successional species that are ultimately replaced by the climax or "steady state" species, i.e., those species that can replace themselves indefinitely under constant environmental conditions.

Disturbance refers to discrete events that alter the function, composition, or population structure of a plant community and some aspect of its physical environment. Within the context of traditional succession theory, disturbance was considered a reset mechanism from which succession could begin anew. More recently, disturbance has become regarded as a major cause of vegetation change that is often more effectual than the slower process of succession.

Disturbance can occur in many different natural and anthropogenic forms. Fire, windstorms, insect outbreaks, landslides, and volcanic eruptions are common examples of natural disturbance agents. Because natural disturbances have occurred over evolutionary time, plant species are adapted to those occurring most frequently within their biogeographic range.

The co-evolution of vegetation and disturbance helps explain why human activities such as logging, agriculture, and overgrazing frequently result in the extreme alteration of natural vegetation. Human disturbance is generally episodic, more frequent, more specific in intent, and otherwise distinct in its characteristics when compared with natural disturbance. Ironically, the restraint of natural disturbances such as fire suppression may result in vegetation change as well. Fire suppression, for example, may ultimately result in more severe fires due to increasing fuel loads or increased risk of insect or pathogen infestation.

Finally, we should note that historical circumstances including climatic and geologic change have shaped the regional vegetation as they acted upon preexisting floras. Accordingly, the vegetation of the Pacific Northwest is the result of evolutionary processes that began during the eve of the late Mesozoic (Table 1).

THE ANCESTRAL FLORA AND VEGETATION OF THE PACIFIC NORTHWEST

Many taxa (families, genera, and species) that comprise the current vegetation of the Pacific Northwest are descendants of Late-Mesozoic and Tertiary floras that once covered large areas of the earth's surface. The histories of these floras are important

EPOCH	Y.B.P.*	CLIMATE	OROGENESIS	VEGETATION
HOLOCENE	10,000	Neoglacials Altithermal		Modern
PLEISTOCENE	2,000,000	Cold/Warm (Glacials/Interglacials)	San Francisco Pks.	Modern
PLIOCENE	13,000,000	Cool–Cold	North Cascades/ Greatest uplift of Sierras	Modern (Desert/Alpine)
MIOCENE	25,000,000	Moderate	South Cascades/Rejuvenation- Rocky Mountains	Temperate (Grasslands)
OLIGOCENE	36,000,000	Warm–Rapid Cooling	Initial Sierra uplift	Temperate
EOCENE	58,000,000	Moderate–Warm	Laramide Revolution	Sub-Tropical/Tropical
PALEOCENE	63,000,000	Moderate–Warm	Rocky Moutains/Great Basin	Sub-Tropical/Tropical

* Denotes approximate beginning of Epoch, years before present.

Table 1. Cenozoic Climate, Orogenesis, and Vegetation in Western North America.

because they illustrate the dramatic change that has occurred to the regional vegetation as migrations and extinctions followed changes in climate and physiography. From a geographic perspective, the existence of these earlier "geofloras" helps explain why many genera and species that occur in the Pacific Northwest also occur in east Asia, Mexico, and Central America.

Climatic cooling approximately 65 million years ago marked the beginning of the Tertiary period and development of three regional geofloras, the Arcto-Tertiary Geoflora, the Madro-Tertiary Geoflora, and the Neo-Tropical Geoflora. These geofloras roughly represented the vegetation that evolved in temperate, dry subtropical, and tropical environments respectively. Most of the mesic (moist) temperate forest plants found in the Pacific Northwest today are descendants of the Arcto-Tertiary Geoflora that covered much of North America and eastern Asia. The extent of this flora is known from the occurrence of plant fossils and plant genera common to both areas. In contrast, many xeric (arid) species found in the Pacific Northwest today are related to the Madro-Tertiary Geoflora that evolved in the arid environments of the Sierra Madre Mountains of Mexico (Figure 1).

Tertiary Development of Vegetation in Western North America

Following the Eocene epoch, global climate began to cool and become more arid in western North America, partly in response to the initial uplift of the Sierra Nevada and Coast Range (Table 1). The rainshadows resulting from the rise of these mountain barriers caused a further increase in regional aridity, allowing the Madro-Tertiary Geoflora to expand its range and replace the Arcto-Tertiary plants in the Great Basin, Colombia Plateau, and western Oregon. Later uplift and development of the Cascades caused the isolation of portions of this flora, resulting in some evolutionary divergence and speciation.

During the Miocene, as aridity became widespread, grasslands became an important part of the North American landscape. Continued cooling during the Pliocene along with continued regional uplift, promoted the development of desert and alpine vegetation in regions of rain shadow and high elevation respectively. It was not until the Pliocene that modern vegetation types became fully developed (Table 1).

Quaternary Climate Change and Vegetation

During the late Pliocene, global climate began to cool dramatically, culminating in the ice ages of the Pleistocene. The net effect of climatic cooling on vegetation was a contraction in the range of more temperate species while taxa associated with higher latitudes were forced to migrate south in front of the advancing continental ice sheets. At the same time, taxa occupying higher elevations in the Pacific Northwest moved down slope as conditions deteriorated ahead of growing mountain glaciers.

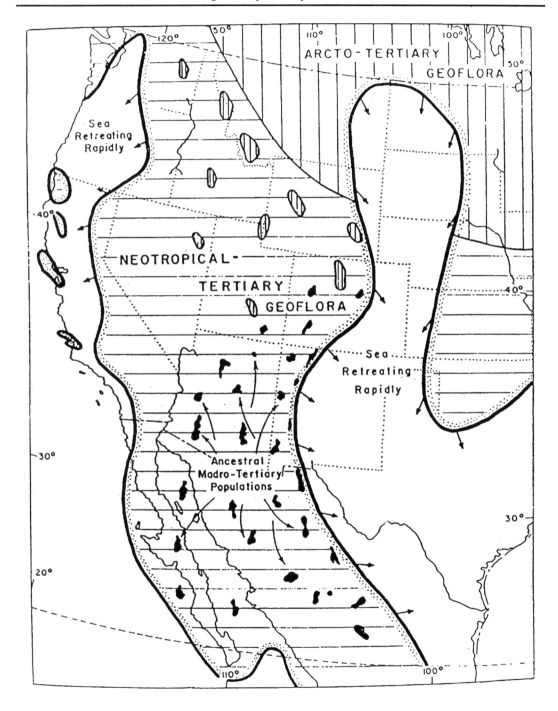

Figure 1. Inferred distributions of the Arcto-Tertiary, Neo-tropical and ancestral Madro-Tertiary Geofloras in pre-Eocene time (From: Axelrod 1958).

Among the more notable changes in the vegetation of the Pacific Northwest during the Quaternary was the arrival of many species that comprise today's evergreen forests. Previously, many of these species had been restricted to higher latitudes. The Quaternary was also a period of southward migration by many circumpolar species along the axes of the Cascades and Rocky Mountains and a time of regional extinction for many less cold-tolerant broadleaf trees and shrubs.

Alternating cool and warm periods during the Quaternary further served to create new and unique plant communities by hastening the intermixing of plant species from different regions. Today, evidence of this intermixing is most pronounced in areas of high species richness such as the Klamath Mountains in southern Oregon and the Iron Mountain area in Oregon's western Cascades. The large number of species in these areas resulted from their central geographic location that allowed plants from floristically distinct areas such as California, the Great Basin, and the Coast Ranges to overlap, and their high habitat diversity.

CONTEMPORARY VEGETATION TYPES

Vegetation can be classified according to several criteria including physiognomy (life forms and structure), floristics (individual or groups of taxa), climax species, geographic location, or site characteristics. Because vegetation is inherently complex, any of these criteria or their combination will have their disadvantages. For our purposes, we will classify the vegetation of the Pacific Northwest using a combination of criteria that focus on important geographic and ecological differences.

At the most general level, the vegetation of the Pacific Northwest is classified according to four vegetation groups as determined by regional climate: (1) Forests, (2) Chaparral, (3) Steppe, and (4) Alpine. Within each group, more specific zone complexes or zones are recognized based upon the dominant or "climax" species or plant form (Figure 2).

Forests

Among the most notable characteristics of the Pacific Northwest vegetation are the great size, productivity, and longevity of the mesic conifer forests that seemingly dominate the character of the regional landscape. In contrast, the dry, interior conifer forests of eastern Oregon and Washington, Idaho and western Montana, are expansive continuations of the northern coniferous forests that extend south along the Rocky Mountains into the American Southwest. Relative to these forests, broadleaf forests and open woodlands are geographically limited to areas of frequent disturbance and pronounced summer drought. In southwest Oregon, broadleaf trees also occur as members of mixed conifer and broadleaf forests.

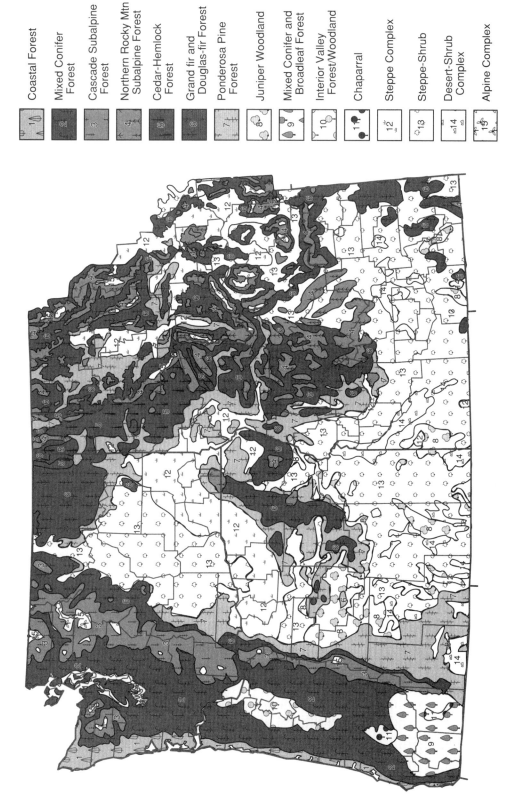

Figure 2. Vegetation Zones and Zone Complexes in the Pacific Northwest (Modified from Kuchler 1964, Franklin and Dryness 1973, and Frenkel 1985).

CONIFEROUS FORESTS

Conifers possess several adaptive traits that allow them to dominate the regional forests of the Pacific Northwest. They are especially well adapted to the forests of western Oregon and Washington where tree growth can proceed without constraint during much of the year. In these forests, large leaf areas, needle shaped leaves, and conical shaped crowns maximize the interception of diffuse sunlight associated with high cloud cover, fog, and the low angle of incident radiation during the winter. In the subalpine forests of the Cascade, Olympic, and Northern Rocky Mountains, the conical crown structure of conifers efficiently sheds heavy snow loads and resists strong winds that can accompany intense winter storms.

The large size of the conifers, especially among the coastal forests, also contributes to the local ecology beyond their considerable lifespan. Among the ecological roles of dead trees are their retention and slow release of nutrients as they decompose. Downed trees also serve as nurse logs for seedlings by providing elevated "safe sites" above competing understory plants of the forest floor.

Coastal Forest Zone

Coastal forests of the Pacific Northwest occupy a narrow fog belt that extends along the shoreline of the Pacific Ocean from southern Alaska to northern California. This area receives copious rainfall (150 to 500 cm per year) and maintains a moderate temperature regime throughout the year with a mean annual temperature of about 16°C.

Sitka spruce *(Picea sitchensis),* a seral species, is the dominant tree in many of these forests by virtue of its great size (up to 5 m in diameter and 90 m tall) and rapid growth. The "climax" species consist of the highly shade tolerant western hemlock *(Tsuga heterophylla),* the long-lived western red cedar *(Thuja plicata),* and Alaska-cedar *(Chamaecyparis nootkatensis).*

Other important tree species of the coastal forests include Douglas-fir *(Pseudotsuga menziesii),* red alder *(Alnus rubra)* in disturbed areas, and shore pine, a subspecies of lodgepole pine *(P. contorta var. contorta)* that occurs on acidic, nutrient-poor soils or in areas of poor drainage. Big leaf maple *(Acer macrophyllum)* may also be a conspicuous component of these forests as it is in the maritime rainforest of the Olympic Peninsular. Although most of these species occur as seral members of the forest community, their presence has increased over the past 150 years due to logging activities.

Mixed Conifer Forest Zone

Inland and at higher elevations than the coastal forests, are the mixed conifer forests of the Coast Range and the western Cascades. These forests are the most widespread in the Pacific Northwest and are well known for their ecologic and economic importance.

Based on its size and growth potential, Douglas-fir is the dominant tree species in these forests. Ecologically, however, these forests are commonly associated with western hemlock, the ubiquitous steady state species. Western red cedar is also an important "climax" species in these forests, which may also include noble fir *(A. procera)* at higher elevations, and incense cedar *(Calcedrus decurrens)*, sugar pine *(P. lambertiana)*, and ponderosa pine *(P. ponderosa)* in southern Oregon.

The dominance of Douglas-fir in these forests can be partially attributed to its greater adaptation to the drier summer conditions of these inland forests. Other factors that contribute to its dominance include its longevity, high seed production, high seed dispersal capacity, and rapid seedling growth following disturbance by fire and logging.

Cascade Subalpine Forest Zone

At higher elevations ($\approx$900 to 1300 m) on the west slopes of the Cascades and in the Olympic Mountains, there is a compositional shift within the mixed conifer forests. Here, the forests possess traits of both the Mixed Conifer Forest Zone and the true Subalpine Forest Zone that occurs at higher elevations. Floristically, these forests share many taxa with the mixed conifer forests but their environmental characteristics are more similar to the subalpine zone. Both western and mountain hemlock *(T. mertensiana)* are present in this transitional zone along with Pacific silver fir *(A. amabilis)*.

At elevations ranging between 1300 and 2000 m, the forests of the western Cascades and Olympics become a true subalpine forest dominated by mountain hemlock. At lower elevations, Douglas-fir, and Pacific silver fir are intermixed with mountain hemlock but give way to subalpine fir (*A. lasiocarpa*) and lodgepole pine (on drier sites) at higher elevations. The subalpine parkland and subalpine/alpine ecotone include subalpine fir and whitebark pine *(P. albicaulis)* on drier sites. Alpine larch *(Larix lyalli)*, a deciduous conifer, occurs at the higher elevations in the Northern Cascades of Washington and Northern Rocky Mountains.

Northern Rocky Mountain Subalpine Forest Zone

The subalpine forests of the Northern Rocky Mountains represent a continuation of the northern coniferous or boreal forests to which they are floristically and structurally similar. Differences between these subalpine forests and those of the Cascade and Olympic Mountains can be attributed to its more continental climate and lower snowpack.

The subalpine forests of the Rocky Mountains are remarkably consistent in their composition from northern Canada to the southwest United States. The climax tree species throughout most of these forests are subalpine fir and Engelmann spruce *(Picea engelmannii)*. In the Pacific Northwest, alpine larch and whitebark pine, which occur at or near treeline, may extend into these forests along rocky and windswept ridges. At the lower elevations the seral species in these forests may include Douglas-fir, lodgepole pine *(P. contorta var. latifolia),* and aspen *(Populus tremuloides).*

Western Red Cedar Zone

The forests of this zone are situated at elevations below 1500 m in the moist valley bottoms of northern Idaho and northwestern Montana. Dominated by western red cedar and western hemlock, these forests are similar to those of the Mixed Conifer Forest Zone in the western Cascades but may also include western white pine and grand fir as subdominant species on drier sites. Although western hemlock is more shade tolerant and its seedlings are more abundant than those of western red cedar, the greater longevity of the latter ($\approx$1000 yrs) allows these species to coexist as "climax" species.

Grand Fir and Douglas-Fir Zone

Among the interior highlands of the Pacific Northwest, grand fir *(A. grandis)* and Douglas-fir dominate a mixed forest zone at elevations between 1100 to more than 2000 m. These forests are typically found in a mesic zone below the subalpine forest zone associated with the Cascades or Northern Rocky Mountains and above the drier zone of the ponderosa pine. Toward the drier continental interior range of these forests, Douglas-fir, again displaying its high adaptability, becomes the dominant tree species.

Ponderosa pine, western larch (L. *occidentalis),* and lodgepole pine are common associates within these forests. These species, along with Douglas-fir, typically compose the early stages of forest succession.

Ponderosa Pine Forest Zone

The drier forests of the eastern Cascades and interior highlands of the Pacific Northwest are dominated by open stands of ponderosa pine. These forests, which have been severely altered by logging, are the product of semi-arid conditions and a high natural fire frequency.

The primary compositional change in these forests is among the understory plants. In the drier forests of Oregon, these understory plants are predominantly shrubs, such as bitterbrush *(Purshia tridentata).* Grasses, including Idaho fescue *(Fescue idahoensis),* are predominant in the more moist forests of Idaho and Montana. In certain areas, notably the pumice region east of the Cascades near Crater Lake in central Oregon, lodgepole pine is the dominant tree species by virtue of its superior adaptation to low nutrient and fine textured soils.

Additional tree species that may occur in "climax" ponderosa pine forests are western juniper, Oregon oak *(Quercus garryana),* and aspen. Oregon oak is restricted to small areas east of the Cascades in northern Oregon and central and southern

Washington. Aspen is found throughout the range of these forests in riparian and poorly drained sites whereas western juniper is found on the most xeric sites.

Juniper Woodland Zone

The juniper woodlands found east of the Cascades and among the upper reaches of the intermountain basins are the driest of the tree-dominated vegetation types in the Pacific Northwest. In general, they occupy areas of intermediate moisture between the Shrub-Steppe Zone and the Ponderosa Pine Forest Zone. Floristically, they represent a northern extension of the Great Basin Pinon Pine-Juniper Zone of Nevada and Utah.

Western juniper *(Juniperus occidentalis)* is the dominant tree species of this zone which may include infrequent ponderosa pine and aspen in moist habitats along streams or at higher elevations. Big sagebrush *(Artemisia* tridentata) and several grass species, including Idaho fescue, bluebunch wheatgrass *(Agropyron spicatum),* and Sandberg's bluegrass *(Poa sandbergii)* are common understory components of these woodlands.

MIXED CONIFER AND BROADLEAF FORESTS

Broadleaf trees are rarely the dominant members of forest communities in the Pacific Northwest. The exception being the oak forests and woodlands of interior valleys of Oregon and the open plains south of Puget Sound in western Washington. Nonetheless, the broadleaf tree species and shrubs inhabiting the Pacific Northwest represent an important floristic component with a diverse history.

Broadleaf species found on mesic sites such as maple *(Acer spp.)* and Pacific dogwood *(Cornus nuttallii),* are relatively shade tolerant and well adapted to their existence among their coniferous companions. Other species such as red alder, willow *(Salix spp.),* birch *(Betula spp.),* Oregon ash *(Fraxinus latifolia),* cottonwood and aspen *(Populus spp.)* are primarily shade-intolerant, pioneer species that invade disturbed sites in riparian and mesic habitats. These species survive by outgrowing conifer seedlings and temporarily dominating the forest canopy. They are short-lived however, and will be replaced by the conifers within 50 to 150 years without repeated disturbance.

Broadleaf trees and shrubs found on more xeric sites including oak *(Quercus spp.),* ceanothus *(Ceanothus spp.),* Pacific madrone *(Arbutus menziesii),* and poison oak *(Rhus diversiloba),* are well adapted to both drought and a high frequency of fire. Their adaptations include expansive root systems, sclerophyllous ("hardened") leaves to decrease transpired water loss, a high capacity for sprouting following fire, and among some oaks, a thick protective bark for protection during ground fires.

Mixed Conifer and Broadleaf Forest Zone

The forests of the Klamath Mountains of southern Oregon and northern California are unique in their diversity and distinct composition. This unique character is the product of an intermingling of northern and southern taxa during the Pleistocene and a complex geology that contributes to a variety of soil types and steep environmental gradients.

These forests typically consist of a combination of conifers and broadleaf trees resulting in a two-storied structure. Conifers usually comprise the upper tree story of these forests with Douglas-fir being the most frequently encountered. Other common conifers include white fir *(A. concolor)*, sugar pine, Shasta fir *(A. magnifica var. shastensis)*, ponderosa pine, and Jeffrey pine *(P. jeffreyi)*. The lower forest story includes several broadleaf trees including tan oak *(Lithocarpus densiflorus)*, Pacific madrone, golden chinkapin *(Castonopsis chrysophylla)*, and three species of true oak; black oak *(Q. kelloggii)*, canyon live oak *(Q. chrysolepis)*, and Sadler oak *(Q. sadleriana)*.

Because of the diverse blend of species, the succession in these forests can be quite complex. Besides a high species diversity, these forests have a complex disturbance history of fire, logging, and grazing. Hence, the relative importance of resprouting by some broadleaf species versus the rapid initial growth of the conifers is an important successional consideration in these forests. Because of these factors, Douglas-fir, white fir, and tan oak are generally considered the major climax species although several other species are considered "minor climax" depending on local conditions.

Interior Valley Forest/Woodland Zone

The interior valley vegetation of western Oregon and Washington are remnants of a former grasslands and savannas that predate the era of Euro-American settlement. These areas were previously maintained as a "disturbance climax" by frequent fires set by local Native Americans to improve game conditions. Today these areas are characterized by closed Oregon oak forests and woodlands which developed following the arrival white settlers and the beginning of a period of lower fire frequency.

The oak forests of the Willamette Valley and lowlands south of Puget Sound are currently being replaced by more shade-tolerant species such as Douglas-fir, big leaf maple, grand fir, or mazzard cherry *(Prunus avium)*, an exotic introduced from Europe. The eventual dominance of these forests by a particular species probably depends on moisture availability. Douglas-fir is the most likely steady state species on drier sites, big leaf maple the steady state species on moist sites, and grand fir on intermediate sites or with Douglas-fir. The mazzard cherry could hypothetically become the climax species in areas where it occurs in dense stands where it could shade out the less shade-tolerant Douglas-fir.

South of the Willamette Valley in Oregon, several additional drought-adapted species become members of these forests and woodlands, including black oak and Pacific madrone. Shrubs and grasses are also important components to these vegetation types. The best-known among the former is the infamous poison oak.

Chaparral

Chaparral is geographically restricted in the Pacific Northwest to portions of the Rogue and Umpqua valleys in southern Oregon. Composed primarily of Madro-Tertiary descendants, this xeric shrub vegetation is prominent in areas having a Mediterranean climate, including southern California and portions of Arizona and Mexico.

Chaparral Zone

The chaparral in the Rogue and Umpqua valleys is generally best developed on south-facing slopes with shallow soils. Often, fire is an important factor in maintaining many of these shrublands as a disturbance climax. Adaptation to serpentine soils that have potentially toxic levels of magnesium, nickel, and chromium and low levels of macronutrients, may also contribute to the persistence of this vegetation type at some sites (see Chapter 5).

Chaparral vegetation is well suited to its seasonally dry environment by virtue of its sclerophyllous leaves, an ability to vegetatively resprout following fire, and variety of root systems to optimize water uptake depending on local soil conditions. Adaptations to the low nutrient status of chaparral soils may also include the ability of some plants, such as ceanothus *(Ceanothus spp.)*, to fix nitrogen.

The primary chaparral species in the Pacific Northwest are narrow-leaved buckbrush *(C. cuneatus)*, white-leaved manzanita *(Arctostaphylos viscida)*, and hoary manzanita *(A. canescens)*. In more moist or less recently burned areas, these shrubs are mixed with several tree species, including Oregon and black oak, tan oak, golden chinkapin, ponderosa pine, and madrone.

Steppe

Steppe vegetation is characterized by its discontinuous ground cover in arid and semi-arid environments. In the Pacific Northwest, steppe vegetation typically includes several species of perennial bunch grasses and sagebrush *(Artemisia spp.)*. Perhaps because of its barren appearance, this vegetation type is often called desert or high desert. Historically, however, this vegetation type is more closely related to high latitude vegetation types than it is to those of the hot, subtropical deserts.

Although similar in appearance, the steppe areas of eastern Oregon and Washington and southwestern Idaho can be divided into three zones or zonal complexes. These include the: (1) Steppe Zone Complex, consisting primarily of perennial grassland communities without sagebrush, (2) Shrub Steppe Zone, consisting of communities with both perennial grasses and sagebrush, and (3) Desert Shrub Zone Complex, consisting primarily of salt tolerant shrub communities.

Steppe Zone Complex

Most of the native steppe in the Pacific Northwest has been altered by burning, grazing, and the mechanical disturbances associated with dryland farming. Still, some regionally significant steppe areas exist in the Palouse Plateau of southeastern Washington and western Idaho and in north-central Oregon. Low moisture availability and the high matric potential of fine textured soils serve to exclude trees from these areas.

Before Euro-American settlement, the native species in this area were a mixture of perennial and annual grasses. These species originated in the mixed grass prairie of the northern Great Plains and annual grasslands of California respectively. In spite of its Mediterranean climate, about 80% of the flora is composed of cool-season grasses derived from the mixed-grass prairie region of the northern Great Plains.

Grass species common to the lesser disturbed steppe habitats include bluebunch wheatgrass, Idaho fescue, Great Basin wildrye *(Elymus condensatus)* and in more moist sites, Sandberg's bluegrass. Today, sagebrush and cheatgrass *(Bromus spp.)* are abundant due to heavy grazing pressures.

Steppe-Shrub Zone

Steppe-shrub vegetation occupies large portions of eastern Oregon and Washington and southwestern Idaho. These areas are more arid than that of the adjacent steppe and consist of a mixture of perennial bunchgrasses and sagebrush. In some areas, especially those associated with rock outcrops, small western juniper may be intermixed with grasses and shrubs.

The dominance of sagebrush in these communities is facilitated by several biological adaptations such as seasonal leaf dimorphism. Seasonal leaf dimorphism refers to the development of two different sets of leaves that allow the plant to maximize its photosynthetic capacity during periods of varying moisture stress. Sagebrush, for example, develop large ephemeral leaves during the early spring that persist until the onset of the summer drought; during the late spring, a second set of smaller leaves develops and will remain in place through the following winter.

Sagebrush also possesses unique root systems that consist of both a fibrous and a tap root component. This adaptation allows the plant to use both surface water and water stored deep within the soil. Other characteristics, such as chemical defenses that inhibit herbivory and seedling establishment by competing plants further ensures the dominance of sagebrush in this arid landscape.

Perennial grasses are only marginally stable members of the shrub-steppe community because of their greater reliance on precipitation for moisture. They do, however, recover better than sagebrush following fire which helps ensure their presence in these communities. Ironically, exotic annual grasses, especially cheatgrass *(Bromus tectorum),* now outcompete the native perennials following fire. Unfortunately, the resulting shift in species composition has resulted in the accelerated degradation of the steppe-shrub community.

The pivotal adaptation of the exotic annual grasses in the Steppe-Shrub Zone is their ability to complete their life-cycle during the spring and early summer when sufficient moisture is available. During the summer, these vast areas of dried grass provide continuous ground cover and fuel for more frequent and hotter fires. As a result, the new fire regime eliminates both the remaining perennial grasses and sagebrush from the plant community and subsequently promotes soil erosion.

Desert-Shrub Zone Complex

Desert-shrub is the most xeric vegetation type in the Pacific Northwest. This vegetation zone is generally restricted to areas of internal drainage within the Great Basin physiographic province.

Many plants that occupy the Desert-Shrub Zone are halophytes, plants adapted to high levels of saline and/or alkaline soils. These plants can exclude salt during water absorption or internally isolate the salts and excrete them.

Habitat types in the Desert-Shrub Zone are often delineated by the depth of the permanent water table or salinity. Among upland areas, where the water table is below 1 m depth throughout the year, saltbush (Cheonopodiaceae) is the dominant plant type. Although extremely salt tolerant, these shrubs can only use water suspended above the saturated zone. Greasewood *(Sarcobatus vermiculatus)* is the dominant shrub in lowland habitats and receives its moisture from a regularly saturated root zone. Marshlands and meadows with standing surface water during much of the year are typically covered by grasses and grasslike plants.

Many ecological problems that confront the Steppe-Shrub Zone also confront the Desert-Shrub Zone. Unfortunately, interest in this vegetation zone is relatively recent and only then following the incursion of exotics, grazing, and the recent arrival of fire. Soil compaction by grazing and vehicles is currently becoming a widespread problem. Soil compaction disrupts the thin biological crust of these soils causing increased soil erosion and decreases in soil nitrogen. Soil compaction also inhibits seeds from germinating in compacted areas.

Alpine

The geographic extent of alpine vegetation in Oregon and Washington is comparatively minor, especially south of the Northern Cascades in Washington. In these areas, true alpine vegetation is confined to small "islands" located at high elevations on the larger volcanos and in the Olympic Mountains. In contrast, the larger contiguous alpine regions of the Northern Cascades and Northern Rocky Mountains possess well-developed alpine vegetation. Floristically, both areas have a strong affinity with the Arctic tundra and the alpine areas of the central Rocky Mountains.

Alpine Zone Complex

Distinctive to the higher elevations of the Pacific Northwest are the extensive parklands of small tree islands distributed among larger meadow communities. The development of this unique forest-alpine ecotone is probably related to broad continuous slopes and the persistence of the winter snowpack into the growing season. These parklands are especially well developed in the volcanic portions of the Cascades and Olympics due to their proximity to maritime air masses.

Plants characteristic of alpine vegetation are predominantly herbaceous, grasslike, or small shrubs. Lichens and mosses become increasingly important at higher elevations. Although not rich in species, the composition of alpine communities is highly variable. At the micro scale changes in the community composition of alpine vegetation are often gradual and associated with local variations in soil type, moisture conditions, and duration of snowpack. At the regional scale, high community diversity is probably due to the recent accessibility of these areas following the Pleistocene glaciations.

Perhaps the most notable structural characteristic of alpine plants is their short stature. Beginning with the krummholtz ("crooked wood") trees at the upper edge of the forest-alpine ecotone, alpine vegetation becomes shorter with increasing elevation. This decrease in stature is the consequence of several factors, most notably, the interaction between snowpack and wind. As wind speeds increase at higher elevations and near ridge tops, snowpack decreases, thus eliminating it as a protective and insulating layer. Plants are thus exposed to lower air temperatures, snow abrasion, and subsequent desiccation.

VEGETATION, BIODIVERSITY, AND CONSERVATION

Biodiversity is a living heritage. In the Pacific Northwest, this heritage is evident in a regional vegetation that links the present landscape with ancient geofloras and tens of millions of years of climatic and geologic change. During the past one hundred and fifty years we have further witnessed a dramatic change in this living heritage, a change that continues today at an unprecedented rate.

Paramount among the contemporary changes in the regional vegetation are those associated with human activities. Some effects of these activities, such as logging, grazing, fire suppression, and soil compaction, are discussed briefly in the preceding text but the majority are not. It is left to the reader to consider the consequences that will be born of these activities. In doing so, however, the reader must consider how organisms that have evolved over millions of years will respond to sudden change, a change so rapid that it has occurred within a small fraction of the lifespan for many species.

Because organisms do not live in isolation, consideration of plant diversity must also implicate animal diversity. In this regard, the diversity of animals in the Pacific

Northwest is notable. For example, more than 1000 species of invertebrates have been found in a single old-growth Douglas-fir stand. Some estimates suggest that the number of species eventually found in these forests may exceed that found in tropical rainforests once the below ground members of the community are cataloged.

The importance of a single species cannot be evaluated in isolation, however, since each species contributes to the overall function and health of its community. The highly publicized issue of the northern spotted owl, for example, is not simply a question of the "value" of a single organism, but rather the northern spotted owl's role as an indicator of the overall health of its community. Unfortunately, as plant communities continue to be degraded, the number of endangered species will continue to rise and today's controversy over the northern spotted owl will become tomorrow's conflict over the marble murrelet.

The importance of biodiversity often reaches beyond the realm of ecology and into the social and economic well-being of humans. Among the socioeconomic advantages of maintaining biodiversity is the potential for developing new medical drugs. One such example is the Pacific yew *(Taxus brevifolia)* which grows among the regional old-growth forests of the Pacific Northwest. This species has attracted international attention since the drug taxol, derived from its bark, has proven effective against several types of cancer. Consequently, a formerly unimportant species must now be reevaluated as a natural resource.

As the landscape of the Pacific Northwest continues to change during the twenty-first century, it will not resemble the vegetation of today any more than today's vegetation resembles that encountered by Lewis and Clark. We can also be reasonably certain that, without increased restrictions of our own activities, this living heritage will become greatly diminished with time.

REFERENCES

Arno, S.F. and Hammerly, R.P. 1977. *Northwest Trees.* The Mountaineers, Seattle, Washington. 222 p.

Axelrod, A.I. 1958. Evolution of the Madro-Tertiary Geoflora. *Botanical Review* 24:433–509.

Barbour, M.G. and Billings, W.D. eds. 1988. North American Terrestrial Vegetation. Cambridge University Press, New York, New York.

Detling, L. E. 1968. Historical background of the flora of the Pacific Northwest. *University of Oregon Museum of Natural History Bulletin 13.* 57 p.

Franklin, J.F. and Dyrness, C.T. 1984. *Natural Vegetation of Oregon and Washington.* USDA Forest Service General Technical Report PNW-8. 417 p.

Habeck, J.R. 1987. Present-Day Vegetation in the Northern Rocky Mountains. *Annals of the Missouri Botanical Garden* 74:804–840.

Hansen, H.P. 1947. Postglacial Forest Succession, Climate, and Chronology in the Pacific Northwest. *Transactions of the American Philosophical Society* 37:1–130.

Heusser, C.J. 1960. Late-Pleistocene environments of North Pacific North America. *American Geographic Society Special Publication* 35. 308 p.

Kuchler, A.W. 1964. Potential natural vegetation of the conterminous United States (map and manual). *American Geographical Society Special Publication* 36:1–116.

Sudworth, G.B. 1967. *Forest Trees of the Pacific Slope.* Dover Publications, New York, N.Y. 455 p.

Vale, T.R. 1982. *Plants and People: Vegetation Change in North America.* Resource Publications in Geography Series, Association of American Geographers, Washington, D.C. 88 p.

Waring, R.H. and Franklin, J.F. 1979. Evergreen Coniferous Forests of the Pacific Northwest. *Science* 204:1380–1386.

Wolfe, J.A. 1968. Neogene floristic and vegetational history of the Pacific Northwest. *Madrono* 20:83–110.

____. 1978. A paleobotanical interpretation of Tertiary climates in the Northern Hemisphere. *American Science* 66:694–703.

LIST OF FIGURES

Figure 1. Inferred distributions of the Arcto-Tertiary, Neo-tropical and ancestral Madro-Tertiary Geofloras in pre-Eocene time (From: Axelrod 1958).

Figure 2. Vegetation Zones and Zone Complexes in the Pacific Northwest (Modified from Kuchler 1964, Franklin and Dryness 1973, and Frenkel 1985).

SOILS OF THE PACIFIC NORTHWEST — Chapter 5

Keith S. Hadley
Western Oregon State College

INTRODUCTION

The soils of the Pacific Northwest are as varied as the geology, land forms, vegetation, and climate that are responsible for their development. Although considered by most to be a natural component of the region's landscape, many soils of the Pacific Northwest have been modified by human activities. In this regard, soils represent a cultural feature of the regional landscape as well as a natural resource and important component of the region's economy.

This chapter describes the nature, distribution, and use of soils in the Pacific Northwest. It begins with an introduction to basic soil concepts and the factors and processes responsible for their formation. This is followed by a brief discussion of soil taxonomy and a description of the soils found in the Pacific Northwest. The chapter concludes with an overview of the uses of soils in the Pacific Northwest and a regional perspective on soil degradation and conservation.

SOILS

Generally defined, a soil is a natural body consisting of layers or horizons of mineral and organic constituents that differ from their parent material in their morphological, physical and chemical properties, and their biological characteristics. Like all components of a natural landscape, soils can be thought of as dynamic systems that are continually changing in response to external environmental factors and internal or pedogenic processes. Overall, external factors such as climate and vegetation define the conditions of soil development since they govern the type and rate of internal pedogenic processes. Pedogenic processes more directly determine the evolutionary characteristics of a soil, including its color, number and type of horizons, texture, structure, and other properties.

To better understand the relationship between soils and environmental conditions, Hans Jenny (1941) developed an often-cited factorial model of soil development. This model identifies five major external soil-forming factors that govern the rate and pathways of soil development. These factors include climate, parent material, relief, biota, and time. In theory, each of these soil-forming factors is considered independent. This means their influence on soil development can be identified provided each of the other factors is held constant. For example, using Jenny's approach, we can compare soils that have developed under the influence of the same climate, vegetation, parent material, and age but occur in different topographic situations. By holding the first four variables constant, we can then determine the role of topography in the development of those soils. One advantage of Jenny's factorial approach is its provision of a conceptual framework that allows us to ask meaningful questions about the development of a particular soil.

EXTERNAL SOIL FORMING FACTORS

The development and distribution of soils in the Pacific Northwest are closely related to climate, geology, and vegetation, as these factors control the chemistry, physics, and biology of a soil. Relief, or the topographic expression of the landscape, is also important because it influences erosion and deposition rates. Time defines the duration of internal soil processes and thus, soil development.

Climate

Climate is the most important environmental factor governing soil development on a global or regional scale. Climate defines how much energy is available for the soil system and the amount of moisture available for biological activity, weathering processes, and the transport of materials through the soil profile. Temperature also plays an important role in defining the type and extent of soil development because it determines the rate of chemical reactions and thus the decomposition rates of organic and mineral constituents of the soil. Temperature also determines the state in which water exists in the soil (liquid, gas, or solid), the rate of soil moisture loss due to evapotranspiration, and the level of biological activity.

In the Pacific Northwest, soil types generally mirror regional moisture and temperature patterns. This is most apparent in the longitudinal variation of soils associated with the windward and leeward sides of moisture barriers such as the Olympic Mountains, the coast ranges of Oregon and Washington, the Cascade Range, and the Northern Rocky Mountains. Like vegetation, soil types vary along elevational gradients in response to local changes in temperature and moisture.

Parent Material

"Parent material" refers to any mineral or organic substance in which a soil develops. Because soils retain many properties of their parent material, they inherit certain chemical and physical characteristics. This is especially true for young soils and soils developing in dry and cool environments. Parent material thus defines, among other things, the chemical composition, nutrient status, and clay mineralogy of a soil. The parent material also contributes to the texture and structure of a soil.

Soil texture refers to the percentages of sand, silt, and clay-sized particles in the soil and is important in determining the porosity and permeability of the soil. Porosity and permeability are important because they control the soil's capacity to hold moisture and gases and the rate at which they can move through the soil profile.

"Structure" refers to the aggregate arrangement of soil particles, which also influences permeability and the movement of gases and solutions through the soil. Both texture and structure are important in determining the suitability of a soil for plant growth and for the identification of locally important pedogenic processes.

Because of their influence on soil development, soil types frequently coincide with parent material boundaries and exhibit distinct geographic patterns. In the Pacific Northwest, the major parent materials for soils exhibit a west to east distributional pattern similar to that of the regional climate. A variety of marine sediments, basalt, and a minor amount of intrusive materials serve as parent material in the coast ranges of Oregon and Washington and in the Olympic Mountains. In the Puget Sound lowlands and Willamette Valley, these materials are replaced by younger sediments derived from a variety of sources. Soils in the Puget Sound lowlands are derived from marine sediments and glacial and fluvial or stream processes. Fluvial and glaciofluvial deposits are also important parent materials in the Willamette Valley as are lacustrine or lake deposits resulting from a series of floods emanating from glacial Lake Missoula during the Pleistocene (see Chapter 2).

Further inland, the western Cascades of Oregon and Washington are underlain by flow basalt, as is much of the Columbia Intermontane and portions of the Basin and Range geomorphic provinces. Each of these areas also includes extensive tracts of Tertiary sediments and Quaternary deposits, including the wind deposited loess of the Palouse.

The Northern and Middle Rocky Mountain geomorphic provinces are geologically complex and provide a diverse array of parent materials for soils, most notably Paleozoic metasediments, Mesozoic marine sediments, and intrusive granitic materials. Volcanic material along the crest and eastern flank of the high Cascades in Oregon and southern Washington and northern Idaho provides silica-rich soil parent materials including andesite, dacite, and tephra. Volcanic ash from the eruption of Mt. Mazama in Oregon serves as the parent material for some soils in northwestern Montana.

Biota

Vegetation governs soil development through a variety of processes. Foremost, vegetation provides a major recycling pathway for soil nutrients. In addition, organic matter derived from plant litter and root decay 1) adds nutrients to the soil, 2) modifies soil texture and structure, and 3) provides a variety of chemical compounds that hasten chemical weathering processes and the transport of minerals through the soil profile. Organic matter further increases the moisture-holding capacity of the soil and serves as the energy source for most soil organisms.

Plants are also important in atmospheric gas exchange of carbon dioxide, oxygen, and water vapor and their modification of local microclimates. Root respiration is a primary source of soil carbon dioxide and thus plays a major role in the development of carbonic acid, an important weathering agent. Vegetation is also important in controlling water and wind erosion by binding the soil in place and decreasing the velocity of water runoff.

Animals also play an important role in soil development and degradation. Soil microbes perform a myriad of essential soil processes, including the decomposition of organic matter and the subsequent mineralization of soil nutrients. Bacteria and mycorrhizae fungi form important symbiotic relationships with plants that improve plant access to nitrogen and other mineral nutrients. Invertebrates contribute to the breakdown of organic matter and are important mixing agents. Burrowing vertebrates also contribute to the mixing of soils, and their tunnels can be an important factor in contributing to the through-flow of water. Grazing animals compact and scarify soils which may lead to increased erosion by decreasing permeability and removal of plant cover.

As with climate, the soils and vegetation of the Pacific Northwest share similar geographic distributions. One consequence of this shared distribution is the type of erosion that occurs in humid versus arid environments. In the humid, high biomass environments west of the Cascades and at higher elevations inland, surface erosion is comparatively low because of the protective covering of the vegetation. In these areas, surface runoff generally occurs as saturated overland flow after the field capacity of the soil has been exceeded. Since vegetation intercepts precipitation before it impacts the soil, it decreases the effects of rainsplash erosion and increases the "residence time" of the moisture, allowing it to infiltrate the soil.

In the arid environments of the Columbia Intermontane and Basin and Range geomorphic provinces, sparse vegetation cover promotes rainsplash erosion and allows precipitation rates to exceed infiltration rates. This results in a non-saturated or Horton overland flow, which results in high erosion rates and the development of rills and gullies common in arid environments. One repercussion of the large-scale removal of vegetation in humid areas is an increase in Horton overland flow and soil erosion.

Relief

Relief or topographic variation influences the rate of soil development by increasing erosion and deposition rates. In either case, soils occupying slopes have a less well-developed profile as the soil is being either continually removed or buried, depending on its topographic position. As a result, soils that occupy a mid-slope position will be morphologically "young" and shallow compared with those occupying more stable landscape positions. Soils in a toeslope position may also be "young" but over-thickened due to the accumulation of sediments. These soils typically have higher soil moisture, more dense vegetation cover, and higher nutrient status than their mid-slope and hilltop counterparts.

Relief also has an important influence on the local climate. All slopes result in an angular change of solar incidence and thus influence how much insolation a slope receives. This can dramatically alter the local vegetation and slope stability. The direction of the slope or aspect can also influence soil development (Table 1). Aspect is important in all mid-latitude regions and is especially notable in the Pacific Northwest where grassland or steppe vegetation is transitional with forest vegetation. In these areas, south-facing slopes are often covered by drought-resistant grasses and shrubs which contrast markedly with tree covered north-facing slopes.

South-Facing Slopes	North-Facing Slopes
High insolation values	Low insolation values
High soil temperatures	Low soil temperatures
Low soil moisture	High soil moisture
Low snow accumulation	High snow accumulation
Drier growing season	More moist growing season
Lower vegetation cover	More dense vegetation cover
Low organic matter	High organic matter
Low soil fertility	High soil fertility
Thinner soils	Thicker soils
Higher surface erosion	Higher mass-movement

Table 1. The relative influence of aspect on soil formation in the Pacific Northwest.

Time

Time defines the duration of pedogenic processes. Assuming a constant rate of pedogenesis, we would expect older soils to possess different morphological, physical, and chemical characteristics than younger soils. For example, older soils exhibit greater rock and mineral weathering than younger soils as well as changes in clay mineralogy through clay alteration and recrystallization. These and other changes in soils over time allow pedologists to date or assign relative ages to soils.

The most commonly noted field characteristics used to determine the relative ages of soils are 1) increased reddening of the soil caused by the oxidation of iron and clay accumulation and 2) development of a clay-enriched subsurface or B horizon (Figure 1). Among arid land soils, calcium carbonate ($CaCO_3$) accumulation is often a good indicator of relative age. Other evidence of soil age includes the degree of mineral etching, clay mineralogy, and trends in iron, aluminum, and phosphorus concentrations. Absolute soil ages can be determined under circumstances that permit the use of radio-isotopic methods, paleomagnetism, chemical dating techniques, or dendrochronology. The relative and absolute dating of soils is important for several reasons, including the determination of rates of soil development and erosion and dating geomorphic features such as fault lines or flood deposits or aging cultural features associated with archaeological sites.

Several additional concepts related to soil age are noteworthy. First, soil age is better characterized by the geologic time scale than an ecological time scale. Although some soils may be very young (tens to hundreds of years old), most soils are hundreds to thousands of years old. Some surface soils in the Pacific Northwest are tens of thousands of years old, whereas paleosols may be hundreds of thousands to millions of years old. Second, older soils are often polygenetic, that is, their longevity has allowed them to experience pedogenesis under a variety of environmental conditions generally associated with climate change. This means that certain soil characteristics may not be in sync with the current environment, but that these soils contain pedogenic information that may provide clues about past environmental conditions. Finally, the age of soils and their rate of development is not without social implications. Because many soil properties develop over comparatively long periods, soil degradation can result in prolonged periods of low agricultural, rangeland, and forest productivity.

INTERNAL SOIL-FORMING PROCESSES

While external factors are important in determining the development of soils over large areas, internal pedogenic processes are responsible for the unique characteristics of soils. Pedogenesis involves many pathways and includes processes that may take place simultaneously or in sequence and may act mutually to reinforce or contradict each other.

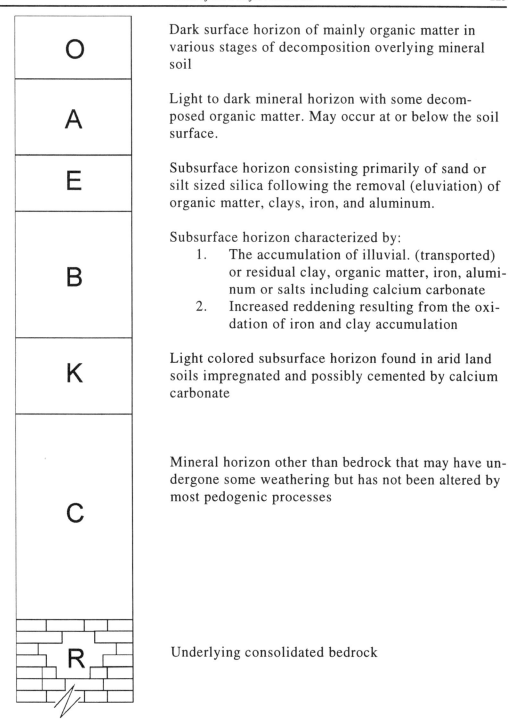

O	Dark surface horizon of mainly organic matter in various stages of decomposition overlying mineral soil
A	Light to dark mineral horizon with some decomposed organic matter. May occur at or below the soil surface.
E	Subsurface horizon consisting primarily of sand or silt sized silica following the removal (eluviation) of organic matter, clays, iron, and aluminum.
B	Subsurface horizon characterized by: 1. The accumulation of illuvial. (transported) or residual clay, organic matter, iron, aluminum or salts including calcium carbonate 2. Increased reddening resulting from the oxidation of iron and clay accumulation
K	Light colored subsurface horizon found in arid land soils impregnated and possibly cemented by calcium carbonate
C	Mineral horizon other than bedrock that may have undergone some weathering but has not been altered by most pedogenic processes
R	Underlying consolidated bedrock

Figure 1.

Pedogenic processes are best thought of as a combination of chemical, physical, and biological subprocesses and reactions that alter the characteristics of a soil (Table 2). The outcomes of these subprocesses or reactions are reflected in the gains and losses to the soil body and the translocation and transformation of materials within the soil. The outcomes of these pedogenic processes are thus manifested in the depth of the soil, number and type of soil horizons, soil color, texture, structure, clay mineralogy, and chemical composition. Each process also imparts a unique and measurable "signature" on the soil that reflects its developmental history. The sum of these "signatures" when compared with unaltered parent material represents soil development or the maturity of the soil.

Soils of the Pacific Northwest are not unique in being subject to a large array of pedogenic processes. However, some of these processes are more notable than others in their contribution to the development of regional soil types. These processes include podzolization, calcification, salinization, gleization, melanization, and braunification/rubifaction/ ferrugination (Table 2).

SOIL TAXONOMY

Soil taxonomy refers to the classification of soils based on observable, measurable, and quantifiable properties. In a formal sense, *Soil Taxonomy* refers to a hierarchical soil classification system developed by the Soil Conservation Service in 1975.

Soil Taxonomy provides for the classification of soils into six increasingly specific categories. From most inclusive to least inclusive these taxonomic levels are 1) Order, 2) Suborder, 3) Great Group, 4) Subgroup, 5) Family, and 6) Series. Each of these taxonomic levels is useful for classifying soils for specific purposes or at different geographic scales. For example, the soil series would be the most useful taxonomic level for assessing the specific properties of a soil and how that soil would respond to a particular management strategy. For our purposes we will focus on the soil orders present in the Pacific Northwest.

Soil Orders

Ten of the eleven soil orders identified in *Soil Taxonomy* are found in the Pacific Northwest (Table 4). Only Oxisols, which typically occur in tropical environments, are absent. Below I describe the basic characteristics of each soil order and their regional distributions. Because several soil orders may be present within a given area, they are mapped as soil complexes with the most common soil orders noted for each area (Figure 2).

Type	Description
Eluviation	Movement of material out of a horizon
Illuviation	Movement of material into a horizon
Leaching	Removal of material from a soil via solution
Enrichment	Addition of material to a soil
Erosion	Detachment and removal of surface material
Cumulization	Hydrologic and eolian addition of soil particles
Calcification	Accumulation of calcium carbonate to a soil horizon(s)
Salinization	Accumulation of soluble salts such as sulfates and chlorides
Alkalization	Accumulation of sodium ions on cation exchange sites
Lessivage	Mechanical transport of silts and clay-sized particles to illuvial horizons
Pedoturbation	Physical and biological mixing of the soil
Podzolization	Chemical movement of iron, aluminum, and organic matter
Decomposition	Chemical breakdown of mineral and biological materials
Melanization	Darkening of material through the addition of organic material
Littering	Accumulation of surface organic material
Mineralization	Release of minerals through the decomposition of organic material
Braunification/Rubifaction/Ferrugination	Browning or reddening caused by the oxidation of iron and the accumulation of iron oxides
Gleization	Reduction of iron under anaerobic conditions

Table 2. Some important pedogenic processes for soils of the Pacific Northwest (modified from Buol, Hoke, and McCracken 1989).

Taxonomic Level	Characteristics
Order	Orders are distinguished by the presence or absence of major diagnostic horizons including epipedons or surface horizons.
Suborders	Suborders are based on soil properties influenced by soil moisture, soil temperature, and natural vegetation.
Great Groups	Great Groups are identified based on the presence, absence or combination of diagnostic subsurface horizons.
Subgroups	Subgroups reflect deviations within a Great Group based on those characteristics that best define them.
Families	Families are based on similarities in physical and chemical properties that affect how the soil could be best managed. This would include differences in soil depth, texture, mineralogy, and temperature.
Series	The most specific and least inclusive classification, series are named after a particular location.

Table 3. Categories of Soil Taxonomy. Categories are listed from most inclusive to least inclusive.

Alfisols

Alfisols are common soils in areas of medium to high precipitation and forest vegetation. In the Pacific Northwest, this includes many forest soils in the coast ranges of Oregon and Washington, the Olympic Mountains, and the inland mountain ranges including the Northern and Central Rocky Mountains. Alfisols can also be present in open woodland and grassland environments where precipitation is sufficient to leach base cations from the soil. Examples of such areas include portions of the Willamette Valley and Puget Sound lowlands.

Figure 2.

Name	Major Characteristics
Alfisols	Ped**alf**er; forest or grassland soils with a moderate abundance of nutrients
Andisols	**And**esite; soils developed from volcanic ejecta
Aridisols	**Arid** soils associated with desert or steppe vegetation
Entisols	Rec**ent** soils with little profile development
Histosols	**Histos** (Greek "tissue"); soils with a high percentage (>30%) of organic matter
In**cept**isols	In**cept**ion of soil development; some diagnostic soil features
Mollisols	**Moll**ic (soft); dark colored surface horizon generally associated with grassland vegetation and high nutrient status
Spodosols	**Spodos** (Greek "wood ash"); soils that develop in acidic environments that include an "ashy" layer (E horizon) and the accumulation of oxidized iron, aluminum, and humus
Vertisols	**Vert**o (Latin "turn"); soils with high concentrations of expandable clays
Ultisols	**Ult**imate soils; old soils or soils with low concentrations of nutrients.

Table 4. Soil Orders of the Pacific Northwest and Their Major Characteristics.

Podzolization is the distinguishing pedogenic process associated with Alfisols. This process is responsible for the development of a weak eluvial E horizon and a clay-enriched or argillic B horizon common in these soils (Figure 1). E horizons are light colored, sandy layers that result from the removal of iron, aluminum and organic matter from the upper portion of the soil profile. Clay-rich argillic horizons develop through two processes, the *in situ* weathering of the parent material or the accumulation of clay transported through the soil profile by water.

The abundance of precipitation and the acidic nature of the organic compounds associated with tree litter contribute to making these moderately acidic soils. Although Alfisols have a moderately high saturation of base ions (base saturation > 35 percent) and nutrient status, they are subject to the loss of cations such as calcium, magnesium, potassium, and phosphorus from the soil.

Andisols

Andisols are more common in the Pacific Northwest than in any other region of the United States. Regionally they are associated with the high Cascades of Oregon and Washington and their eastern slopes and portions of the Northern Rocky Mountains in Idaho and Montana.

Andisols are soils dominated by silica-rich minerals derived from recent volcanic ejecta. These minerals include allophane and imogolite (amorphous, hydrated aluminosilicates of various composition) or aluminum-organic matter complexes.

Overall, Andisols are weakly developed soils that were until recently classified as Inceptisols. Like Inceptisols, Andisols show no translocation of clays through the soil profile. The distinguishing morphological characteristics of Andisols include a dark color and low bulk density. One unique characteristic of Andisols is their potential to be either fertile or infertile depending on local pH conditions.

Aridisols

Based on their geographic distribution, Aridisols are one of the more common soils in the Pacific Northwest. These soils dominate the steppe-covered landscapes of the Columbia Intermontane and western portion of the Basin and Range geomorphic provinces in eastern Oregon and Washington and southern Idaho.

Aridisols are light colored soils that develop in environments where evaporation exceeds precipitation and where prolonged droughts are common. Calcification is common in these soils where the quantity, concentration, and depth of calcium carbonate accumulation depend on the amount of precipitation and the depth of wetting within the soil profile. Where calcium carbonate has accumulated over tens of thousands of years, a thick indurated or cemented K horizon may be present near the soil surface (Figure 1).

Salinization, the accumulation of soluble salts such as sodium sulfate (gypsum) and sodium chloride (table salt), is another pedogenic process common in some Aridisols. Salinization is similar to calcification but restricted to hyper-arid environments with negative moisture balance, i.e., where more moisture is lost through evapotranspiration than is gained through precipitation. This process also involves the accumulation of soluble salts in areas where overland flow occurs during some part of the year. In the Pacific Northwest, salinization is primarily restricted to the Basin and Range geomorphic province where water accumulates in internally drained basins during the rainy season and evaporates during the summer.

Salinization can also be human-induced in irrigated areas. It results from a build-up of salts at or near the soil surface due to the "wicking effect" of capillarity and can decrease the agricultural potential of a soil. In extreme cases, salinization results in the formation of a white crust on the soil caused by the accumulation of sodium chloride or gypsum. The black crust found in some Aridisols indicates the accumulation of sodium carbonate (black alkali) through the process of alkalization.

Morphologically, Aridisols are typically characterized by a weakly developed A horizon over an argillic horizon with some accumulation of calcium carbonate or other more soluble salts. Soils having high concentrations of both clay and calcium carbonate represent polygenetic soils that developed during the alternating moist and dry periods of the Quaternary.

Entisols

Entisols are very young soils that lack most horizons. Any soil development is restricted to a weakly developed A horizon over a weakly developed C horizon. Entisols are common in areas that have experienced recent soil disturbance, including 1) areas of recent deposits caused by flooding, landslides, recent lava flows, glaciation, or non-active sand dunes or 2) recently eroded areas. Because the distinguishing characteristics of Entisols are their young age and weak development, they can occur in any climate regime or environmental setting. The most expansive area of Entisols in the Pacific Northwest occurs on the Pumice Plateau of Central Oregon where recent accumulations of volcanic ejecta cover a weakly dissected landscape.

Uses of Entisols are generally restricted to recreation, wildlife preservation, grazing, and military test sites. However, where Entisols develop as the result of flooding, they may be productive agricultural soils.

Histosols

Histosols occupy the least area of any soil order in the Pacific Northwest. These soils are restricted to areas having high water tables and poor drainage. Regionally, they occur in waterlogged areas near estuaries along the Oregon and Washington coast, marshes in the Basin and Range geomorphic provinces, bogs and fens at higher elevations, and in flood plains. The largest areas of Histosols in the Pacific Northwest are associated with marshes near the margins of pluvial lakes in Basin and Range geomorphic province.

The accumulation of organic matter in Histosols is the result of slow decomposition rates under anaerobic and saturated conditions during a prolonged portion of the year. The saturated condition of these soils also results in gleying, the development of a bluish gray coloring and a mottled or spotted appearance caused by the reduction of iron. Where Histosols are properly drained, they can be productive agricultural soils for acid-tolerant crops including vegetables such as potatoes and berries. The productivity of these soils is related to their high nutrient content, low bulk density, and ability to hold soil moisture.

Inceptisols

Inceptisols are young soils with weak to moderate soil development. Although similar to Entisols in occupying recently stabilized portions of the landscape, Inceptisols are more widespread. Inceptisols are common soils in mountainous areas throughout the Pacific Northwest and are common on flood plains and in alpine environments where their "young" age reflects the slow rates of chemical weathering.

Although Inceptisols are weakly developed soils, they do exhibit some profile development, including the presence of incipient B horizons. Typically, the type of soil development for Inceptisols can be predicted, suggesting that they represent a developmental or early stage of pedogenesis. In some environments, however, soil disturbance is sufficiently frequent to ensure that soil development does not reach pedologic maturity.

Because many Inceptisols lack the soil development needed to support agriculture, they are often used for grazing stock, forestry, and recreation. However, where they have developed as the result of cumulative processes such as flooding, Inceptisols may be productive agricultural soils.

Mollisols

Mollisols are generally considered the most agriculturally productive soils because of their high nutrient status and favorable structure. These soils are common in subhumid and semi-arid grassland environments that possess seasonal water deficits. Mollisols are widespread in the Pacific Northwest and are a conspicuous component of the grassland environments. They are common among the soils of the Palouse in eastern Washington and Oregon and western Idaho, the northeastern portion of the Basin and Range Province, the western margins of the Central Rocky Mountain Province, the western foothills of Montana, and portions of the upland areas of northeastern Oregon and the Willamette Valley.

Mollisols are characterized by a dark brown epipedon or surface horizon resulting from the process of melanization. Melanization, or the darkening of a soil, is the result of the decomposition of the dense fine roots below the soil surface. This high below-ground biomass is associated with native perennial grasses and produces a highly tillable soil structure well suited for dry-land farming.

The chemical and physical characteristics of Mollisols are notable. These soils possess high base saturation (> 50 percent), moderate to high pH, and a high cation exchange capacity. The high base saturation is the result of the accumulation of calcium (calcification), phosphorous, and potassium that serve as macronutrients for most plants.

Spodosols

Spodosols are the result of intense podzolization and chelation. These processes result in the translocation of iron, aluminum, and humus through the soil profile. Consequently, Spodosol development requires abundant precipitation, the presence of organic acids, and a permeable, acidic parent material. Spodosols develop best on granitic or sandy parent materials because of their high silica content and high permeability. These conditions are most common in the Pacific Northwest along the

Oregon and Washington coast in areas of paleodune deposits or along old marine terraces and in the Puget Sound area. Spodosols can also occur in humid forest environments near the coast or at higher inland elevations. However, Spodosols are comparatively rare at these sites because they lack the unique climatic, vegetative, and geologic conditions needed for their development.

Although Spodosols and Alfisols both develop under the process of podzolization, the intensity of this process is much greater in Spodosols. As a result, Spodosols display better developed E horizons and a diagnostic spodic B horizon having low clay concentrations but high concentrations of iron, aluminum, and organic matter. Spodosols also differ from Alfisols in having a lower pH, lower base saturation, lower cation exchange capacities, and lower fertility, as many plant macronutrients are leached from the soil.

Ultisols

Ultisols are highly weathered soils that reflect their comparatively great age or the high susceptibility of the parent material to weathering. They are an important soil order in humid areas underlain by basalt such as the coast ranges of Oregon and Washington and the western Cascades and in areas that have undergone pedogenesis over long periods.

Ultisols are characterized by their red color caused by the oxidation (braunification/rubifaction/ferrugination) of iron and a clayey B horizon. Typically, cations have been leached from these soils resulting in low base saturation, low cation exchange capacity, and low fertility. Ultisols are, however, productive forest soils and suitable for most types of land use except intensive agriculture.

Vertisols

A unique quality of Vertisols is their dynamic nature. These soils have high concentrations of montmorillonite, a clay that has a great expansion and contraction potential. Because of the dynamic nature of these clays, the motion of these soils is sufficient to prevent the development of horizons.

Vertisols occur in wet/dry climates where large cracks in the soil develop during the dry season and small mounds or gilgai form during the rainy season. Vertisols are similar to Entisols and Inceptisols in that they can occur in a variety of environments and lack strong horizonation. However, their distribution is restricted to areas having parent materials that consist of high concentrations of expanding clays, especially montmorillonite, a common weathering product of basalt. Within the Pacific Northwest, Vertisols can be found in the Willamette Valley, the Puget Sound lowlands, and in isolated areas of the Columbia Intermontane Province where the surface geology consists of marine shales.

SPECIAL SOIL TYPES

Of the many unique soils found in the Pacific Northwest, serpentine soils and paleosols merit special mention. Serpentine soils are distinctive because of their role in plant ecology and evolution. Paleosols or "fossil soils" are distinguished by their utility as stratigraphic units and the information they provide about past environmental conditions.

Serpentine Soils

Serpentine soils are common in areas near continental plate boundaries where ultramafic rocks such as serpentinite, dunite, and peridotite are components of the local geology. In the Pacific Northwest, this includes approximately 450 square miles in Oregon and 200 square miles in Washington.

The distinctive nature of serpentine soils results from their unique chemical and physical properties. Chemically, serpentine soils are limited in their abundance of macronutrients, including calcium, nitrogen, phosphate, and potassium. Conversely, they often possess near-toxic concentrations of magnesium and heavy metals such as nickel, chromium, and cobalt. Thus, the most significant peculiarity of serpentine soils is their influence on vegetation.

Floras occurring on serpentine soils are often species rich and include many highly endemic or locally evolved plants. This high level of endemism is the result of selective pressure presented by the extreme properties of these soils. Serpentine soils also promote distinctive changes in plant physiognomies as compared to adjacent non-serpentine soils. This is commonly expressed in a shift from trees to chaparral or chaparral to grassland vegetation where serpentine soils contact other soil types. In extreme cases of heavy metal toxicity, vegetation may be absent on serpentine soils.

Paleosols

Paleosols are soils that developed under past environmental conditions. They include buried, relict, and exhumed soils, but must possess pedogenic properties that are out of phase with current environmental conditions. For example, paleosols might include traits that suggest they formed in wetter, drier, warmer or cooler conditions than those that currently exist at a given location. The study of paleosols is important because it provides a method for understanding prehistoric environmental conditions and allows for the relative dating of land forms and geologic features. For example, the characteristics of a paleosol are often used to learn the environmental conditions that existed at archaeological sites during periods of prehistoric occupation. Paleosols are also useful for determining relative ages of geologic deposits and the frequency of tectonic activity along fault lines.

Paleosols of various ages have been identified in several areas of the Pacific Northwest. Some older paleosols include Tertiary (Eocene) and younger soils identified by Retallack (1981) in central Oregon. These paleosols represent a variety of soil orders and are identified by several pedogenic properties such as soil horizons, textural changes, soil structure, calcium carbonate accumulation, plant litter, root casts, and rooted plants. The different types of paleosols in this area represent a variety of environmental conditions ranging from humid tropical to arid climates.

Pliocene, Pleistocene, and Holocene paleosols are common in several areas of the Pacific Northwest, including the Snake River Plain in Idaho, the Puget Sound lowlands, the Cascades of Washington, the Palouse, and the Willamette Valley of Oregon. The series of paleosols described for the Snake River Plain is Pliocene (ca. 3 million years) to Pleistocene in age and believed to be developed on glacially derived materials. Associated with these soils are Pliocene and Pleistocene faunal remains that, along with the development of different aged paleosols, suggest several periods of environmental change during the past three million years.

More recent paleosols in Western Washington and the Washington Cascades are developed on glacial till, outwash, or stream deposits and suggest Pleistocene changes in climate and vegetation. In contrast, the paleosols of the more arid Palouse are found in a series of loess deposits where they developed between episodes of eolian deposition. Near Washtucna, Washington, 21 different paleosols have been identified in 26 meters of loess exposure. These soils provide localized evidence of Pleistocene environmental change and flooding caused by the outbreak of glacial Lake Missoula. Pleistocene and younger paleosols are also present in the Willamette Valley where they have been used to date geomorphic surfaces and flood deposits associated with glacial Lake Missoula.

THE SOIL RESOURCE

Although generally considered a component of the natural environment, the soils of the Pacific Northwest are also one of the region's most important natural resources and economic assets. Soils are explicitly linked to the region's agriculture and forest industry as noted elsewhere in this book. What is often overlooked is their importance to fisheries, water resources, and urban systems. Furthermore, soils in the Pacific Northwest are also used for many non-agricultural purposes, including waste disposal, construction, and recreation. They are also the primary concern in mine reclamation and land use.

Soil Erosion and Degradation

Soil erosion has been called the "quiet crisis" because it is not widely perceived nor well understood by the public. Although this "quiet crisis" is global in scale, it is well evidenced nationally and in the Pacific Northwest. Recent studies have shown that in the United States, 90 percent of the croplands and 54 percent of grazing lands, including large portions of the Pacific Northwest, are being eroded at a greater than sustainable rate. Regionally, approximately 40 percent of the Palouse soils have become unsuitable for agriculture over the past 100 years.

The "quiet crisis" in the Pacific Northwest is, unfortunately, not a recent phenomenon. In a 1930 soils inventory of Oregon, Washington, and Idaho, the Soil Conservation Service found that 4.5 million acres in arid and semi-arid areas had been permanently removed from agricultural production due to erosion. The study further found that six million additional acres had lost 50 percent of their topsoil. According to the same report, three million acres of range land were without forage plants due to erosion, with an additional seven million acres degraded beyond full recovery. The remaining 48 million acres of range land showed lesser stages of erosion. In a more recent study, Rasmussen et al. (1989) have documented continued decreases in soil organic matter, nitrogen, and biological quality since 1931.

Although much of the soil loss in the Pacific Northwest is due to water and wind erosion, there are other forms of soil loss and soil degradation. Soil mass movement is responsible for large volumes of soil erosion, especially in areas having unstable slopes (see Photo 4 in Chapter 2) and abundant precipitation. Like water and wind erosion, soil loss caused by mass movement is highly influenced by human activities, especially clear-cut logging and road building. These activities increase slope failure by decreasing evapotranspiration rates, altering drainage patterns, reducing root strength of trees, and redistributing the mass of slopes during construction.

Urbanization is another process causing the loss of prime agricultural soils. In Oregon, recent estimates suggest that 25 square miles are lost each year through urbanization. From a regional perspective, this process is most evident in the Puget Sound lowlands and the Willamette Valley of Oregon. However, this form of soil loss is widespread throughout the region.

Other common sources of soil degradation in the Pacific Northwest include heavy metal contamination resulting from sewage applications to crops and nuclear waste contamination. Pesticide and herbicide applications can also have detrimental effects on soils and soil organisms.

A SUMMARY PERSPECTIVE

Soils are a unique component of the regional landscape and provide a natural link for integrating the region's physical geography and natural resources. Although dynamic in their response to the external environment and internally complex, the soils of the Pacific Northwest provide firm geographic mooring to our natural and cultural heritage.

To appreciate this regional heritage fully, it is important that we seek a better understanding of the relationship between soils and society. Although soils clearly represent a natural resource and provide the basis for much of our region's economic well-being, they represent far more in terms of creating a sustainable society. Soils store our water, purify our waste products, and act as a global carbon sink that helps control our climate. Furthermore, they give us the means to better understand ourselves as soils have historically been a powerful force in shaping the cultural, economic, and political geography of the region. For these and other reasons, we should share a concern for the future of this regional heritage. It is paramount that we recognize the implications of soil degradation and that we seek to develop a regional land ethic that embodies our regional values.

REFERENCES

Agricultural Experiment Stations of the Western States Land-Grant Universities and Colleges. *Soils of the Western United States*. Pullman, Washington: Washington State University, 1964.

Balster, C.A. and Parsons, R.B. *Geomorphology and Soils, Willamette Valley, Oregon*. Special Report, Agricultural Experiment Station, Corvallis, Oregon: Oregon State University, 1968.

Barker, R.J.; McDole, R.E.; and Logan, G.H. *Idaho Soils Atlas*. Moscow, Idaho: University of Idaho Press, 1983.

Birkeland, P.W. *Soils and Geomorphology*. New York: Oxford University Press, 1984.

Birkeland, P.W.; Crandell, D.R.; and Richmond, G.M. "Status of Correlation of Quaternary Stratigraphic Units in the Western Conterminous United States." *Quaternary Research,* 1(1971):208–227.

Brady, N.C. and Weil, R.R. *The Nature and Property of Soils*. 11th ed. Upper Saddle River, New Jersey: Prentice Hall, 1996.

Brooks, R.R. *Serpentine and Its Vegetation: A Multidisciplinary Approach*. Portland, Oregon: Dioscorides Press, 1987.

Buol, S.W.; Hole, F.D.; and McCracken, R.J. *Soil Genesis and Classification*. 3rd. ed. Ames, Iowa: Iowa State University Press, 1989.

Franklin, J.F. and Dyrness, C.T. *Natural Vegetation of Oregon and Washington*. USDA, Forest Service General Technical Report, PNW-8. Portland, Oregon: 1973.

Jenny, H. *Factors of Soil Formation*. New York: McGraw-Hill, 1941.

Jenny, H. *The Soil Resource*. New York: Springer-Verlag, 1980.

McDowell, P.F. "Quaternary Stratigraphy and Geomorphic Surfaces of the Willamette Valley, Oregon," p.156–164. In: Morrison, R.B., ed. *Quaternary Nonglacial Geology: Conterminous U.S.* The Geology of North America Vol. K-2. Boulder, Colorado: Geological Society of America, 1991.

Montagne, C.; Munn, L.C.; Nielsen, G.A.; Rogers, J.W.; and Hunter, H.E. *Soils of Montana*. Bulletin 744. Bozeman, Montana: Montana Agricultural Experiment Station, Montana State University, 1982. Soils map 1:1,000,000.

Pimentel, D.; Harvey, C.; Resosudarmo, P.; Sinclair, K.; Kurz, D.; McNair, M.; Crist, S.; Shpritz, L.; Fitton, L.; Saffouri, R.; and Blair, R. "Environmental and Economic Costs of Soil Erosion and Conservation Benefits." *Science,* 267(1995):1117–1123.

Rasmussen, P.; Smiley, R.; and Collins, H. *Long-term Management Effects on Soil Productivity and Crop Yield in Arid Regions in Eastern Oregon.* Agricultural Experiment Station, Special Report 675, Corvallis, Oregon: 1989.

Retallack, G.J. "A Field Guide to Mid-Tertiary Paleosols and Paleoclimatic Changes in the High Desert of Central Oregon—Part 1." *Oregon Geology,* 53(1991):51–59.

Retallack, G.J. "A Field Guide to Mid-Tertiary Paleosols and Paleoclimatic Changes in the High Desert of Central Oregon—Part 2." *Oregon Geology,* 53(1991):75–80.

Rowalt, E.M. *Soil and Water Conservation in the Pacific Northwest.* Farmers' Bulletin 1773. Washington, D.C.: U.S. Government Printing Office, 1937.

Sidle, R.C. *Slope Stability on Forest Land.* PNW Bulletin-209. Corvallis, Oregon: Oregon State University Extension Service, 1980.

Singer, M.J. and Munns, D.N. *Soils: An Introduction.* 3rd ed. Upper Saddle River, New Jersey: Prentice Hall, 1996.

Soil Conservation Service. *General Soil Map: Idaho.* Moscow, Idaho: 1984. Soils map 1:1,000,000.

Soil Conservation Service. *General Soil Map: Oregon.* Corvallis, Oregon: 1985. Soils map 1:500,000.

Soil Survey Staff. *Soil Taxonomy.* Washington, D.C.: United States Government Printing Office, 1975.

Swanston, D.N. "Effect of Geology on Soil Mass Movement Activity in the Pacific Northwest," p. 89–115. In: *Forest Soils and Land Use.* Youngberg, C.T., ed. Proceedings of the Fifth North American Forest Soils Conference, Fort Collins, Colorado: Colorado State University. 1978.

USDA. *Summary Report:1992 National Resources Inventory.* Natural Resources Conservation Service, Washington, D.C.: Government Printing Office. 1994.

Veseth, R. and Montagne, C. *Geologic Parent Materials of Montana Soils.* Bulletin 721. Bozeman, Montana: Montana Agricultural Experiment Station. 1980.

LIST OF FIGURES

Figure 1. Hypothetical soil profile showing six master soil horizons. These horizons may occur in different combinations, and some horizons may be absent from a particular soil. Note that the relative depth of these horizons may also vary among different soils.

Figure 2. Soils of the Pacific Northwest. Each soil unit represents a combination of soil orders based on their relative importance.

NATURAL RESOURCES

FISHING

Chapter 6

James G. Ashbaugh
Portland State University

Fishing has always played an important role in the economy of the Pacific Northwest. The North Pacific is a productive fishery for many cold-water species. Hundreds of miles of shoreline around Puget Sound and along the Washington and Oregon coasts provide a variety of habitats for fish (Figure 1) and the region's rivers serve as spawning areas for five species of Pacific salmon.

Pacific Northwest fishing can be divided into four parts:

1. The first is offshore fishing and consists of the catch of halibut, tuna, salmon, and a variety of groundfish. Among the groundfishes are three groups of fish, called rockfish, flatfish, and roundfish (see Table 1). Because halibut is a groundfish and is regulated by international treaty, it is dealt with separately in this discussion.
2. The second category consists of anadromous fish caught in the rivers and offshore. These anadromous fish migrate up rivers from the sea to breed in fresh water and consist mostly of the five species of salmon, the sea-run rainbow trout or steelhead, shad, and sturgeon.
3. The third category consists of the various shellfish, including oysters, crab, shrimp, and clams.
4. The fourth group consists of farmed and ranched salmon, farmed trout, tilapia, and catfish.

For many years, the major catch landing at Pacific Northwest ports was of salmon, halibut, tuna, and shellfish such as Dungeness crab. Groundfish were plentiful, but until World War II little market existed for them, and while production did increase during the war, it decreased after the war ended. It has been only in the last 40 or 50 years that markets have been created for fish such as the Pacific Whiting (hake), several varieties of sole and rockfish, and the Pacific shrimp. For example, landing of rockfish rose from 2.5 million pounds in 1941 to 43 million pounds in 1945. Because of reduced post-war demand, rockfish then declined to about 15 million pounds in the 1950s. During the period from 1985–1989, rockfish landings again reached the 40-million-pound level.

Figure 1. Major Washington and Oregon Fishing Ports.

SPECIES
Roundfish
Ling Cod
Pacific Cod
Pacific Whiting
Sablefish
Rockfish
Pacific Ocean Perch
Shortbelly
Widow
Other Rockfish
Bocaccio
Canary
Chilipepper
Yellowtail
Remaining Rockfish
Flatfish
Dover Sole
English Sole
Petrale Sole
Other Flatfish (Except Arrowtooth Flounder)
Other Fish
Jack Mackerel
Others

Table 1. Groundfish.
Source: Status of the Pacific Coast Groundfish Fishery through 1995 and Recommended Acceptable Biological Catch for 1996. Pacific Fishery Management Council, Portland, Sept. 1995.

GROUNDFISH

Groundfish are made up of a wide variety of bottom-feeding fish, including roundfish such as Ling Cod, Pacific Whiting, Sablefish, Spiny Dogfish Shark; rockfish such as Pacific Ocean Perch, Bocaccao, and Yellowtail; flatfish including Dover and English Sole; and a variety of other fish including Jack Mackerel.

Landings from the Pacific Coast groundfish fishery adjacent to Washington, Oregon, and California were relatively stable until the early 1970s, with an average of about 66 million pounds. By 1982, landings had reached 255 million pounds. Pacific Ocean Perch was depleted in the late 1960s by foreign fishing ships.

In 1976, the Fishery Conservation and Management Act was passed. Known as the Magnuson Act, it addressed the problem of overfishing of Pacific Ocean Perch and the concern over the future of groundfish by creating fishery management councils. The Pacific Fishery Management Council consists of the states of California, Oregon, Washington, and Idaho and has authority over the fisheries on the Pacific Ocean from 3 to 200 miles from shore, an area known as the Exclusive Economic Zone. The council was required to develop a fishery management plan based on the best scientific information available that would prevent overfishing while achieving, on a continuing basis, the optimum yield from each fishery.

Since the early 1980s, landings of several species have reached or exceeded maximum production levels. The International North Pacific Fishery Commission has divided the marine waters off the Pacific Coast into statistical areas. The areas related to the Pacific Northwest are the Columbia and Vancouver areas (Figure 2). The largest and most rapid growth occurred in the Columbia area. One of the criticisms of the Magnuson Act was that it had no mechanism for regulating the number of commercial fishing boats. If vessels were not allowed in certain fisheries or found certain fisheries to be uneconomical, they moved into the Exclusive Economic Zone and fished for groundfish. Also, new technology, improved electronic navigation and fish-finding equipment, has enhanced the ability to catch fish (see Conclusion). In 1982, a report by the Pacific Fisheries Management Council said that with the exception of Pacific Whiting and Shortbelly Rockfish, there was little room for expansion of the groundfish industry.

Before the passage of the Magnuson Act, large numbers of groundfish were caught in the North Pacific by foreign fleets, which would both catch and process the fish. Fish caught by foreign vessels were not recorded for the Pacific Northwest Fishery. Since the passage of the Magnuson Act, foreign fishing in the 3 to 200 mile Exclusive Economic Zone is regulated, with the states being responsible for the first 3 miles. In addition, the Act provided for a joint-venture fishery where United States trawl vessels deliver their catch to processing vessels at sea. Processing at sea is now done almost entirely by United States-owned ships with Whiting now processed at Northwest Pacific Ports. Shore-based landings of Whiting increased since 1978 but amounted to less than 5 percent of the total foreign and domestic catch of Whiting

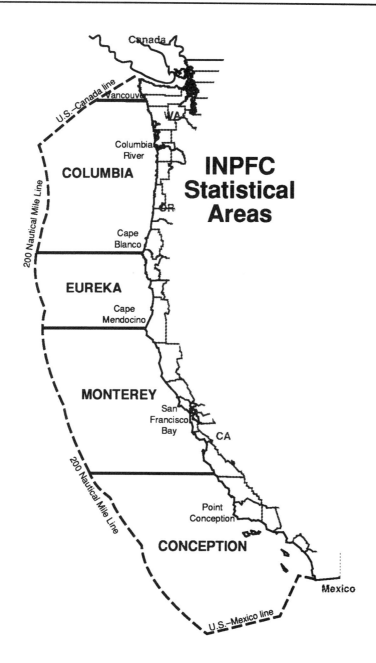

Figure 2. INPFC statistical area in the U.S. EEZ seaward of Washington, Oregon, and California.
Source: Status of the Pacific Coast Groundfish Fishery through 1991 and Recommended Acceptable Biological Catch for 1992. Pacific Fishery Management Council, Portland, Oct. 1991.

Year	Foreign Fishery (mt)	Joint Venture (mt)	U.S.-Processed (mt)[a]	Total Landings (mt)[b]	Quota or Harvest Guideline (mt)	Quota Landed (percent)
1978	96,827	856	689	98,372	130,000	76
1979	114,910	8,834	937	124,681	198,900	63
1980	44,023	27,537	793	72,353	175,000	41
1981	70,366	43,557	838	114,761	175,000	66
1982	7,089	67,465	1,024	75,578	175,500	43
1983	0	72,100	1,051	73,151	175,500	42
1984	14,772	78,889	2,721	96,382	175,500	55
1985	49,853	31,692	3,894	85,439	175,000	49
1986	69,861	81,639	3,463	154,963	295,800	52
1987	49,656	105,997	4,795	160,448	195,000	82
1988	18,041	135,781	6,876	160,698	232,000	69
1989	0	203,578	7,418	210,996	225,000	94
1990	0	170,972	12,828	183,800	196,000	94
1991	0	0	217,505	217,505	228,000	94
1992	0	0	208,575	208,575	208,575	100
1993[b]	0	0	141,222	141,222	142,000	99
1994[b]	0	0	252,729	252,729	260,000	97
1995[b]			176,571	176,571	178,400	99

a/ U.S. processing was entirely shore based through 1989. Since 1990, domestic at-sea processing vessels have operated in the whiting fishery.
b/ Preliminary.

Table 2. Landings and quotas for Pacific whiting (including discards in the foreign and joint venture fisheries).
Source: Status of the Pacific Coast Groundfish Fishery Through 1996 and Recommended Acceptable Biological Catches for 1997. Pacific Fishery Management Council.

each year from 1978 to 1990 (Table 2). For example, in 1990 approximately 376 million pounds of Pacific Whiting were caught. The total landed in Oregon was 5,058,341 pounds and in Washington 6,837,000 pounds. The average value of the landed Whiting in 1990 was less than 10 cents a pound, with the total catch valued at about $30 million. By comparison, in the state of Washington alone, all salmon caught in 1990 totaled 39,781,000 pounds, with a total value of $51,954,000.

Since 1990 there has been a marked change in the Pacific Whiting fishery. In 1991 virtually all fish were caught by U.S. vessels where they were processed, taken to mother ships to be processed or landed and processed at ports such as Newport, Oregon. In 1994 over 250,000 metric tons (a metric ton is 2,204.6 pounds)were caught in both categories in Oregon, Washington and California. Oregon is the leader in Whiting fishery, with Washington second and California a distant Third. (Table 2.)

When shore processors became alarmed that the operations at sea would take so much whiting that they would be unable to operate, a plan was established for 1994–96. The plan reserved 40 percent of the yearly harvest guideline established by the Pacific Fishery Management Council for shore processing after the first 60 percent was taken in competition between shore-based and at-sea processors.

In 1995 a total of 176,571 million metric tons of Whiting was caught in both sectors. The decline between 1994–95 was due to a reduction in the harvest guidelines. Harvest guidelines allowed a harvest of almost 200,00 metric tons in 1996. (Table 2.)

Though the initial price of Whiting is low, it is processed into surimi which is used to make products such as artificial crab and lobster. The value of surimi is several times that of the unprocessed Whiting (see Changing Food Habits).

The Spiny Dogfish Shark (*Squalus acanthias*) is commercially abundant off the coasts of Washington and British Columbia. It is ovoviviparous, producing litters of from one to twenty pups but averaging only six to eight fully formed pups after a gestation period of 18 to 24 months. Males mature in 14 years and females in 24 years. The life span is 30 to 50 years, although some reach 100 years of age.

During the 1940s, the Spiny Dogfish Shark was fished heavily for its liver, a rich source of vitamin A. More than 250,000 metric tons was caught during the 1940s, with a total estimated stock of between 300,000 to 500,000 metric tons. Development of synthetic vitamin A destroyed this market. Sustained yield now is estimated at eight to ten thousand tons annually. The use of the Spiny Dogfish Shark for food in the United States started about 1975. The fish is also exported to Europe because production from the Atlantic Ocean has declined. There is some indication of overfishing of this species.

Halibut (*Hippoglossus stenolepsis*), a deep sea flounder, a major United States food fish, was originally abundant off the coasts of Washington and Oregon. With the inauguration of refrigerator cars and fast transcontinental rail in the latter part of the 19th century, it was shipped in increasing quantities to markets in the eastern United States when the Atlantic halibut fishing industry was in steep decline—a decline from which it has not yet recovered.

Most halibut reproduce when they are 12 to 15 years old. The parent fish spawn from December to March along the continental shelf. After hatching, they work their way to shallow water where they spend their early lives. As they age, they move to deeper water. Females grow to a length of $8\frac{1}{2}$ feet and may weigh 500 pounds. The males are smaller. Halibut live to be from 25 to 35 years old, and are classified according to size: "medium," up to 60 pounds, and "large" or "whales," over 60 pounds. Halibut are caught on long bottom lines baited with herring, at depths from 50 to 175 fathoms.

Early Pacific Northwest halibut fishing was on the southern coastal area within 500 miles of home ports. As the catch declined, fishermen ranged north into Alaskan waters. A peak in production was reached in 1915 with 63 million pounds. Several signs of depletion appeared. The southern area off the Oregon and Washington coasts had smaller and smaller catches. Another indication of depletion was that catches contained larger and larger percentages of smaller fish under ten pounds, and the numbers of fish caught per line was decreasing.

In 1924 the United States and Canada created the International Fisheries Commission to study the halibut and recommend restoration and conservation methods. In 1953, the Commission's name was changed to the International Pacific Halibut Commission. One action taken by the commission was to have a closed period during the winter spawning season. Also, nursery areas were set aside where fishing was prohibited and a limit on the catch was established. The halibut fishing area was divided into ten sections, with individual regulations for each section. Washington, Oregon, and California are in Section 2A. Section 2B is British Columbia, which is off limits to United States fishermen. The remaining sections extend along the coast of Alaska and include the Aleutian Island chain (Figure 3).

Catch quotas and discrete fishing seasons are established for each of the sections. Section 2A, historically the most overfished area, has the smallest quota. The allocations are divided among commercial, sport, and treaty users. In 1990 the total allowable catch for all users was 520 thousand pounds. Treaty users are members of 12 Native American tribes from western Washington who are allocated a total of 130 thousand pounds of halibut each year for subsistence and ceremonial use. The non-treaty portion of the catch, 390 thousand pounds, was subject to an allocation of 50 percent commercial and 50 percent sport, with approximately 195 thousand pounds each.

The commercial season is short throughout the commission area. For example, in 1990 in section 2A, the non-Native American commercial catch was taken in four 12-hour fishing periods on July 10, 30, August 27, and September 11. The tribe's season lasted 26 days from March 1 to 27, 1990. The sport season varies within 2A but is as long as two months off the north Washington coast.

The main objective of the commission is to provide a sustained yield of halibut for the North Pacific Ocean. By establishing quotas and controlling the season, it was able to stabilize the fishery. In the 1960s, halibut bycatch, fish caught inadvertently while fishing for other fish, became a problem. Foreign trawls were a major part of the problem. In 1965, the halibut bycatch had increased to 21 million pounds. Even when released, many of the halibut die. Regulation of drift net fishing should result in a reduction of the bycatch.

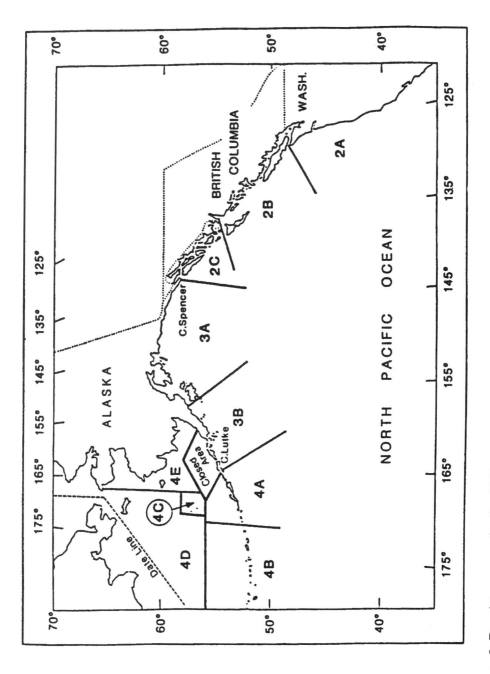

Figure 3. Regulatory areas for the 1990 commercial halibut fishery.
Source: International Pacific Halibut Commission Annual Report, 1990, Seattle, 1991.

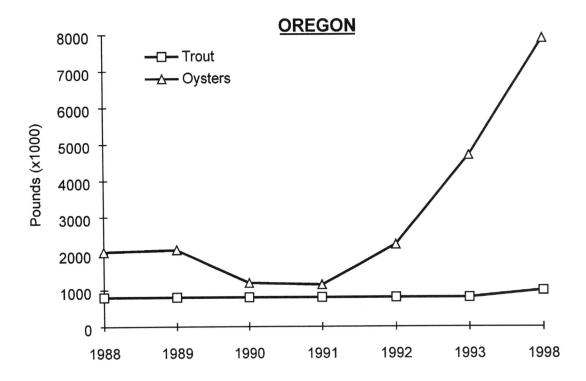

Table 3A.

Halibut landings in the Pacific Northwest exceed the allowed catch in zone 2A because both United States and Canadian fishermen bring halibut caught in other Zones to United States ports. For example, in 1990 325 thousand pounds was the authorized quota for zone 2A, but landings at Pacific Northwest ports totaled almost five million pounds. Although important for the Pacific Northwest, this amount is minor when compared with the total of 56 million pounds landed at all ports in Alaska. The economic value of halibut in the Pacific Northwest has also declined: in the period from 1948 to 1952, it ranked third in value after Pacific salmon and flounders (flatfish including the various soles).

In 1990 in Washington, halibut accounted for 5.8 percent of the value of all fish; in Oregon, the percentage was 1.3 percent. In 1995 both states accounted for 0.7 percent.

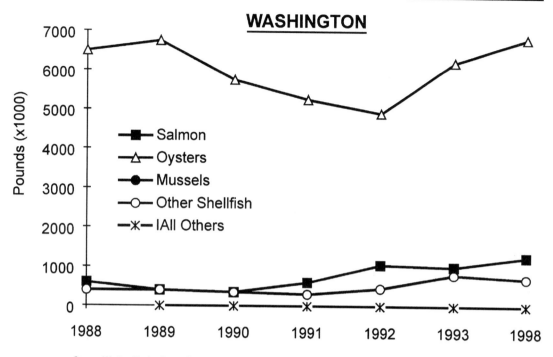

Source: *Western Region Aquaculture Industry Situation and Outlook Report*, Vol 3., Sept 1995.

Table 3B.

SHELLFISH

The name "shellfish" is applied to a wide variety of fish including oyster, clams, shrimp, and crabs. Collectively, in 1993 shellfish accounted for 38 percent of all ocean fish by value in Oregon and 45 percent in Washington. The state of Washington is a leading United States producer of oysters, which are farmed under carefully controlled conditions. The main variety is the Pacific oyster (*Crassostrea gigas*) brought from Japan in 1900. Oyster seed used to come from Japan in bundles of clutch (oyster shells on which the seed was attached) but is now produced in the Pacific Northwest. As many as four million oyster seeds can be held in the palm of the hand.

Farmers in the Pacific Northwest own their own oyster beds. They plant and harvest oysters as they would an agricultural crop. In Washington, five areas dominate in oyster production. In rank of production, they are Willapa Harbor, South Puget Sound, Grays Harbor, and central and northern Puget Sound (map). Oregon production is concentrated at Tillamook, Yaquina, Coos, and Netarts bays (Figure 1).

Washington has by far the largest production, producing more than 60 million pounds in 1993, mostly by aquaculture. Oregon also produced over 4.5 million pounds (Tables 3A and 3B).

The Olympia (*Ostica luriola*), a native oyster, was brought to the brink of extinction by overfishing and pollution. In 1938 the state of Washington produced 299 thousand gallons while in 1993, only 415 gallons were produced. The Kumamoto, a variety of Pacific oyster, is smaller than other Pacific oysters and is prized by those preferring oyster on the half shell.

The Triploid is a Pacific oyster that was developed as an all-season oyster: it can be eaten in months without an "R" in the name. Triploids have three sets of chromosomes instead of two and are sterile. Pacific oysters normally convert 80 percent of their body weight into sperm and eggs during the summer spawning period (the months without an "R" in their name: May, June July, and August), which renders them unsatisfactory for eating. Because the Triploid is sterile, it does not change during the summer. The glycogen that is normally used to provide energy for the production of gametes (eggs and sperm) is not lost. Triploids are produced by various methods including temperature, high-pressure, and chemical shocks. They can be marketed throughout the year, increasing the profitability to the oyster farmer.

Crab

Washington and Oregon are major producers of Dungeness crab (*Cancer magister*), with almost all of the Dungeness crab in the United States caught in Alaska and these two states. They are caught mainly in the ocean although Puget Sound is a significant producer. The crab are caught in a welded steel frame covered by handwoven wire mesh with chopped herring for bait. Dungeness crab production has large fluctuations from year to year. As can be seen in Table 4, in 1989, Washington and Oregon produced more than 33 million pounds. In 1990 only 26 million pounds were caught. In 1986 Washington and Oregon produced about 12 million pounds; in 1977, another good year, the two states produced about 35 million pounds. Production in 1993 totalled almost 29 million pounds.

Clams

The habitat for clams is similar to that for oysters. Bays and estuaries are major centers of production. In recent years, the clam industry has suffered from poor natural seed sets of spawn. As in the case of oysters, clam production is a form of aquaculture in the state of Washington. Oregon statutes do not provide for the culture of clams. Clam seed is grown in hatcheries in large numbers and planted in beds. Seven kinds of clams are produced by aquaculture. The principal clam is the Manilla introduced from Asia, the second is the native littleneck.

The decline of the razor clam (*Siliqua patula*) illustrates the impact of overfishing and the degradation of ocean beaches. This clam lives in sand on ocean beaches and

	1981	1982	1983	1984	1985
Oregon	6,712,248	7,550,695	7,934,349	7,758,923	10,740,071
Washington	3,949,000	3,998,000	6,560,000	4,744,000	4,979,000
Total	10,611,248	11,548,695	14,494,349	12,502,923	15,719,071
	1986	1987	1988	1989	1990
Oregon	6,588,472	8,351,876	11,276,973	13,564,259	14,544,833
Washington	5,322,000	7,476,000	18,085,000	20,007,000	12,089,000
Total	11,910,472	15,827,876	29,361,973	33,571,259	26,643,833
	1991	1992	1993		
Oregon	4,923,571	11,908,102	10,456,154		
Washington	5,944,000	15,395,000	18,261,000		
Total	10,867,571	27,303,102	28,822,154		

Table 4. Dungeness Crab—Oregon-Washington, 1981–1993. Pounds caught
Source: 1993 Pounds and Values of Fish and Shellfish Landed in Oregon, Oregon Department of Fish and Wildlife, March 1996, 1993 Fisheries Statistical Report, Deparment of Fisheries, State of Washington, Spring 1996.

is a very rapid burrower. The commercial production in Washington state was 355 thousand pounds in 1978. In 1990, commercial production had decreased to 27 thousand pounds. In Washington none have been reported since 1992. Oregon, in 1976, produced 117 thousand pounds and 13 thousand pounds in 1990. In 1993 only 62 pounds were recorded.

Washington is the leading state in the recreational digging of razor clams. In 1979, more than 13 million clams were dug; by 1990, 440,000 were dug. The main areas of production extend from Long Beach on the south to beyond Copalis to the north. By 1973, the limit was reduced from 24 to 15 per day. Several factors are associated with the decline. Overfishing is certainly one factor. In 1979, there were 967,000 digger trips. The digging season was reduced for conservation reasons and because of parasite infestation of clams. Oil spills are also believed to have had an adverse effect on razor clams. There were only 32 thousand digger trips in 1990.

The geoduck (*Panope generosa*) is produced in Washington from wild beds. Weighing one to four pounds, it is the largest clam in the Pacific Northwest with a commercial harvest of about five million pounds per year. Hatchery-produced seed may increase commercial production in the future. Oregon has no production of the geoduck.

Shrimp

Both Oregon and Washington produce the Pacific (pink) shrimp. Oregon is the most important producer of the two states. In 1993 its production was 27 million pounds, while Washington's production was approximately 16 million pounds. Shrimp are caught by using trawls, a large cone-shaped net that is towed through the water. Efforts to product the Pacific Shrimp by aquaculture have, as yet, not been successful.

Mussels

Two kinds of mussels are produced in the Pacific Northwest: cultured (*Mytilus Tropulus*) and wild (*Mytilus Californiaus*). Washington is the major producer of cultured mussels. In 1993 it produced over 702 thousand pounds of Blue or Bay Mussels and 703 thousand pounds of wild mussels. In the same year Oregon's catch of wild mussels totaled 51 thousand pounds with only minimal production of cultured mussels.

Scallops

At one time, scallops were considered to have great potential along the Oregon coast. When, in 1981, fishermen took 16,853,845 pounds in Oregon, it appeared as though the state had a new fishery, but subsequent years proved otherwise. Overfishing resulted in a steady decline, and by 1990 the production from Oregon was only 1,781 pounds (Table 5). In 1993 this had rebounded to 286 thousand pounds, still far below the 3,329,234 pounds caught in 1984.

SALMON

Salmon was an important resource in the Pacific Northwest long before the entry of Europeans in the 16th century. Native populations throughout the region depended on the fish as a source of food and as a commodity for trade. After being dried and pulverized, salmon could be kept for several months. Indians also depended on a variety of other aquatic life.

Euro-white populations found fish, especially salmon, to be a valuable addition to their supply of plant foods, domestic animals, and wild game. It was obvious to early day settlers that salmon in seemingly inexhaustible numbers were an extremely valuable commercial resource. Canneries were built along the Columbia River, around Puget Sound, and at various locations along the coast. In 1880, 82 percent of all United States canned salmon came from the Columbia River. Fishing methods capable of landing huge quantities of fish were employed; however, by 1896, catches were declining, and it became apparent that salmon were not an inexhaustible resource.

Year	Pounds
1979	3,434 pounds
1980	
1981	16,853,845
1982	1,487,941
1983	2,648,965
1984	3,329,234
1985	819,030
1986	105,523
1987	13,590
1988	29,226
1989	
1990	1,781
1991	1,021
1992	1,944
1993	286,194

Table 5. Scallops, Weathervane, Oregon.
Source: 1993—Pounds and Values of Fish and Shellfish Landed in Oregon, March 1996.

Historically, the salmon has dominated the Pacific Northwest fishery. For a number of reasons this is no longer true. For example, Oregon, which shares the Columbia River fishery with Washington and was a leading United States producer, produced only about 2.8 million pounds of salmon in 1995. In value, this was approximately 13 percent of Oregon's fishery production. For Washington, salmon still leads other fish production with approximately 34 percent of the total value of the fishery in 1993 (Table 6).

The Pacific Northwest has five species of salmon: the Chinook or King (*Oncorhynchus tshawytscha*); Pink or Humpback (*Oncorhynchus gorbuscha*); Chum or Dog (*Oncorhynchus keta*); Coho or Silver (*Oncorhynchus kisutch*); and Sockeye, Red, or Blueback (*Oncorhynchus nerka*).

The Chinook, the largest of the salmon, is also one of the most prized economically. It ranges from California to Alaska but historically was found in the greatest abundance on the Columbia River. In the latter part of the 19th century, the Columbia River alone produced more Chinook salmon than are now produced in California, Oregon, Washington, and Alaska combined. A report published in 1887 said that the

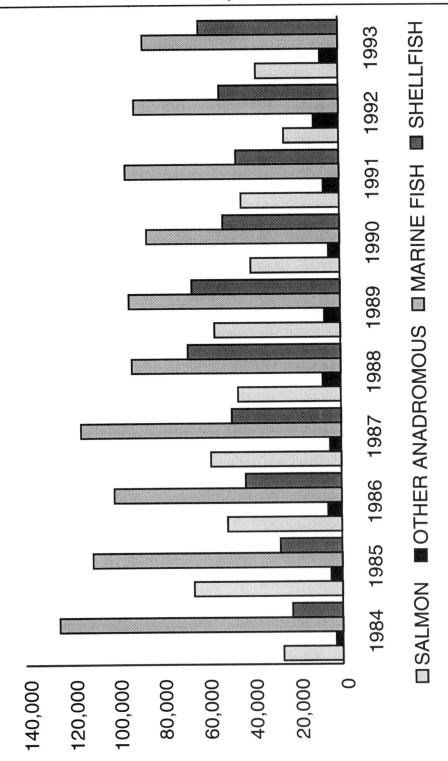

Table 6. 1984–93 Commercial Harvest by Species Group (round weight in thousands)
Source: 1993 Fisheries Statistical Report, Dept. of Fisheries, State of Washington.

quinat (Chinook) salmon, from its great size and abundance, was more valuable than all other fishes on the Pacific Coast combined. In the early 1880s, two million spring and summer Chinook were landed each year on the Columbia River. At an average of 25 pounds a piece, this was equal to 50 million pounds. In 1995 the total United States production of Chinook salmon was 24.3 million pounds, of which only 604,600 came from the Columbia River (Table 7).

The three runs of Chinook salmon on the Columbia River are in spring, summer, and fall. Each is distinctive because of the differences in spawning area. The spring salmon return after two to six years at sea between March and May, with spawning occurring in the early fall. Migration to salt water occurs in the spring after the salmon have lived for about a year in fresh water. Much of the spring run is from hatcheries. Others spawn naturally in the tributaries of the Columbia River system. Historically, the largest run of spring Chinook was spawned in the Salmon River of Idaho, a tributary of the Snake River. The construction of dams on the Snake River above the mouth of the Salmon River has interfered with their downstream passage (Figure 4). In 1991 the Snake River wild portion of the spring Chinook salmon run and the Snake River summer Chinook were proposed for listing as threatened under the Endangered Species Act.

Summer Chinook migrate upriver during June and July. The construction of Grand Coulee Dam cut off about 1,200 miles of spawning area (Figure 5). These fish, called "June Hogs," were the prime salmon on the Columbia River. Production peaked in the early 1880s, when two million fish were landed annually. In order to compensate for the loss of the spawning areas above Grand Coulee Dam, the federal government embarked on a program to transplant salmon runs from their native streams to new distant spawning areas, a practice called supplementation. The fish were trapped at Rock Island Dam and hauled by truck to new streams that seemed best adapted to their needs. This program continued for nine years. Willamette River Dams also reduced spawning areas (Figure 5).

The fall Chinook return to the rivers in August and September and spawn in the fall. Several months after hatching in the early spring, the fish migrate to the sea. The main spawning area historically was the Columbia River in the same area where Bonneville, The Dalles, John Day, and McNary dams have been constructed.

In the past, hatcheries have been quite successful in propagating fall Chinook. The majority of hatcheries are below Bonneville Dam or between Bonneville and the Dalles Dam on the Columbia River or on Puget Sound.

At one time as many as three million Sockeye salmon spawned on the Columbia River system, which also contained lakes essential for the natural propagation of this species. After spawning they moved to lakes where they resided one to three years before going to sea. Today, however, blockage by dams on the Columbia and Snake rivers has reduced the original surface acreage of the Columbia Basin Sockeye nursery lakes by 96 percent. In 1992 the Snake River Sockeye was placed on the endangered species list (Figure 5).

Year	Chinook[3/]	Coho	Sockeye	Chum	Steelhead	Total[4/]
1866	272.0	0.0	0.0	0.0	0.0	272.0
1867	1,224.0	0.0	0.0	0.0	0.0	1,224.0
1868	1,904.0	0.0	0.0	0.0	0.0	1,904.0
1869	6,800.0	0.0	0.0	0.0	0.0	6,800.0
1870	10,200.0	0.0	0.0	0.0	0.0	10,200.0
1871	13,600.0	0.0	0.0	0.0	0.0	13,600.0
1872	17,000.0	0.0	0.0	0.0	0.0	17,000.0
1873	17,000.0	0.0	0.0	0.0	0.0	17,000.0
1874	23,800.0	0.0	0.0	0.0	0.0	23,800.0
1875	25,500.0	0.0	0.0	0.0	0.0	25,500.0
1876	30,600.0	0.0	0.0	0.0	0.0	30,600.0
1877	25,840.0	0.0	0.0	0.0	0.0	25,840.0
1878	31,280.0	0.0	0.0	0.0	0.0	31,280.0
1879	32,640.0	0.0	0.0	0.0	0.0	32,640.0
1880	36,040.0	0.0	0.0	0.0	0.0	36,040.0
1881	37,400.0	0.0	0.0	0.0	0.0	37,400.0
1882	36,808.4	0.0	0.0	0.0	0.0	36,808.4
1883	42,799.2	0.0	0.0	0.0	0.0	42,799.2
1884	42,160.0	0.0	0.0	0.0	0.0	42,160.0
1885	37,658.4	0.0	0.0	0.0	0.0	37,658.4
1886	30,498.0	0.0	0.0	0.0	0.0	30,498.0
1887	24,208.0	0.0	0.0	0.0	0.0	24,208.0
1888	25,328.4	0.0	0.0	0.0	0.0	25,328.4
1889	18,135.4	0.0	1,210.2	0.0	1,726.6	21,072.2
1890	22,821.1	0.0	3,899.5	0.0	2,912.1	29,632.6
1891	24,065.7	0.0	1,052.8	0.0	2,010.4	27,128.8
1892	23,410.2	284.0	4,525.2	0.0	4,919.7	33,139.0
1893	19,636.6	1,979.3	2,071.2	157.1	4,435.4	28,279.6
1894	23,875.2	2,907.5	2,979.4	0.0	3,564.7	33,326.8
1895	30,253.8	6,772.9	1,225.0	1,529.5	3,378.1	43,159.3
1896	25,224.1	2,999.3	1,154.8	0.0	3,377.1	32,755.4
1897	29,867.2	4,137.8	882.1	0.0	3,137.9	38,025.0
1898	23,180.5	4,449.3	4,533.6	0.0	1,786.8	33,950.2
1899	18,771.0	2,013.3	1,629.9	773.8	815.6	24,003.6

Table 7. Columbia River Commerical Landings (in Thousands of Pounds) of Salmon and Steelhead, 1866–1994. 1/
Source: Status Report—Columbia River Fish Runs and Fisheries

Year	Chinook[3/]	Coho	Sockeye	Chum	Steelhead	Total[4/]
1900	19,245.2	3,054.9	895.0	1,203.3	1,400.6	25,799.0
1901[2/]	—	—	—	—	—	29,832.4
1902	23,033.7	716.2	1,158.5	707.3	584.3	26,200.0
1903	27,917.3	828.3	570.0	680.0	493.1	30,488.7
1904	31,782.5	2,125.3	877.9	1,407.1	671.0	36,863.9
1905	33,028.7	1,824.2	528.2	1,751.1	667.9	37,800.1
1906	29,970.7	2,818.3	531.5	1,890.5	442.0	35,653.1
1907	24,250.4	2,159.5	374.3	1,533.8	402.6	28,720.6
1908	19,742.5	2,137.4	583.5	1,148.1	729.4	24,340.9
1909	17,118.9	2,868.1	1,704.2	1,668.9	1,175.2	24,535.3
1910	25,325.6	4,686.7	423.9	4,524.6	369.6	35,330.4
1911	36,602.1	5,400.3	407.2	3,636.0	584.4	49,480.0
1912	21,388.0	2,165.3	558.3	1,271.5	2,147.2	27,530.2
1913	19,384.5	2,785.9	758.3	904.6	2,167.9	26,556.2
1914	25,409.1	4,744.3	2,401.1	3,351.4	1,907.6	38,501.3
1915	32,126.8	2,266.8	371.2	5,884.0	2,690.2	43,838.7
1916	31,992.9	3,541.7	257.7	5,288.1	1,580.9	42,746.3
1917	29,521.9	4,372.3	541.8	3,648.8	2,233.1	40,448.0
1918	29,249.1	6,673.9	2,572.6	2,029.5	3,022.6	44,125.4
1919	30,325.3	6,169.5	494.2	5,133.5	1,899.9	44,934.5
1920	31,094.3	1,837.6	178.0	1,277.9	1,165.9	36,311.5
1921	21,551.7	2,337.9	411.1	327.8	1,021.1	26,712.5
1922	17,914.7	6,149.7	2,090.5	601.4	2,162.8	30,152.7
1923	21,578.3	6,965.1	2,605.0	1,734.5	2,684.3	35,667.3
1924	22,365.2	7,796.4	500.9	3,926.9	3,192.8	38,167.1
1925	26,660.0	7,936.6	384.2	3,795.2	2,907.2	42,333.4
1926	21,241.0	6,605.7	1,478.0	2,234.0	3,843.1	35,566.7
1927	24,010.7	5,209.5	408.3	4,654.5	3,147.3	37,688.4
1928	18,149.3	3,722.9	327.4	8,496.8	2,160.2	33,127.1
1929	18,151.1	6,701.1	684.9	3,714.1	2,870.1	32,321.3
1930	20,078.6	7,736.9	668.0	773.2	2,404.1	31,923.4
1931	21,378.4	2,714.2	280.5	239.2	2,126.0	27,031.8
1932	16,000.9	4,096.5	190.1	1,173.7	1,431.8	23,330.2
1933	19,528.4	2,701.6	470.6	1,659.1	1,958.3	26,846.8
1934	18,787.5	4,774.7	467.1	1,662.9	1,919.2	27,901.9

Table 7. Continued.

Year	Chinook[3/]	Coho	Sockeye	Chum	Steelhead	Total[4/]
1935	15,266.4	7,108.3	88.5	1,053.7	1,472.2	25,756.0
1936	16,213.6	2,495.8	668.9	2,080.6	1,940.8	23,528.6
1937	18,653.6	1,841.6	335.1	1,909.8	1,933.4	24,673.5
1938	12,418.5	2,311.0	424.6	1,915.4	1,764.4	18,833.9
1939	13,498.8	1,529.7	269.8	1,174.4	1,438.5	17,911.2
1940	13,516.1	1,373.2	361.9	1,253.5	2,815.4	19,320.1
1941	23,238.5	1,045.0	505.7	4,149.8	2,663.7	31,602.7
1942	18,679.1	644.5	192.4	5,191.1	1,839.1	26,546.2
1943	11,426.5	706.3	146.1	959.9	1,514.5	14,753.3
1944	14,059.6	1,533.3	54.8	275.4	1,720.1	17,643.2
1945	12,972.1	1,835.5	8.7	588.8	1,963.5	17,368.6
1946	14,277.8	1,059.6	128.5	886.6	1,725.6	18,078.1
1947	17,302.7	1,498.1	718.3	496.2	1,648.7	21,664.0
1948	17,352.3	1,174.7	95.8	1,044.8	1,579.0	21,246.6
1949	10,768.5	899.2	24.0	545.0	814.0	13,050.7
1950	10,421.7	1,048.0	169.2	700.2	945.2	13,284.3
1951	10,036.3	968.0	169.4	532.3	1,207.2	12,913.2
1952	7,271.1	1,074.0	608.7	308.6	1,461.9	10,724.3
1953	6,966.6	457.5	146.2	249.2	1,898.3	9,717.8
1954	5,312.7	303.4	243.4	320.0	1,450.8	7,630.3
1955	8,581.9	598.8	200.4	125.6	1,320.0	10,826.7
1956	8,178.5	460.0	287.1	45.7	815.0	9,786.3
1957	5,918.9	390.7	240.2	32.1	741.0	7,322.9
1958	6,434.0	167.6	723.5	89.3	700.0	8,114.4
1959	4,594.3	119.7	635.8	42.9	628.5	6,021.2
1960	3,928.0	159.1	394.1	15.3	657.4	5,153.9
1961	4,160.2	382.6	158.0	17.3	612.3	5,330.4
1962	5,467.3	600.0	51.7	48.1	715.3	6,882.4
1963	4,346.1	501.1	48.8	15.3	972.9	5,884.2
1964	4,484.0	1,963.5	68.2	23.9	421.0	6,960.6
1965	6,142.9	1,901.8	22.9	6.1	510.1	8,583.8
1966	3,612.2	4,389.1	17.2	11.0	393.0	8,422.5
1967	4,974.1	3,817.9	195.1	9.7	445.6	9,442.4
1968	4,097.1	962.2	89.6	3.4	433.9	5,586.2
1969	5,775.9	1,663.3	104.5	4.0	495.0	8,042.7

Table 7. Continued.

Year	Chinook[3]	Coho	Sockeye	Chum	Steelhead	Total[4]
1970	6,461.7	5,745.6	55.7	8.0	311.8	12,582.8
1971	5,967.2	2,277.9	285.6	5.9	467.5	9,004.1
1972	5,684.6	1,239.4	275.6	16.0	667.1	7,882.7
1973	8,552.4	1,904.8	15.9	18.0	634.1	11,125.2
1974	3,637.8	2,432.7	0.2	10.7	185.2	6,266.6
1975	6,586.5	1,581.4	0.0	5.7	69.5	8,243.1
1976	5,586.7	1,328.6	0.5	16.9	86.6	7,019.3
1977	4,688.4	316.6	0.5	2.3	425.7	5,433.5
1978	3,674.4	1,096.9	0.1	20.4	249.2	5,041.0
1979	3,226.3	1,096.8	<0.1	1.6	68.6	4,393.3
1980	3,072.3	1,122.5	<0.1	3.1	65.6	4,263.5
1981	1,739.4	473.2	<0.1	18.5	98.0	2,329.1
1982	3,036.1	1,600.2	0.5	22.3	96.5	4,755.6
1983	1,038.9	45.4	6.5	2.0	156.7	1,249.5
1984	2,069.0	1,621.3	110.5	22.1	908.4	4,731.3
1985	2,646.7	1,674.7	287.4	7.5	766.2	5,382.5
1986	4,729.0	6,820.1	23.8	20.2	683.8	12,276.9
1987	9,016.9	1,313.4	256.5	14.1	753.8	11,354.7
1988	10,539.1	2,683.5	181.1	30.2	764.8	14,198.7
1989	6,120.3	2,683.1	0.1	17.3	591.0	9,411.8
1990	3,095.3	501.2	<0.1	9.4	331.0	3,936.9
1991	1,994.5	2,730.3	<0.1	4.2	307.4	5,036.4
1992	952.3	303.8	<0.1	6.9	465.2	1,728.2
1993	877.8	271.4	<0.1	0.5	263.0	1,412.7
1994	604.6	505.0	0.0	0.4	110.2	1,220.2

1/ Sources: 1866–1936—Craig and Hacker (1940), 1937—Cleaver (1951), 1938–70—FCO and WDF (1971), and 1971–present—ODFW and WDFW landings records.

2/ Landings by species unavailable.

3/ Chinook total for 1938–56 differ slightly from those calculated by summing individual landing tables in this report.

4/ From 1911–36, totals include landings of unknown salmonid species.

Table 7. Continued.

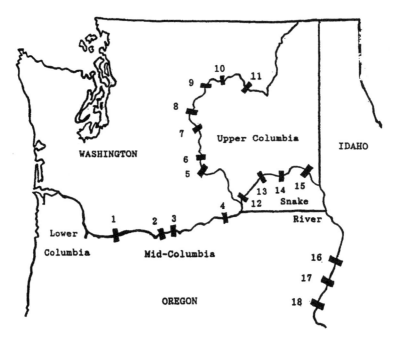

Major Dams on the Columbia and Snake Rivers

	Dam	Years in Service	Miles to Mouth	Gross Head (Feet)	Miles of Reservoir	Operator	Adult Fish Passage
1.	Bonneville	1938	146	65	45	Corps of Engineers	Yes
2.	The Dalles	1957	192	85	31	Corps of Engineers	Yes
3.	John Day	1968	216	105	76	Corps of Engineers	Yes
4.	McNary	1953	292	75	61	Corps of Engineers	Yes
5.	Priest Rapids	1959	397	82	18	Grant County PUD	Yes
6.	Wanapum	1963	416	84	38	Grant County PUD	Yes
7.	Rock Island	1933	453	54	21	Chelan County PUD	Yes
8.	Rocky Reach	1961	474	93	42	Chelan County PUD	Yes
9.	Wells	1967	515	72	30	Douglas County PUD	Yes
10.	Chief Joseph	1955	545	177	51	Corps of Engineers	No
11.	Grand Coulee	1941	597	343	151	Bureau of Reclamation	No
12.	Ice Harbor	1961	334	100	32	Corps of Engineers	Yes
13.	Lower Monumental	1969	366	100	29	Corps of Engineers	Yes
14.	Little Goose	1970	395	100	37	Corps of Engineers	Yes
15.	Lower Granite	1975	432	98	39	Corps of Engineers	Yes
16.	Hells Canyon	1967	571	210	22	Idaho Power Company	No
17.	Oxbow	1961	597	120	12	Idaho Power Company	No
18.	Brownlee	1958	609	272	57	Idaho Power Company	No

Figure 4. Location and Completion Dates of Columbia and Snake River Dams That Affected Anadromous Fish.
Source: Status Report—Columbia River Fish Runs and Fisheries 1938–91, Washington Dept. of Fisheries and Oregon Dept. of Fish and Wildlife, July 1992.

COLUMBIA RIVER BASIN

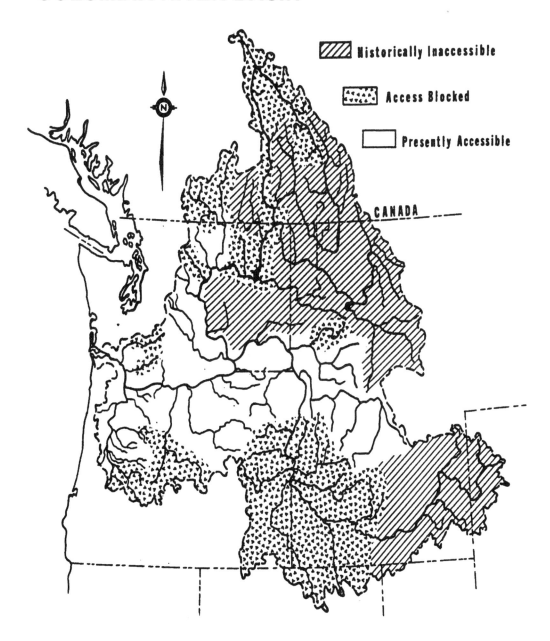

Figure 5. The Columbia River System Showing Area Accessible to Salmon and Steelhead.
Source: Status Report—Columbia River Fish Runs and Fisheries 1938–91, Washington Dept. of Fisheries and Oregon Dept. of Fish and Wildlife, July 1992.

	Dam	Stream	Year in Service	Miles to Mouth	Operator
1.	Willamette Falls	Main Willamette	1904	26.6	Portland General Electric
2.	River Mill	Clackamas	1911	48.1	Portland General Electric
3.	Faraday	Clackamas	1906	51.0	Portland General Electric
4.	North Fork	Clackamas	1958	55.9	Portland General Electric
5.	Big Cliff	N. Santiam	1953	166.1	Corps of Engineers
6.	Detroit	N. Santiam	1953	168.9	Corps of Engineers
7.	Foster	S. Santiam	1967	145.7	Corps of Engineers
8.	Green Peter	S. Santiam	1967	153.7	Corps of Engineers
9.	Leaburg	McKenzie	1930	210.6	Eugene Water & Electric
10.	Blue River	Blue River	1968	230.5	Corps of Engineers
11.	Cougar	S. Fork McKenzie	1964	236.0	Corps of Engineers
12.	Fern Ridge	Long Tom River	1941	171.6	Corps of Engineers
13.	Fall Creek	Fall Creek	1965	206.0	Corps of Engineers
14.	Dexter	Mid. Fork Willamette	1953	203.8	Corps of Engineers
15.	Lookout Point	Mid. Fork Willamette	1953	206.9	Corps of Engineers
16.	Hills Creek	Mid. Fork Willamette	1961	232.5	Corps of Engineers
17.	Dorena	Row River	1949	2153	Corps of Engineers
18.	Cottage Grove	Coast Fork Willamette	1942	216.7	Corps of Engineers

Figure 6. Location and Completion Dates of Willamette Basin Dams.
Source: *Status Report—Columbia River Fish Runs and Fisheries 1938–1994,* Washington Dept. of Fish and Wildlife and Oregon Dept. Fish and Wildlife, August 1995.

Another important Sockeye salmon area is the Fraser River system in British Columbia which has lakes essential to the successful natural propagation of the Sockeye. Because the fish move through United States territory on the way to the Fraser, the United States and Canada, by treaty, agree to equally share Sockeye production. In 1913, construction of the Canadian Pacific Railroad caused the massive Hells Gate slide on the Fraser, which resulted in a decline of the fishery until the construction of fishways allowed the salmon to move to their spawning areas.

Varying degrees of success have been achieved with the reproduction and maintenance of other species of salmon but not the Sockeye. The difference is that the life cycle of Sockeye requires lakes. Puget Sound makes Washington a major Pacific Northwest producer of Sockeye because of the many lakes in spawning areas around the Sound.

The Coho or Silver salmon returns to fresh water from August through November, with spawning occurring from October through December. The first year of their life cycle is spent in fresh water. Almost all production is from hatcheries; indeed natural production is less than 10 percent of the run. The commercial catch reached a peak of eight million pounds in 1925. Of all commercial hatchery programs, the Coho has been the most successful. Most of the hatcheries are located downstream from Bonneville Dam on the Columbia and on streams entering Puget Sound. The smolt, therefore, are not subject to the problems faced by both naturally spawned and hatchery fish above the dams—problems such as reductions in the velocity of flow on the dammed sections of the river, downstream passage blocked by the dams which forces fish through turbines, and thermal pollution (see Environmental Degradation).

Chum salmon, at one time, were abundant in the Pacific Northwest. They return to fresh water in October and November and spawn in November and December. Chum come from small tributaries of the Columbia below Bonneville Dam and from small streams around the Puget Sound. Their spawning areas have been badly damaged by poor forestry and agricultural practices and spreading urbanization.

Pink salmon are very important in the state of Washington. In odd-numbered years, in numbers of salmon and pounds, Pink salmon are the leading salmon landed in Washington. They run every odd-numbered year: 1987, 1989, etc. They are not economically significant in the state of Oregon. Pink salmon spawn near the sea in the fall and the young move to the sea in the spring. In 1993, 22 percent of the production by species was Pink. Others were Chum at 27 percent, Coho at 5 percent, and Sockeye, the leader, at 38 percent. Most pink are caught in Puget Sound. Of the rivers emptying into the Sound, the Skagit is by far the most productive.

Salmon is clearly the most important fish to the economy of the Pacific Northwest. However, Washington and Oregon together in 1995 produced only about 2.5 percent of the United States' commercial catch. California produced less than 1 percent, and Alaska the rest. In 1995 Oregon, Washington and California no longer dominated the Chinook fishery. Alaska's 12.9 million pounds of Chinook was more than the combined Oregon and Washington total of almost 4.8 million pounds.

Washington has been more successful than Oregon in maintaining salmon stock. Oregon has no area that even approaches Puget Sound in salmon productivity. The Columbia River, once so productive, has declined because of the problems associated with dams, including the elimination of spawning areas (see Figures 4, 5 and 6).

Another indication of the decline of the salmon is the length of the open commercial season on the Columbia River below Bonneville Dam. In 1938, it was 272 days, in 1994, 20 days (Table 8). For many years, Native American tribes in the Pacific Northwest claimed that various state regulations violated treaty agreements. In the last 20 years, a series of court decisions in favor of the Indians provided them with 50 percent of the salmon catch in the Pacific Northwest. This is clearly shown by the following: in 1974 Indians in the state of Washington caught 890 thousand salmon compared to a commercial catch of 5,704,000; in 1989, the commercial catch was 8,581,000, and the Indian catch was 4,624,121. The Native Americans also share in the management of the fishery. On the Columbia River system, four tribes make up the Columbia River Intertribal Fish Commission (Figure 7).

ENVIRONMENTAL DEGRADATION

Overfishing has reduced the fish population substantially at all sites. In making their way to spawning beds, fish have to overcome rapids and move through narrow areas where they can easily be caught using a variety of nets and, in early years, fish wheels. Large numbers of salmon, for example, were taken in Puget Sound when they made their way between islands and the mainland.

Long before the building of dams, anadromous fish numbers were depleted when spawning areas were destroyed or greatly modified. For example, early logging practices were very destructive. The practice of "cut out and get out" caused excessive runoff, erosion, and siltation of spawning areas. Streams left unshaded by trees were subject to increases in water temperature. Streams too small to carry logs were diverted into flumes which carried enough water to move the logs downhill single file. In other cases, streams were dammed, creating a log-holding pond. When the dam was removed, the flood waters would carry the logs downstream.

In other places, the removal of vegetation increased runoff and reduced infiltration. This resulted in a reduction of stream-bank storage. Because the Pacific Northwest has a dry summer climate, spawning streams became dry or had substantial reductions in flow.

Before large-scale trapping and agricultural development, the beaver (*Castor canadenses*) helped maintain summer stream flow. Their numerous dams stored enough water in the drainage areas to provide conditions favorable for spawning by anadromous fish. The beaver population, of course, has been greatly reduced because of loss of habitat, especially in agricultural areas. Renewed interest in the preservation and restoration of wetlands should enhance habitat for the beaver and possibly restore lost spawning areas.

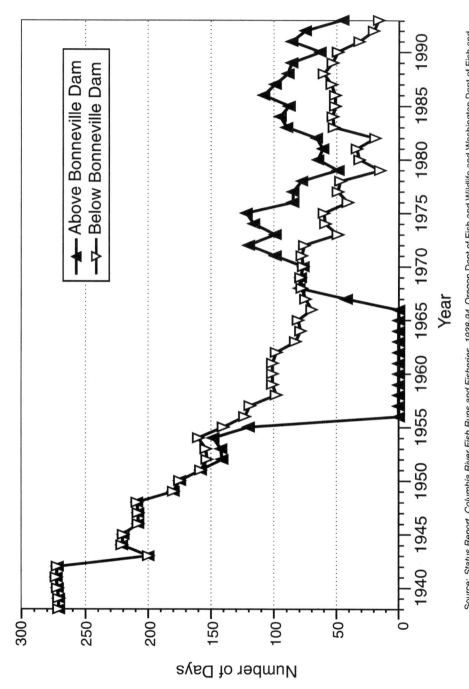

Table 8. Columbia River Number of Days Open to Commercial Fishing, 1938–94.

Source: Status Report—Columbia River Fish Runs and Fisheries 1938–94, Washington Dept. of Fisheries and Oregon Dept. of Fish and Wildlife, August 1995.

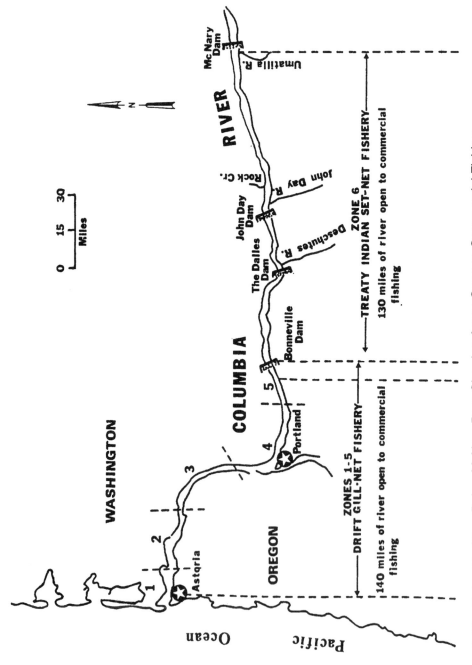

Figure 7. The Columbia River Below McNary Dam Showing Areas Open to Commercial Fishing.
Source: Status Report—Columbia River Fish Runs and Fisheries 1938–91, Washington Dept. of Fisheries and Oregon Dept. of Fish and Wildlife, July 1992.

Spawning areas for anadromous fish are not confined to the areas west of the Cascade Mountains. Mountainous areas in eastern Oregon, Washington, Idaho, and western Montana receive orographic precipitation, especially as winter snow. These snow packs may provide enough water to maintain stream flow during the region's dry summers and fall.

Total spawning areas for anadromous fish covered thousands of miles of streams of different sizes. Even before the building of scores of dams on these streams, they had been severely degraded as spawning areas. Diversion by farms for irrigation quite literally dried up stream courses. The doctrine of prior appropriation in water rights permitted those with water rights to take all the water from a stream. For example, the Yakima River, a major tributary to the Columbia River, is believed to have once had about 500 thousand salmon. Since 1890, when irrigation controlled the stream flow, the fish runs suffered decline. Presently, about 5,000 fish spawn in the Yakima River, only one percent of historic runs.

The grazing of cattle and sheep has also had adverse effects on streams, with overgrazing resulting in accelerated runoff and erosion. This has led to reduced stream flow, sedimentation of spawning beds, and increased water temperatures. Streams in many areas are unfenced, and livestock trample stream beds and foul the water with urine and excrement.

In parts of the Pacific Northwest, mining operations have also led to stream degradation. Dredging for placer deposits has resulted in the destruction of the mined areas and, through increased sedimentation, of other areas as well. The leaching of minerals from mining spoils has, in many cases, damaged the habitat by changing stream chemistry.

However, the construction of dams is now considered to be the leading cause of the decline of the salmon. Grand Coulee Dam on the Columbia River completely stopped the migration of fish which had once spawned more than 1,200 miles from the sea into British Columbia. In 1979 there was a total of 2,560 dams over 10 feet high in Washington, Oregon, and Idaho. Most do not have fish facilities.

Many problems are associated with dams. Originally, the most obvious seemed to be getting the fish upstream past the dams. Large amounts of money were spent by public and private bodies to overcome this problem. Fish ladders were the most common solution and, in most cases, performed satisfactorily. Where they did not work, for example at Pelton Dam on the Deschutes River, fish had to be carried in tanks by truck around the dam.

Another problem associated with dams on the Columbia was nitrogen super-saturation. Water spilled at each dam trapped nitrogen. The cumulative effect from all the dams produced enough nitrogen to cause the adult salmon to die. Some changes were made in the bases of spillways, but the major reduction in nitrogen super-saturation occurred because of less spilling of water. When power is generated, water goes through penstocks to turbines, not over the spills. As the demand for power has steadily increased, less and less water is being spilled at each dam.

The major problems with dams, however, are related to the movement of fish downstream. Salmon moving downstream are killed at each dam when they are drawn into the penstocks and the power turbines. Changes in pressure from the top to bottom of the dam, moving turbine blades, and the shearing action of turbine water all contribute to injury and death. Also, after passing through the turbine, fish become stunned and disoriented and are vulnerable to predators such as squawfish, which are found at the base of dams. From 10 percent to 15 percent of the salmon die at each and every dam.

Another problem facing salmon relates to the use of the river for flood control and the generation of electricity. The runoff of the Columbia is stored in reservoirs in Canada and the United States to be used during periods of low flow and for flood control. Unfortunately for the salmon, water is being stored in the spring when it is most needed to carry the young salmon to the sea. The combination of reduced flows and large pools behind each dam which reduce stream velocity slows the fish on their way to the ocean. The increased travel time results in the fish having problems adjusting to the change from fresh to salt water. One problem is "smoltification," the physiological change that occurs in fish as they adapt to the ocean. Another problem is caused by higher water temperatures occurring in the dammed parts of the river which may also result in changing water chemistry and thus increasing susceptibility of fish to disease. The extended time period also increases the fishes' exposure to birds and predatory fish.

The United States Army Corps of Engineers felt that fish could be barged downriver to below Bonneville Dam and started a program to do so in 1977. The program has been very controversial because the mortality rate of the fish is often high. Some feel that this could be alleviated by holding the salmon longer in specially prepared areas before releasing them. Figure 8 shows the fish transport route.

To reduce the travel time for the fish the Northwest Power Planning Council created a water budget. This would release water downstream during the spring (April 15–June 15) and simulate the spring freshet. However, water would still pass through the turbines causing increased fish mortality. To avoid turbine-caused mortality it was proposed that water be spilled at each dam. A large volume of water would be lost as far as the generation of electricity is concerned because it would not pass through the turbines. Also water would be released from storage in the spring, reducing the amount of water available to generate electricity during the high demand period in winter. Spilling would also cause reservoir drawdown and is opposed by irrigators and barge companies who fear that it would interfere with their activities. In addition, spilling may cause nitrogen super-saturation. To solve this problem bypass facilities at each mainstream dam would be constructed.

Puget Sound, historically Washington's most productive fishery, also suffers from problems of degradation. These are especially significant for the resident fish population. For years the Sound was the sewage dump for surrounding populations. For many urban areas including Seattle, the salt water of the Sound offered an easy and inexpensive way to take care of sewage. Eventually it was realized that human and industrial wastes were exacting a terrible toll on water quality and marine life. In addition, nonpoint sources from agricultural areas added to the problem.

Industrialized areas suffered the most damage. Estimates of cleanup costs for highly polluted areas such as Tacoma's Commencement Bay and Seattle's Elliott Bay run into hundreds of millions of dollars and these projects pose formidable technical problems. The discovery of cancerous growths in some resident fish is indicative of the urgency of the problem.

Urban areas surrounding the Sound must now provide secondary sewage treatment. This biological process will reduce some problems of disease. The recycling and pre-treatment of industrial waste has reduced the flow of toxins and heavy metals. In contrast, Victoria, B.C., across the Strait of Juan de Fuca, continues to discharge untreated waste into the ocean.

Among the leading proponents of water cleanup in Puget Sound are the producers of shellfish, especially oysters and clams. Because these shellfish feed by filtering water through their systems, the industry cannot tolerate diseased or toxic clams and oysters.

SALMON RANCHING

Salmon are raised by two forms of aquaculture: ranching and farming. Salmon ranching is carried on in Oregon, but in Washington, only Native American tribes may practice ranching. Private hatcheries raise the salmon until they can be released to the ocean. When mature, the fish return from the ocean to the hatchery, where they are killed and processed. Legislation was passed in 1971 for Chum salmon ranching and was amended in 1973 to include Chinook and Coho. A 1979 amendment included Pink salmon. In 1988, 12 Oregon hatcheries had permits for ranching; the three largest, at Coos Bay, Newport, and south of Gold Beach, have had mixed success. Chum salmon, for example, have little resistance to bacterial disease in the hatcheries. Environmental degradation has reduced the carrying capacities of the estuaries. Returns have been in the 1 to 1.5 percent range. Returns in the 3 to 4 percent range are needed to make operations profitable. In 1988, 2,719 Chums returned to the hatcheries, yielding 21,838 pounds. In 1990, only 321 returned, totaling 2,674 pounds.

More success has been achieved with Chinook and Coho salmon. However, their numbers have been in decline in recent years. In 1986, 66,363 Chinook adults returned. By 1990, this had declined to 6,910. Coho have suffered an even sharper decline from 445,103 to 35,592. The average ocean catch plus the average return to private hatcheries for Coho from 1978 to 1987 was 2.24 percent. Between 1985 to 1987, it increased to 4.55 percent. Ranches have also contributed to commercial and recreational fishing. The average ocean catch for Coho between 1985 to 1987 was 662,000 fish. Of these, 16 percent originated from ranching hatcheries, about the same

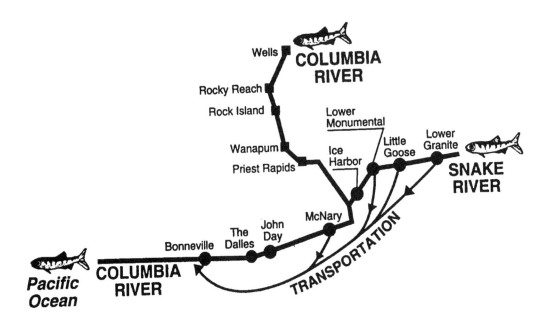

Figure 8. Fish Transport Route
Source: Salmon Passage Notes, U.S. Army Corps of Engineers—North Pacific Division, Sept. 1995.

percent from natural spawners, and 448,000, the largest number, from state and federal hatcheries. None were reported in 1993

A number of theories have been suggested for this decline. One theory blames the El Niño effect which raises water temperature and adversely affects salmon. Another theory is that hatchery fish have a lower survival rate than native fish. Overfishing by both foreign drift-net fleets and United States fishermen is more than a theory. The Pacific Fisheries Management Council, which regulates catches and seasons, is believed to have overestimated the number of Coho and to have set limits too high. Several nations practice drift-net fishing in the Pacific. Japan fishes for squid and uses nets almost identical to nets used to catch salmon. Multilateral agreements between the United States, Japan, Taiwan, and Korea to control drift-net fishing may help restore some of the runs. Tightened regulations on catch and season for United States fishermen are a result of substantial declines in the return of native and hatchery Cohos.

FISH FARMING

Salmon farming which is practiced in at least 13 countries and has increased world production of salmon, has been suggested as a means to bolster salmon production in the Pacific Northwest. The state of Washington seems particularly well adapted to salmon farming. The large protected waters of Puget Sound and Grays Harbor appear to be ideal for net-pen culture salmon farming. In the farming operation, year-old smolts are purchased from a hatchery, and put into pens that are actually floating frames supporting large enclosures of nylon netting. For the next 1 1/2 to 2 years, the farmer feeds them pellets produced by commercial feed mills, wet feed from whole fish, or fish waste. Fish may be given vaccines to prevent disease. At the end of the grow-out period, the salmon weigh from three quarters to nine pounds. They are then removed from pens, killed, and processed. A typical salmon farm will have 16 pens, each about 40 feet square. The farm will cover about 2.5 acres and produce about 50 thousand fish a year.

The growth of net-pen farming has been slow in Washington. While start-up mistakes were a major problem, today, fragmented government policies and public opposition from property owners inhibit the industry's growth. Each of the 11 counties bordering Puget Sound has authority to establish its own criteria for aquaculture permits. Potential threats to water quality and concern over the visual impact of floating pens have certainly slowed growth.

Salmon farms currently raise Atlantic salmon and Coho. Atlantic salmon accounts for 85 percent of the world's farmed salmon production. Norway, the world's largest producer of farmed salmon, developed the techniques to grow Atlantic salmon. Atlantic salmon grow faster, have a better survival rate in net pens, and keep better after being processed. They therefore command a higher market price than Pacific salmon.

In 1993 Washington produced 9.4 million pounds of farmed salmon (Table 9). Commercial landings in that year totaled 36,411,000 pounds. Oregon has not been important for net-pen farming because of a lack of sites along the coast. Most of the major world farming areas have sheltered, deeply indented, or fiord coastlines. In Washington the value of aquaculture salmon exceeded that of the commercial catch.

THE DECLINE OF "WILD" SALMON

Of great concern in the Pacific Northwest is the decline of the wild stocks of salmon; presently, about 75 percent of the runs on the Columbia River are from hatcheries. Wild fish have a survival rate twice that of hatchery fish. More wild fish will survive the years in the ocean than hatchery fish. Wild fish adapt to the changing conditions of their stream of birth. The fittest stock survive and pass on their genetic characteristics to the next generation. Interbreeding with hatchery fish lowers wild fish production and hybridization reduces disease resistance. Some consider wild fish as insurance. If hatcheries fail because of disease and genetic inbreeding, wild fish will be needed to avert total collapse of the industry.

In 1939 Congress passed the Mitchell Act to address the problems of fish passage and propagation resulting from the construction of the dams on the Columbia River system. Millions of dollars have been spent on fish ladders, power-house bypasses, hatcheries, and the transport of fish by barge. Despite all of this, the decline in salmon caught on the Columbia continues. In 1911 nearly 50 million pounds of salmon were canned and otherwise processed (Table 10). In 1994, only 1.2 million pounds were caught in the Columbia River. Especially alarming is the decline of the coastal Coho. Stream sedimentation caused by poor logging practices, chemicals from farms and intensified land use, and overfishing has put the entire coastal wild Coho fishery at risk. Several groups are seeking protection under the Endangered Species Act for 39 different runs of Coho from 39 streams throughout Oregon.

The role that hatcheries play in the propagation of salmon has been increasingly debated in recent years. Arguments against hatcheries focus on the susceptibility of hatchery fish to disease, and on the fact that there is too much reliance on hatcheries to replace natural spawning blocked by dams and severely degraded by destructive grazing, mining, irrigation, and navigation improvements. Critics also point out that crossbreeding of native and hatchery stocks weaken fish and that native fish mixed with hatchery stocks are being overfished. Also, natural spawners die in the streams and replenish nutrients in addition to supplying wildlife with food.

Those who defend hatcheries say that few natural spawning areas still exist and that hatcheries are the only way to maintain the supply of fish.

Both sides make strong arguments, but the sharp decline in recent years in salmon runs indicates that an unbiased analysis of all methods of propagation is long overdue.

THE AMERICAN SHAD

The American Shad (*Alosa sapidissima*), an anadromous fish native to the Atlantic Coast, was introduced to the San Francisco Bay area in 1871 and to the Columbia River in 1885. It adapted very well to its new environment and is now an important fish in the Pacific Northwest, especially in the Columbia River. The 1990 count at the Dalles Dam was more than 3.7 million. As of 1994, it stands as a record. The 1994 run of 2 million was the lowest since 1987. Shad average between three to five pounds and up to 2 1/2 feet in length. They are considered to be an excellent sports fish but, because of their boniness, have not become important commercially. Also, because they run at the same time as spring and summer Chinook, Sockeye, and summer steelhead, which are in decline, the various restrictions and regulations designed to reduce the incidental catch of the declining salmon and steelhead run have hurt the Shad gill-net fishery. For many years, Winchester Bay on the southern Oregon coast was the major landing area for Shad. A superior marketing location on the lower Columbia River resulted in a shift to the Astoria area in the mid-1970s.

SPECIES	TYPE	NUMBER	POUNDS	VALUE
Coho	Market	123,956	89,872	$167,328
	Juvenile	200,000	620	$0
	Eggs	8,845,000	5,363	$278,454
	Total	9,168,956	95,855	$445,782
Rainbow/ Steelhead Trout	Market	733,374	481,775	$1,000,002
	Juvenile	435,215	37,011	$97,258
	Eggs	261,603,000	65,402	$3,262,094
	Total	262,771,589	584,188	$4,359,354
Atlantic Salmon	Market	1,279,746	9,305,457	$23,537,868
	Juvenile	443,612	72,277	$638,394
	Eggs	4,837,000	1,259	$290,098
	Total	6,560,358	9,378,993	$24,466,360
Other Finfish	Market	752	2,088	$5,197
	Juvenile	0	0	$0
	Eggs	0	0	$0
	Total	752	2,088	$5,197
Total Finfish	Market	2,137,828	9,879,192	$24,710,395
	Juvenile	1,078,827	109,908	$735,652
	Eggs	275,285,000	72,024	$3,830,646
	Total	278,501,655	10,061,124	$29,276,693

Table 9. Aquaculture Finfish Harvest, 1993.
Source: 1993 Fisheries Statistical Report. Dept. of Fisheries, State of Washington.

WHITE STURGEON

The fishery for White Sturgeon (*Acipenser transmontanus*) is divided into two major sections, or zones. Below Bonneville Dam, sturgeon are freely anadromous and move to and from Washington and Oregon rivers and bays. This is the most stable and productive area and accounts for most of the White Sturgeon caught. Above

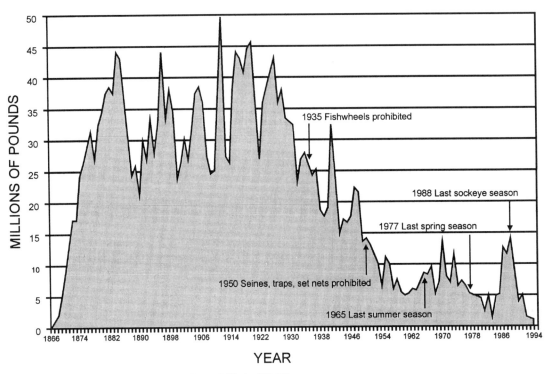

Source: State Report Columbia Ricer Fish Runs and Fisheries 1938-1994.

Table 10. Commercial Landings of Salmon and Steelhead from the Columbia River in Pounds, 1866–1994.
Source: Status Report, Columbia River Fish Runs and Fisheries 1938–1994.

Bonneville, sturgeon are confined in each reservoir because no provisions were made for their movement when the dams were constructed. Facilities designed for salmon cannot accommodate sturgeon, which may weigh hundreds of pounds and may be several feet in length. The general area above Bonneville Dam in the Columbia and the Snake Rivers has fewer sturgeon and is less productive, a condition that is referred to as "depressed."

Below Bonneville Dam, overfishing during the mid-1980s severely reduced the stock of White Sturgeon. In 1986, 696,000 pounds were landed, about 30 percent of the stock, an amount too high for maintenance. A sustainable yield was considered to be 15 percent of the stock. In 1991, 166,800 pounds were caught, about 7 percent of the stock. The high since 1938 was 757,100 pounds in 1976. This represented 23,400 fish caught in all zones on the Columbia River system. In 1994, 6,400 fish were caught below Bonneville and 1,600 above Bonneville by Native American tribes.

By far the most White Sturgeon are landed by the sports fishery. In 1987 over 60,000 fish were landed below and approximately 5,000 above Bonnevile Dam.

GREEN STURGEON

The Green Sturgeon (*Acipenser medirostris*) is not in as great demand as the White Sturgeon. Green Sturgeon are common in Washington and Oregon coastal bays and are found during the summer months in the Columbia River estuary. They are caught by gill nets. In 1994, 200 Green Sturgeon were caught, totaling 6,400 pounds.

ALBACORE TUNA

Several species of Tuna have been important in the fishery of the Pacific Northwest, but none more so than the Albacore (*Thunus alalaunga*). The Albacore fishery started in 1936 in Oregon. Tonnage landed increased to a peak of 34 million pounds in 1944. The catch in 1993 for Washington and Oregon totaled approximately 9.6 million pounds with Washington contributing 4,880,000 pounds.

Many factors have been responsible for the decrease since the peak in 1944. Perhaps the major factor is environmental: ocean temperatures have not been conducive for tuna in the ocean off the coast of the Pacific Northwest. Also, because the Albacore ranges the entire temperate portion of the Pacific, it may migrate across the Pacific in one year and spawn in mid-ocean. Increased competition among vessels of many nations has certainly been an additional important cause of tuna decline.

TROUT

Idaho is the leading producer of trout by aquaculture with about 90 percent of the United States production. A major aquifer fed by mountain water flowing below the porous volcanics north of the Snake River discharges from the north side of the Snake River at Thousand Springs near the city of Twin Falls. The high quality and temperature of the water are well suited for trout production. In 1993, Idaho produced more than 40 million pounds of trout. In 1990, 25 operations produced over $33 million worth of trout. The method of production is to divert water through raceways before it returns to the Snake River. Water runs continuously at a rate that provides several exchanges each day. Excremental waste from so many fish draining into the Snake River has caused substantial changes in the river's water quality, further exacerbated by the fact that the volume of water has been severely depleted by irrigation.

Idaho also produces catfish, Tilapia, and alligators in areas with geothermal wells that provide year-round water of the proper temperature.

CHANGING FOOD HABITS

The increasing popularity of Japanese cuisine has resulted in the use of previously underutilized marine life and the culture of at least one kind of seaweed.

Sushi is a special rice preparation including various marine products, usually raw. Sea urchin gonads, raw squid, raw shrimp, and boiled octopus are served with rice in sushi bars. Washington has been catching sea urchins since the 1970s and in 1900 landed 3,351,000 pounds at Washington ports. Oregon landed red sea urchins (*Stronaylocentrotus franciscahus*) in 1987. This increased to 9,320,868 pounds in 1990. By 1993 the catch was 2,183,266 pounds.

Sashimi is fish or other seafood thinly sliced and eaten raw. Many types of fish, including tuna, are used for sashimi. It is eaten with soy sauce and wasabi, a horseradish paste. Surimi is a processed fish product. Fish such as Alaska pollock and Pacific Whiting, which have white flesh, can be used. These fishes can be processed into products with different flavors and textures: imitation crab is surimi. One source reports that 100 million pounds of whole fish are needed to produce 20 million pounds of surimi.

A small amount of Porphyra formerly was produced by aquaculture in Puget Sound. It is used to make Nori, a seaweed that sushi is sometimes wrapped in. It has been difficult for growers to obtain permits in the state of Washington, and in 1990, Washington harvested only 2,500 pounds of Porphyra. No production has been recorded since 1990.

Sea cucumber (*Parastichopus*), found in Puget Sound, is used in a number of ways, including being cut up for soup and stew when dried.

CONCLUSION

A natural environment that once supported a wide variety of fish has been inundated with changes that have substantially reduced its ability to sustain these fish. Impediments to the movement of salmon; the modification and destruction of estuaries; sedimentation caused by overgrazing and improper logging practices; the discharge of sewage, toxic waste, and radioactive materials into water; thermal pollution; and severe overfishing have reduced the fishery of the Pacific Northwest to a position far below its potential. Ignorance, greed, apathy, and conflicting values are largely responsible for the present condition of the fishery. The increasing demand for fish in the United States and the rest of the world will probably prevent any meaningful restoration. Demand increases even as some success is achieved in sustaining the yield of some fishes. The addition of previously underutilized fishes can obscure the decline of others (Table 11). Overall, Washington's total production has remained static since 1982. In contrast, Oregon, because of very large landings of Whiting, has increased tonnage landed in the last 10 years.

Many hope that aquaculture will increase the production of fish. There is a potential, especially in Puget Sound and some embayments such as Grays Harbor and

YEAR	SALMON POUNDS	SALMON VALUE	OTHER ANADROMOUS POUNDS	OTHER ANADROMOUS VALUE	MARINE FISH POUNDS	MARINE FISH VALUE	SHELLFISH POUNDS	SHELLFISH VALUE	TOTAL POUNDS	TOTAL VALUE
1978	40,758	$52,751	3,443	$ 1,036	82,153	$21,340	37,827	$15,077	164,181	$ 90,204
1979	52,538	$61,216	2,911	$ 4,010	93,047	$23,248	37,497	$18,840	185,993	$107,314
1980	34,442	$45,972	4,113	$ 1,379	99,022	$19,716	32,948	$17,853	170,525	$ 84,920
1981	47,058	$38,417	3,994	$ 2,408	99,683	$20,967	27,685	$14,378	178,420	$ 76,170
1982	49,513	$45,972	4,060	$ 2,109	108,928	$25,129	23,072	$14,612	185,573	$ 87,822
1983	26,312	$52,096	4,882	$ 3,036	111,505	$25,070	26,007	$21,503	168,706	$101,705
1984	27,784	$18,817	3,679	$ 3,967	125,572	$28,707	24,360	$17,306	181,395	$ 68,797
1985	65,585	$31,402	6,348	$ 8,237	108,499	$28,722	29,686	$20,464	210,118	$ 88,825
1986	49,660	$50,598	7,399	$ 7,724	101,975	$34,427	43,481	$35,788	202,515	$128,537
1987	56,465	$54,513	6,599	$10,588	113,311	$40,172	47,954	$44,015	224,329	$149,288
1988	41,748	$80,804	9,360	$20,622	92,771	$35,608	68,254	$55,481	212,133	$192,515
1989	54,055	$73,286	8,700	$19,368	94,443	$33,143	64,274	$62,168	221,472	$187,965
1990	39,790	$51,955	6,843	$15,677	85,580	$32,140	49,050	$57,042	181,263	$156,814
1991	45,228	$32,066	9,198	$21,523	95,599	$35,801	42,738	$50,007	192,763	$139,397
1992	24,742	$21,613	14,099	$33,688	90,718	$32,451	51,183	$59,933	180,742	$147,685
1993	36,411	$25,240	10,794	$30,007	84,123	$32,300	57,819	$71,488	189,147	$159,035
Five Year Avg	41,113	$51,945	9,640	$22,176	91,822	$33,829	55,100	$56,926	197,675	$164,875
Table Avg	43,712	$47,432	6,375	$10,358	100,187	$29,109	40,401	$33,631	190,675	$120,531

Table 11. Commercial Landings of Foodfish and Shellfish by Species Groups (Round Weight and Value in Thousands)
Source: Dept. of Fisheries—State of Washington.

	QUANTITY (MILLIONS OF POUNDS)	VALUE (MILLIONS OF DOLLARS)
Newport, OR	112,000,000	22,000,000
Astoria, OR	89,000,000	26,000,000
Bellingham, WA	27,000,000	15,000,000
Westport, WA	24,000,000	25,000,000
Coos Bay, Charleston, OR	21,000,000	14,000,000
Waco, Chinook, WA	18,000,000	
Blaine, WA	12,000,000	
Shelton, WA		12,000,000
Anacortes-Iaconner, WA	12,000,000	11,000,000
Seattle, WA		10,000,000

Table 12. Pacific Northwest Ports.
Source: Current Fisheries Statistics, #9500, Fisheries of the United States 1995, U.S. Department of Commerce, NOAA, Silver Spring, MD, July, 1996.

Willapa Bay in Washington, for the production of more shellfish, salmon, and seaweed. However, sharp differences that have not been resolved exist on how these areas are to be used.

In addition, a myriad of private and governmental organizations in the United States and foreign countries clash over methods and strategies to perpetuate the fishery. Often the result is that little is done until it is too late. The destruction of the wild Columbia River salmon is a classic example. An attempt is now underway to reduce the pressure on the ocean fishery. The Pacific and North Pacific Fisheries Management Councils, mandated by the Magnuson Act to plan and carry out fish maintenance, are now limiting permits, a change from the past when the fishery was collectively owned by all U.S. fishermen within the 3 to 200 mile zone. The claim has been made that collective ownership has led to too many boats chasing too few fish. Methods of determining these rights in the future are being discussed.

Fishing is an important economic activity in many urban areas in the Pacific Northwest (Table 12). Recent declines in prices (Whiting selling at three cents per pound) have hurt coastal economies. Table 13 shows the results of a study of the effect of fishing and fish processing on the state of Oregon in 1993. It appears that the Pacific Northwest fishery has passed its peak and will contribute a declining amount to the total economy of the region in the future.

	Astoria and Columbia R.	Tillamook Area	Newport Area	Coos Bay Area	Brookings Area	All Coastal Communities	Total State Level Impact
Oregon Landings	$38.7	$5.2	$32.4	$23.7	$9.9	$109.9	$132.9
Distant Water	$13.7	$1.9	$37.6	$4.8	$0.8	$58.7	$69.7
Total	$52.4	$7.1	$70.0	$28.5	$10.6	$168.6	$202.6
Percent of Total Personal Income /3	8.2%	2.0%	9.9%	2.8%	3.0%	5.5%	0.34%
Percent of Total Earned Income /4	13.3%	3.9%	18.7%	4.9%	6.7%	10.0%	0.52%

1/ Estimates by Hans Radtke (Consulting Economist) and Chris Carter (Staff Economist) with some data provided by the Oregon State University Sea Grant Extension Agents. The personal income impact estimates are calculated with the Commercial Fisheries Economic Assessment Model developed by William Jensen and Hans Radtke for the West Coast Fisheries Development Foundation. Total personal income includes direct, Indirect and Induced effects.

2/ In most cases the economic impact for areas listed includes impacts for smaller ports in the area. The Tillamook area includes Pacific City. The Newport area includes Depo Bay. The Coos Bay area includes Florence, Reedsport and Bandon. The Brookings area includes Prt Orford and Gord Beach.

3/ Computed as total Income generated directly and indirectly by the commercial fishing and processing Industry divided by total Income from all sources. Total income from all sources is based on personal income estimates from the U.S. Department of Commerce, Bureau of Economic Analysis. Personal income is the sum of wages and salaries, other labor income, proprietors' income, personal dividend income, personal interest income and transfer payments less personal contributions for social security. Estimates for Coos Say area are slightly overstated due to Inclusion of Reedsport and Florence impacts in comparison to Coos County personal income.

4/ Computed as total income generated directly and indirectly by the commercial fishing and processing industry divided by toatl earned income as represented by estimated net earnings by place of residence, which does not include personal contributions for social insurance. Earned income does not include transfer payments, dividends, interest or rent.

Table 13. Personal Income Contribution of the Oregon Commercial Fishing and Processing Industry to Oregon Coastal Communities and to Oregon, 1993. (Millions of dollars) 1/ 2/
Source: 1993—Pounds and Valued of Commercially Caught Fish and Shellfish Landed in Oregon Area. Dept. of Fish and Wildlife, March 1996.

REFERENCES

1. Natural Resources Law Institute, *Anadromous Fish Law Memo,* Lewis and Clark Law School, Portland, Oregon: 50 issues from June 1979 to August 1990.
2. Bureau of Indian Affairs in Cooperation with U. S. Fish and Wildlife Service, *Background Information on Indian Fishing Rights in the Pacific Northwest,* Portland, Oregon, Revised January 1976.
3. Eschmeyer, William N., Herald, Earl S., and Hammann, Howard. *A Field Guide to Pacific Coast Fishes of North America,* Houghton-Mifflin Company, Boston 1983.
4. Jordan, David Starr and Gilbert, Charles H. *The Salmon Fishing and Canning Interests of the Pacific Coast in the Fisheries and Fishery Industries of the United States,* Section V, Volume 1, Part XIII by George Brown Goode. United States Commission of Fish and Fisheries, Washington Government Printing Office 1887.
5. Holts, David B. Review of United States West Coast Commercial Shark Fisheries. *Marine Fisheries Review,* Vol. 50, Number 1, 1988, pp. 1–8.
6. Johnson, Frederick G. and Stickner, Robert R. *Fisheries: Harvesting Life from Water,* Kendall/Hunt Publishing Co., Dubuque, Iowa, 1989.
7. Western Regional Aquaculture Consortium, *Waterlines* (Newsletter) Seattle, Washington, University of Washington, School of Fisheries, 1997
8. Columbia River Inter-Tribal Fish Commission, *Wana Chinook Tymoo,* Fall 1991, Portland, Oregon 1991.
9. Washington Department of Fisheries and Oregon Department of Fish and Wildlife. *Status Report: Columbia River Fish Runs and Fisheries 1938–91,* Portland, Oregon, July 1992.
10. Oregon State Planning Board, *A Study of Commercial Fishing Operations on the Columbia River,* Salem, Oregon 1938.
11. Pacific Fishery Management Council, *Status of the Pacific Coast Groundfish Fishery Through 1991 and Recommended Acceptable Biological Catches for 1992,* Portland, Oregon, 1992.
12. Oregon Department of Fish and Wildlife, *1990—Pounds and Values of Fish and Shellfish Landed in Oregon in 1987,* Portland, Oregon 1991.
13. Northwest Power Planning Council, *Columbia River Basin Fish and Wildlife Program,* Portland, Oregon 1987.
14. Nihlsen, Willa, Williams, Jack E. and Lichatowich, James A. Pacific Salmon at the Crossroads: Stocks at Risk from California, Oregon, Idaho and Washington. *Fisheries,* Vol. 16, No. 2, March–April 1991.
15. Rostland, E. *Freshwater Fish and Fishing.* University of California Publications in Geography, Berkeley, CA, 1940.
16. Damkaer, David M. and Dey, Douglas B. Evidence for Fluoride Effects on Salmon Passage at John Day Dam, Columbia River 1982–1986. *North American Journal of Fisheries Management,* Vol. 9, pp. 154–162, 1989.

17. Northwest Power Planning Council, *Amendments to the Columbia River Basin Fish and Wildlife Program (Phase 2)*, Portland, Oregon, December 11, 1991.
18. Pacific Halibut Commission, Pacific Halibut Fishery Regulations, Seattle, Washington 1992.
19. Pitts, John L. The Use of Aquaculture for Enhancement of the Common Property Fishing in Oregon, Washington and Alaska. *NOAA, Technical Report NMFS102, Marine Ranching,* 1991.
20. Fitzgerald, Roger. The Rise of the Pacific Oyster, *Seafood Leader,* September/October 1989, pp. 67–78.
21. Department of Fisheries, State of Washington, *1990 Fisheries Statistical Report,* Olympia, Washington 1991.
22. Chew, Kenneth K. and Toba, Derrick. *Western Region Aquaculture Industry Situation and Outlook Report,* Seattle Regional Aquaculture Center, University of Washington, Seattle, 1993.
23. *The Pacific Halibut: Biology, Fishery and Management,* International Pacific Halibut Commission, Technical Report, No. 22, Seattle, WA 1987.
24. Wood, Les, Anutha, Karen and Peschken, Achim. Aquaculture: Marine Farming of Atlantic Salmon, *Geography,* v. 30, 1990.
25. Rieman, Bruce E. and Beamesderfer, Raymond C. White Sturgeon in the Lower Columbia River: Is the Stock Overexploited. *North American Journal of Fisheries Management,* Vol. 10. pp. 388–396, 1990.
26. Atkinson, Clinton E. The Montlake Laboratory of the Bureau of Commercial Fisheries and its Biological Research, 1931–91. *Marine Fisheries Review,* Vol. 50, No. 4, 1988, pp. 97–110.
27. Pacific Fisheries Management Council, *Review of 1990 Ocean Salmon Fisheries,* Portland, Oregon 1991.
28. U. S. Statutes at Large 1976, *Fishing Conservation and Management Act of 1976,* Public Law 94-165, 94th Congress, U. S. Government Printing Office, Washington, 1978.
29. Grout C. and Margolis L. Eds. *Pacific Salmon Life Histories,* Vancouver, B.C. University of British Columbia Press 1991.
30. Cone, J. and Riddington S. Eds. *The Northwest Salmon Crisis: A Documentary History,* Corvallis, Oregon, Oregon State University Press. 1996.
31. Northwest Power Planning Council, Return to the River, *Restoration of Salmonid Fishes in the Columbia River Ecosystem,* Independent Scientific Group, Pre-Publication Copy, Portland, Oregon 1996.
32. Oregon Dept. of Fish and Wildlife, 1993—*Pounds and Values of Commercially Caught Fish and Shellfish Landed in Oregon,* Portland, Oregon 1996.
33. U.S. Army Corps of Engineers North Pacific Division, *Migration Exceeds Expectations,* Salmon Passage Notes, Portland, Oregon 1995.
34. Pacific International Halibut Commission, *Annual Report,* Seattle, Washington 1994.
35. Pacific States Marine Fisheries Commission, *48th Annual Report for 1995,* Gladstone, Oregon 1995.

36. Pacific Fisheries Management Council, *Review of 1995, Ocean Salmon Fisheries,* Portland, Oregon 1996
37. Pacific Fishery Management Council, *The Status of the Pacific Coast Groundfish Fishing through 1996 and Recommended Acceptable Biological Catches for 1997.* Portland, Oregon 1996.
38. Oregon Dept. of Fish and Wildlife and Washington Dept. of Fish and Wildlife, *Status Report, Columbia River Fish Runs and Fisheries* 1938–1994, Portland, Oregon 1995.

FORESTS

Chapter 7

Teresa L. Bulman and Bettina von Hagen
Portland State University

INTRODUCTION

The Pacific Northwest consists of nearly 200 million acres of land, of which 50 percent is forest land. Noncommercial forests make up about one quarter of the forest land and are either inaccessible, withdrawn from commercial harvest for parks or wilderness, or grow species of low commercial value. Commercial forests, which are timberlands not withdrawn from commercial harvest and capable of growing a minimum of 20 cubic-feet-per-acre per year, represent the majority of forest lands and produce timber products such as lumber and pulpwood, and non-timber products such as mushrooms, Christmas greens, and medicinals. Generally, these forest lands are accessible to logging and contain valuable tree species. Commercial forests are the basic natural resource for the economically important logging industry of the Pacific Northwest (discussed in Chapter 9), providing wages, employment, and commodities, and supporting related industries of shipping, marketing, and manufacturing.

For centuries, forests have provided food, shelter, fuel, and livelihood for people of the Pacific Northwest. Native Americans hunted game and harvested berries on the forested slopes and used timber for fuel and for log shelters. Later settlement in the Pacific Northwest was strongly tied to logging and milling operations, eventually leading to the establishment of a lumber manufacturing industry that supported a major part of the population. Over time, the region's forests have proven to be an extraordinary economic and environmental resource. In addition to timber and non-timber products, forests provide noncommercial and less quantitative values of soil and watershed protection, water supply through storage of winter snowfall, recreation, hunting, fishing, scenic beauty, and wildlife habitat.

Today, the Pacific Northwest's forest resources attract unprecedented national attention, not only because of their value as a resource but also because of the environmental effects of their use and exploitation. Attitudes toward the forests have changed over time, and forest resources have been variously appraised, exploited, destroyed, conserved, and modified, with the result that sweeping changes have reshaped Pacific Northwest forests since the Lewis and Clark expedition in the first decade of the 19th Century. Nearly 150 years of logging have taken their toll on the forest resources of the region and, as the 20th Century draws to a close, the forests

State	Total Forest Land	U.S. Forest Service	BLM	Other Federal	Non-Federal
Idaho	45,325	20,132	11,985	265	12,943
Oregon	52,133	15,583	15,740	511	20,299
Washington	31,076	9,072	307	1,767	19,934
Montana*	75,893	16,744	8,142	2,432	48,574
Total	204,427	61,531	36,174	4,975	101,750

*Figures are for entire state

Table 1. Forest Lands and Ownership in the Pacific Northwest (in thousands of acres).

of the Pacific Northwest are at the center of a regional and national debate that is pitting those who are interested in exploiting the commercial timber value of the forests against those who want to preserve forest areas and enhance the noncommercial values of those resources. Although management issues and logging practices are at the core of this debate, an understanding of the extent, types, and distribution of forest resources is essential to its resolution.

FOREST EXTENT AND OWNERSHIP

The commercial and noncommercial timberland of the Pacific Northwest of today comprises over 200 million acres and extends from the coastal margins to the Continental Divide (Table 1). The diverse ownership of forest resources reflects the rapid settlement of the region, the range of commercial enterprises involved in the settlement process, and the integral part forests played in settlement and early economic development. Today, a complex pattern of proprietorship exists, with ownership divided among federal, state, and local government, as well as private corporate and individual ownership.

During early settlement of the Oregon Territory, primarily by agriculturalists, land was cleared for farms, and sawmills were established to provide local building materials. The areal expansion of settlement contributed to a retreating forest, as patterns of rapid forest depletion emerged around mining and agricultural settlements. In 1846, as part of the Oregon Compromise with Great Britain that settled the region's boundary dispute, the federal government obtained title to the vast domains of the Pacific Northwest. After that time, the general federal policy—embodied in the Oregon Donation Land Act (1850), the Preemption Act (1841), the Homestead Act (1862), the General Mining Law (1872), and the Timber and Stone Act (1878)—was to transfer ownership to homesteaders, miners, and loggers at uniform low prices to encourage settlement and development of the land. Upon completion of the transcontinental railroad along the Columbia River in 1883, new markets were opened to the lumber industry and lands were acquired for their forest resource value. Railroad deeds, wagon road grants, and state land grants for education and other public purposes were also made, as were treaties between the federal government and Native Americans ceding lands to Northwest tribes for hunting, gathering, and spiritual purposes.

Forests were increasingly important in the burgeoning economy of the region. As the value of the timber resource was recognized, individuals and corporations began to acquire large tracts of forest land for logging ventures. With the devolution of federal land to state and private ownership, conflicts over forest use arose, eliciting nationwide public concern, and resulting in the establishment of forest reserves in 1891. By the end of the 19th Century, Romanticism in art and literature was praising the primitive wilderness, while nature observers were noting that unregulated deforestation was producing eroded, nonproductive, and visually scarred landscapes. Ironically, at the same time there was a nationwide regeneration of forests caused by farm abandonment, the initial development of fire management and evolving corporate timber management strategies. The first half of this century saw increasing corporate management of private forests and an increasing government role in control of publicly-owned forests.

The cumulative result of the land acquisitions, transfers, and reservations of the last century and a half is a complex forest ownership pattern throughout the Pacific Northwest and the absence of a unified forest management strategy, creating conflict between different government agencies and between agencies and private owners that further impedes effective management.

Federal agencies own approximately 45 percent of the region's commercial timberland. In Oregon, Idaho, and Western Montana, more than 50 percent of all forest lands are federally owned; in Washington, approximately one-third is federally owned. The largest of the federal owners is the U.S. Forest Service (Forest Service), an arm of the Department of Agriculture, which has responsibility for management of the region's national forests (Figure 1). The Bureau of Land Management (BLM) is also an extensive federal landowner whose holdings include not only large tracts of old-growth forest in Oregon and Washington, but also 80 percent of the Oregon & California lands (O&C lands) in its capacity as the successor agent to the General Land Office. The O&C lands were originally granted in 1869 to the Oregon and California Railroad, which proceeded to default on the requirements of the land grant, resulting in the reconveyance of the lands to the United States. Pursuant to legislation passed in 1937, these lands are separately managed from other timber resources and are not subject to the full range of timber management laws affecting federal lands, such as multiple-use policies.

Private companies own approximately 25 percent of the region's commercial timberland. The largest owners are Weyerhaeuser Company and the Plum Creek

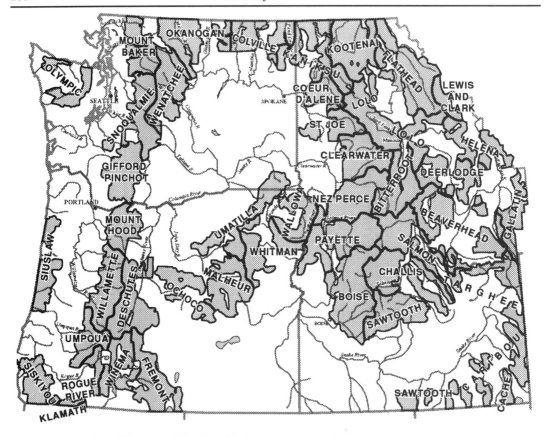

Figure 1. National Forests of the Pacific Northwest.

Timber Company, both of which have extensive logging and milling operations throughout the area.

The balance of forest land, owned by state and local governments and private individuals, is often scattered in small parcels in diverse areas. The State of Oregon alone has over 11 million acres of commercial state and private forest lands and, although representing less than one-half of the commercial forest area of the state, these lands produced more than 50 percent of the state's harvested timber throughout the 1980s.

FOREST ECOSYSTEMS

Forests are a complex association of trees, shrubs, ferns, mosses, fungi, microscopic plant and animal life, birds, reptiles, insects, and large animal species. Combined with soils and climate, the forest environment produces a range of ecological effects with varying growth regimes and productive capacities.

The ecological functions of forests are many and varied. Forests regulate the seasonal flows of water by absorbing, retaining, and eventually releasing snow-melt and rainwater which flow downstream to recharge rivers and aquifers. Forests also affect the global carbon cycle by removing carbon dioxide from and adding oxygen to the air through the process of photosynthesis. In this process, the extensive forests of the Pacific Northwest contribute to natural defenses against global warming. Forests also assist in the global recycling of nitrogen and affect regional climate patterns by influencing wind, air temperature, humidity, precipitation, and evaporation. Reduction of gas, particulate, and noise pollution is also achieved by forest processes.

Forests are the earth's largest reservoir of biological diversity and provide habitat for many wildlife species. Pacific Northwest species include large mammals such as elk, bison, moose, mule deer, pronghorn antelope, bighorn sheep, and black bear; small mammals such as porcupine, red fox, martens, marmots, weasels, and beaver; and birds such as golden eagles, bald eagles, Canada geese, various ducks, and northern spotted owls.

Forests are the natural vegetation of areas with ample soil moisture and a sufficient growing season. Trees are more long-lived than other plants, and their growth rates are relatively slow. However, because of their size, large amounts of nutrients are stored in the tree biomass. The biomass of dead trees is taken up by new growth. Occasional cyclic disturbances, such as fire, disease, insects, and windstorms, can kill nearly all the trees in a stand at once but also release the stored nutrients for a new cycle of growth.

Succession and Disturbance: Ecological succession is the process of evolution of a biotic community, such as a forest, and is an important factor in determining the forest composition and species distribution. Soil develops throughout a forest's ecological succession, derived from the parent material and from organic matter and nutrients in the biotic community. Climatic factors further influence soil development, as well as directly influencing the biotic community through precipitation and temperature regimes. The interaction of soil, climate, and organic matter in the process of ecological succession produces a sequence of tree and other plant species best adapted to the particular soil and climate regime of the region. The biotic community existing at the end of ecological succession is called the climax community and is preceded by a series of biomes called seral stages, such as the coniferous forests of the Pacific Northwest.

In most forests, cyclic disturbances occur that affect the rate and result of ecological succession. Minor disturbances, such as root damage by burrowing animals, wind abrasion, small landslides, or unusual droughts or frosts, can release small amounts of growing space in short terms. Larger disturbances, such as logging activities, fire, large avalanches and landslides, and insect infestations, can dramatically affect forests by killing vegetation and clearing large areas of growing space. Forest composition is strongly influenced by both natural and human disturbances in that such events reduce the life span of existing species and allow opportunistic species to occupy growing space to the exclusion of other species.

Throughout the Pacific Northwest, disturbance has played a significant role in ecological succession. Catastrophic fires, devastating windstorms, and wide-spread insect infestations have periodically destroyed large portions of forested lands. Combined with extensive logging activities, these disturbances have caused many forests to remain in intermediate successional stages rather than developing climax communities.

FOREST TYPES AND GEOGRAPHIC DISTRIBUTION

The extent and topographic variability of the Pacific Northwest provide for a number of regional climates and environmental conditions that affect the character, species composition, and rate of growth of forests. The forest zones of the Pacific Northwest span 16° of longitude and several large north-south-trending mountain ranges. Mean annual precipitation varies from more than 300 cm (120 inches) on the western slopes of the Coast Range to as little as 20 cm (8 inches) in the semi-arid Intermontane region. The orographic barriers of the Cascades and Rocky Mountains result in complex mountain climates, steep precipitation and temperature gradients, with year-round snow fields and glaciers on the highest peaks and rainshadow effects on the leeward slopes. As distance from the Pacific Ocean increases, seasonal temperature ranges also increase, from between 60°F and 40°F in the Coast Ranges to between 65°F and 25°F in the Intermontane and Rocky Mountain regions.

Local variability of elevation, slope, and aspect of surface features produces a complex mosaic of mesoclimates. Diurnal, seasonal, and annual temperature ranges generally increase with altitude. The southern sides of mountain ranges receive greater amounts of solar radiation, and thus are generally warmer and drier. In addition, soils vary in depth, moisture holding capacity, and nutrient value, further affecting forest growth patterns.

The net effect of the diversity of the environmental factors is the development of pronounced vegetation zones, with major variations primarily from west to east. Conifers dominate throughout the forested land of the Pacific Northwest. The dry summers and edaphic conditions of the region result in an absence of hardwood dominants that is unique among temperate forests of the world. Forested areas of the Pacific Northwest can be divided into three major subregions: (1) West Coast forests including Western Washington, the Coast Range of Northern Oregon, and the Cascades; (2) the Intermontane region of Eastern Washington, Eastern Oregon, and much of Southern Idaho; and (3) the Northern Rocky Mountains of Northern Idaho and Western Montana. Within these major subregions lie a range of forest types. (See Chapter 5 for a detailed description of forest types.)

Occupying a narrow belt along the Pacific Coast west of the Coast Range and Olympic Mountains, and extending halfway down the Oregon Coast, are forests influenced by a maritime climate of cool temperatures and high precipitation and dominated by Sitka spruce (*Picea sitchensis*). Extending up to approximately 150 meters (500 feet) above sea level, these productive forests have lush understories with dense shrub communities. Further inland, and extending in a wide belt across the Coast Range to the western slopes of the Cascades, are forests dominated by Douglas fir (*Pseudotsuga menziesii*) and western hemlock (*Tsuga heterophylla*). Here, the forests are characterized by lush understories and dense stands of tall trees. Many of these forests have been extensively altered by fire and logging, with the result that the successional patterns are frequently disturbed and climax vegetations are rare.

Occurring on the West Coast—across the Coast Range into the Willamette and Rogue River Valleys and Puget Trough and onto the western slopes of the Cascades—are mixed stands of coniferous forests occupying a region of strong moisture gradients and resulting in stands of variable structure and composition depending on moisture availability. Within these stands, Douglas fir dominates, often in association with western hemlock, western red cedar, and grand fir. Massive old-growth forests, with trees several hundred years old, are unique forest habitats that are at the center of the controversy over forest management practices.

Southwest Oregon is an area of complex geology and edaphic conditions and a long history of fire, grazing, and logging activities. When combined with the steep temperature and moisture gradients of the region, a complex and variable forest pattern is created of mixed conifers and evergreens. Dominant species include Douglas fir, tanoak (*Lithocarpus densiflorus*), canyon live oak (*Quercus chrysollpis*), and Pacific madrone (*Arbutus menziesii*). On lower interior slopes and in the interior valley oak (*Quercus*) woodlands are the primary forest cover. At higher elevations, various pines, such as sugar pine (*Pinus lambertiana*) and ponderosa pine (*Pinus ponderosa*) dominate.

Hardy species occupy the subalpine forests in an irregular belt on the upper slopes of the Cascades and in the Olympic, Selkirk, Bitterroot, and Rocky Mountains—in areas of heavy snow, severe winter temperatures, and thin soil. The principal species are subalpine fir (*Abies lasiocarpa*) and Engelmann spruce (*Picea engelmannii*).

East of the high elevations of the Cascades, and extending over wide portions of the mountainous Pacific Northwest, are forests dominated by ponderosa pine. Frequently ponderosa pine is as much as 85 percent of the forest volume, creating an environment of open stands with grasses and shrubs suitable for grazing. As a result, these economically important forests are severely harvested and grazed.

In the Northern Rockies, in the intermontane region, and to a small extent in the West Coast and eastern side of the Cascades, lodgepole pine forest dominate. Often consisting of nearly pure, dense stands, these forests reflect the way fire has shaped the forest ecosystems of the Pacific Northwest, particularly in the Northern Rockies where opportunistic fire pioneers, such as lodgepole pine and western larch (*Larix occidentalis*), often replace climax conifers such as Douglas fir.

FOREST MANAGEMENT AND ADMINISTRATION

Because forests vary in their species composition, accessibility, and ownership, a unified management scheme does not exist for the forests of the Pacific Northwest. As a general rule, federal law governs management of federal forest lands while state law regulates forest practices on state, local, and private forest land. To some extent, federal law affects state and private forest lands. For example, provisions of the Clean Water Act govern water pollution related to timber activities. Oregon's Forest Practices Act (1971, amended in 1991) is typical of state regulation in its efforts to limit size of clearcuts, increase reforestation requirements, and require measures for wildlife habitat and stream protection.

A chronology of forest management in the United States indicates that not only have forest practices varied over time, but perceptions about forest values have greatly modified management plans.

The Era of Exploitation: Prior to the arrival of Lewis and Clark in the early 1800s, the forests of the Pacific Northwest were the exclusive domain of Native Americans. Evidence exists that regional gatherers affected forest succession by burning to promote huckleberry growth, and it is likely that there was little attempt to prevent or control naturally occurring fires. Early settlement of the Oregon Territory meant cutting the forests for industry, timber, and urban and agricultural land use. With the development of the nation and the growth of railroads and shipping corridors, the home market for forest products expanded, and exploitation of the forests quickly followed. Wasteful cutting practices, uncontrolled fires (frequently caused by humans), and extensive soil erosion of deforested slopes became commonplace. These destructive practices were most clearly felt in the large and valuable Douglas-fir forests. By the mid-19th Century, the region's wood-producing industry achieved national importance and logging operations and mills dotted much of the Pacific Northwest landscape. By 1905 Washington was the nation's leading wood-producing state.

The Beginnings of Conservation: The widespread acquisition of forest land and the wasteful practices and uninhibited exploitation of forest resources continued into the latter half of the 19th Century when, in 1891, the U. S. Congress authorized the president to withdraw forest lands in the public domain from homesteading and other settlement practices. These areas were to be set aside in timberland reserves (now national forests) and brought under the general management of the federal government through the General Land Office of the Department of the Interior. The Forest Service Organic Act in 1987 established that reserved forest lands were to be used to supply timber, and as a result, forest destruction continued apace. At the same time, concerned citizens were bringing national attention to the exploitation of the nation's forests, increasing public pressure for federal action, and promoting conservation reforms.

The Transfer Act of 1905 transferred management of the forest reserves from the Department of Interior to the newly created U.S. Forest Service in the Department of Agriculture. The first Chief Forester of the Forest Service was Gifford Pinchot, who replaced forest reserves with national forests, created new national forest areas, and promoted forestry as a legitimate profession concerned with the scientific management of forest resources. Though the focus of the early years of the Forest Service was the development of timber sales programs for the national forests, early legislation related to the national forests also contained conservation provisions, including funding for protection of navigable waters and for reforestation projects.

The most significant legislation affecting federal forest lands was enacted after World War II and embodied concepts of exploitation of the timber wealth of the national forests, conservation for noncommercial values, and development of scientific management principles. In 1960 the Multiple-Use Sustained Yield Act was passed, supplementing the Organic Act of 1897 and giving the Forest Service jurisdiction over not only timber but also over watershed protection, wilderness and wildlife habitat, recreation, grazing, and other uses of national forest lands. The act served to establish the principles of multiple use and sustained yield as the functional guidelines of forest management. It also required the Forest Service to act as managers of various and often conflicting uses within the national forests, requiring it in theory to adhere to the Forest Service's stated policy goal of establishing "the greatest good for the greatest number in the long run." However, conflicting interests can be difficult to balance and budgetary constraints often prevent implementation of comprehensive plans.

The National Forest Management Act of 1976 expanded public policy concerns to forest operations on all federal lands. In essence, the act was designed to promote planned harvesting operations and to reduce environmental degradation on harvest sites. The statute restricts allowable cuts and establishes general harvesting restrictions related to site-specific land use, diversity of species, clearcut sizes, and protection of riparian corridors.

Numerous other federal laws affect the forest resources on the public lands of the Pacific Northwest. The most notable are: 1) the Wilderness Act (1964), which provides for the establishment of Wilderness Areas (Table 2) and which bars timber harvesting, mechanized recreation, and mining, in designated wilderness areas on federal lands; 2) the Wild and Scenic Rivers Act (1968), which requires agencies to protect designated segments of rivers flowing through federal lands; 3) the National Environmental Policy Act (1969), which mandates compliance with environmental assessment requirements for timber sales and any other federal programs affecting forest lands; 4) the Endangered Species Act (1973), which mandates protection of habitats vital to endangered species, and which may have a cumulative effect of severely reducing timber harvest programs; and 5) the Clean Water Act (1972), which affects logging practices that can cause water pollution and calls for "best management practices" to control nonpoint source pollution from logging and milling operations. The administrative framework of federal forest management in the Pacific

State	Acreage
Idaho	3,825
Oregon	2,068
Washington	2,522
Western Montana*	3,360
Total	11,775

*Estimated figures for Western Montana

Table 2. Wilderness Areas in National Forests (in thousand acres).

Northwest involves five principal agencies: 1) the Forest Service, which manages the national forests; 2) the Bureau of Land Management, which oversees certain timbers holdings in the region including the Oregon & California lands; 3) the Fish and Wildlife Service, which promotes fish and wildlife habitat protection; 4) the Soil Conservation Service, which evaluates and regulates certain public and private woodlot and forest areas subject to soil erosion; and 5) the Environmental Protection Agency, which enforces national environmental policy.

The wide scope of these agencies and the laws they enforce suggest a substantial and comprehensive management scheme governing the forest resources of the Pacific Northwest. In fact, the authority of these agencies and the interpretation of the laws are continually disputed in administrative and court proceedings and frequent lawsuits charge that the agencies fail to develop comprehensive plans or even to enforce their own regulations. At the heart of most disputes are logging practices and their effect on the environment.

LOGGING PRACTICES

Forests are potentially renewable resources if the rate of harvest does not exceed the rate of regrowth over a long period. However, for decades timber in the Pacific Northwest was harvested at rates far exceeding the rate of regrowth. Clearcutting, the indiscriminate cutting of all trees in an area, has left a fragmented and scarred wilderness, lacking in biodiversity and wildlife corridors, and has led to an intense debate over logging practices on all lands, public and private.

Early forest management consisted almost exclusively of surveying and mapping with limited efforts directed to protection and maintenance. In 1928 the McSweeney-McNary Act authorized and mandated a forest survey of the entire United States, which was not completed until 1949. A re-inventory of forests west of the Continental

Divide was undertaken between 1953 and 1958. Both surveys were primarily concerned with the potential commercial output of forest products, especially timber. More recent inventories have been conducted, but changes in inventory techniques and definitions make comparisons difficult, and the exact amount of forested and logged land is not fully ascertained.

Before the 20th Century, forests were seen as areas of infinite endurance, providing an endless supply of wood and containing little else of value. At the turn of the century, scientists and preservationists, concerned about forest deterioration, managed to convince the federal government that sustained yield was the only reasonable approach to long-term management of forest resources. Sustained yield is the harvesting of species at rates equal to the species' reproduction and growth rates. In theory, under a program of sustained yield a species is neither depleted nor harvested prior to maturity. Difficulties arise in application, however, since not all species mature at the same rate and demand for wood fluctuates continually. Officially, the concept of sustained yield has been federal policy since 1960 although a range of harvesting techniques and reforestation methods are used throughout the forests of the Pacific Northwest.

The principal timber harvesting techniques, each with specific impacts and goals, are shelterwood cutting, selective cutting, and clearcutting. Shelterwood cutting involves thinning, cutting, and removal of poor quality trees, which opens the stand to more light and promotes seedling development. The remaining mature trees provide shelter for the seedlings and can be cut after the seedlings are established. Shelterwood cutting requires small plots, homogeneous species, and large labor inputs.

Selective cutting is used in forests of mixed age or mixed species, where only the most mature and economically valuable trees are harvested. Harvesting specific species can reduce diversity, and thus affect wildlife and other aspects of the forest ecosystems. The high costs associated with selective cutting mean only stands with trees of high value tend to be cut this way.

Clearcutting is the dominant method of tree harvesting in the Pacific Northwest and the rest of the nation, representing about two-thirds of harvested timber in the United States. In clearcutting, all trees are cut, regardless of maturity, size, or species, removing the entire forest canopy. As a result of comprehensive cutting, new, same-aged stands flourish, usually containing only a few of the fastest growing species that require moderate or high amounts of sunlight for germination and seedling growth. Clearcuts may consist of entire stands, strips, or patches. Reforestation can occur naturally from seeds released in the cutting process, but more often, genetically enhanced, nursery-grown seedlings are planted by hand.

Clearcutting is generally preferred by the timber industry because this method of harvesting provides the maximum economic return: the volume of timber harvested per acre is high; fewer roads are required; less planning, less surveying, and less skilled labor is required; and reforestation is relatively easy and controlled. Clearcutting has some environmental advantages in that fewer roads are built than in selective and shelterwood harvesting, thus reducing soil compaction and erosion. In addition, forage and habitat for some herbivore species (including elk and deer) is enhanced, and stands infected by insects or disease can be removed and salvaged for their timber value before they become unmarketable.

On the other hand, clearcutting can lead to severe soil erosion, especially on steep slopes, greater sediment loading of water, greater flood hazards, and increased landslide activity. Soil infiltration capacity is reduced and surface erodibility is increased in heavily disturbed places that contain skid trails, cable log paths, and slash and burn piles. Soil erosion through exposure and compaction impairs regeneration and future biomass productivity, reducing wildlife habitat and disrupting or destroying wildlife corridors. Recreational and scenic values of forests are damaged as well in clearcutting.

The newest forest practice is frequently referred to as "New Forestry" although the term encompasses a range of cutting, clearing, and reforestation regimes. In general, new forestry preserves pieces of the forest habitat by leaving dead snags, living trees of diverse ages and species, fallen logs, and other debris. Slash is burned or left to decay. Such practices retain habitat for wildlife, return nutrients to the soil, and create a more diverse second-growth forest.

All logging methods affect forest ecosystems to some degree. Even minimal clearing can stimulate growth by exposing the ground to sunlight. Cut debris left on the ground decays and releases nutrients. On the other hand, soil erosion is accelerated, as are water and nutrient runoff. The use of heavy equipment and logging roads further damages soil development and increases soil erosion, particularly on the steep slopes found throughout the Pacific Northwest.

After harvesting, reforestation occurs through natural regeneration or seedling planting. Natural regeneration is most effective in harvesting techniques that leave seedlings and seed sources in place. Seedling planting, by machine or hand, is the most common form of reforestation and allows for greater control over tree density and seedling survival. Fertilizers, pesticides, and broadleaf defoliants further impact the species survival and composition. Increasingly, logged forests are being replanted as tree farms of genetically enhanced species of mind-numbing uniformity. Reforestation rates have improved in recent decades, but private and public timber harvests have exceeded sustainable yields for so long that these efforts may be too late.

In the 1970s, logging practices in the Pacific Northwest came under fire because of their severe environmental effects, not only in the forests but off-slope as well. The logging and grazing practices that led to soil erosion have filled river fishing holes with silt and sand, contributing to depletion of Pacific Northwest salmon runs through destruction of spawning areas. The National Forest Management Act of 1976 was intended to address public concerns about logging and other uses of the nation's timberlands. The act envisioned increased public participation in the forest management planning process, of which one result has been frequent and increasing attempts to halt timber sales and logging practices that are perceived as detrimental to the long-term sustainability of the environment.

In 1990 alone, more than one-half of the timber sale volume offered by the Forest Service was subject to appeals by citizens and environmental groups. Conservationists claim that over-logging in Idaho's Targhee National Forest has reduced big-game hunting and hurt local economies by driving timber operators out of business as cuts continue in excess of sustainable yields. The Mt. Hood National Forest, one of Oregon's most popular tourist areas as well as the watershed for Portland's metropolitan water supplies, introduced a new management plan in 1991 that slashed timber sales by nearly half and increased emphasis on protecting wildlife habitat, scenic views, and water quality—and the plan was immediately attacked by environmentalists and industry representatives alike.

One federal judge attributed the increased number of appeals and conflicts over forest management plans to a deliberate and systematic refusal of the Forest Service to comply with environmental laws. The General Accounting Office, the investigative arm of Congress, concluded that the Forest Service and the BLM typically ignore their own experts' advice to protect wildlife, instead favoring mining, logging, and grazing interests. Within the Forest Service itself, an increasingly vocal group of professionals are promoting conservation and better management for wildlife protection and other noncommercial interests. Even when marginal protection is provided in one area, such as the principal habitat of the endangered northern spotted owl habitat on the west side of the Oregon Cascades, the result can be increased pressure for logging elsewhere including logging in private forests already facing critical log shortages due to decades of overcutting.

Another concern related to logging practices on public lands is the financing of timber harvests. The 1992 federal budget included $1.17 billion to finance uses of national forests, including timber ($583.4 million), recreation ($342.1 million), wildlife ($92.9 million), minerals, watershed, and range. The Forest Service also raises money by selling timber and by charging fees for grazing, mining, and camping. Pacific Northwest forests, in part, make money because of the high volume cut. Nationwide, however, timber sales are not always profitable for the government (for example, 65 of the 122 national forests lost money on timber sales in 1990). Part of the financing problem is the failure to correctly estimate the amount of available timber and the failure to price the timber to reflect the true value of the resource and the actual costs of resource management. Additionally, accusations of logging scams, ranging from alleged kickbacks, biased pricing, and fraudulent scaling to brand switches and bid-skewing, hurt the credibility of both loggers and the federal agencies responsible for regulation and oversight.

Federal guidelines for timber harvests and logging practices reflect a time when people perceived forest resources as inexhaustible. If federal agencies are indeed the custodians of the land for the benefit of the public, then interests beyond those of the deeply vested timber industry must be recognized and protected when logging plans and timber sales are developed.

ISSUES IN FOREST MANAGEMENT TODAY

Ecosystem Management: Confronted with diminishing stands of old-growth forests, pleas from lumber-dependent timber communities as well as challenges from environmental groups, Native American leaders, and the scientific community, the Forest Service is attempting to find a new balance in its forest management practices.

The most recent manifestation of the Forest Service's efforts to balance conflicting demands is its 1990 Resource Planning Act. The act identifies major priorities for the Forest Service in the last decade of this century: 1) increased attention to recreation, fisheries, wildlife, and protection of water and soil; 2) production of commodities in environmentally acceptable ways; and 3) improvement of scientific knowledge about natural resources and global resource issues. In addition, the act contains elements of both New Forestry and New Perspectives, the two most significant programs in modern forest practices. Both of these management practices focus on ecosystem management techniques rather than the traditional focus of multiple-use and timber production.

New Forestry, a scientific and ecological approach to forest management was developed in the late 1980s by Forest Service biologists working in the H. J. Andrews Experimental Forest in Oregon. It is rooted in the recognition that various forest ecosystem processes, both above and below ground, are interdependent and essential to maintaining a healthy forest. Thus, for long-term production of commodity items, such as timber, the production practices must simultaneously protect the biological legacy (including biodiversity and succession processes), ecosystem complexity, and the dynamic landscape of the forest. The commodity most affected by this constraint is timber and the clearcut-logging practices associated with it. New Forestry research has indicated that limited cuts, which leave mature trees, snags, and downed debris, are necessary for soil, wildlife habitat, and regrowth and further serve to minimize forest fragmentation. The Forest Service has initiated New Forestry techniques on selected timber harvesting sites in the Pacific Northwest.

A second program adopted by the Forest Service is New Perspectives, an ecosystem management approach currently being developed in several Pacific Northwest forests. Its three main objectives are adoption of ecologically sustainable management practices, multiple-resource use, and preservation and restoration of old growth forests. In practice, New Perspectives will require the development of resource management programs that provide for stewardship of soils, water, air, and biological diversity while at the same time meeting national demand for forest products.

Three major obstacles confront the Forest Service in its transition to New Forestry and New Perspectives management practices: 1) internal resistance to change, which will inhibit implementation of programs within each forest; 2) regional and national

Name of Fire	Year	State	Acres Burned
Yaquima	1846	Oregon	450,000
Nestucca	1860	Oregon	320,000
Silverton	1865	Oregon	1,000,000
Coos	1868	Oregon	300,000
St. Helens	1868	Oregon/Washington	300,000
Columbia	1902	Oregon/Washington	604,000
Great Idaho	1910	Idaho/Montana	2,000,000
Tillamook	1933	Oregon	267,000
Saddle Mountain	1939	Oregon	190,000
Wilson River	1945	Oregon	180,000
Oxbow	1966	Oregon	46,000

Table 3. Historic Major Fires in the Pacific Northwest.

politics, which include harvest levels set by the U.S. Congress and concerns about rural, timber-dependent communities; and 3) the inadequacy of the scientific knowledge base on exactly what "sustainable" levels of timber and other forest product harvests are. Whether the introduction of New Forestry and New Perspectives will lead to a new age of enlightened ecosystem management of national forests will depend on the political will and economic commitment of all forest users.

Fire Management: Fires are an important and natural component of the forest ecosystems of the Pacific Northwest. Some species, such as the lodgepole pine and western larch, have adapted to fire by developing mechanisms for rapid regeneration after fires. Fires promote the health of forests by burning accumulated fuels and aging wood that collect over time and by reducing chances of catastrophic fires.

The disruptive effects of fire on the ecosystem of forests include extensive burning of dead and live trees, decreases in water consumption and transpiration, increased runoff and soil erosion, and increased loss of nutrients. The beneficial effects of fire include the release of nutrients stored in dead biomass and the removal of dead timber susceptible to disease and insect infestation.

Fire frequency, the average number of years between successive fires, is dependent on the rate of fuel accumulation, fuel and soil moisture levels, and ignition sources. Natural fire frequency ranges from 100 to 400 years in the Pacific Northwest. Prior to the adoption of extensive programs of fire suppression in the first half of this century, frequent catastrophic fires engulfed forests throughout the region (Table 3).

Because fires are a normal part of forest ecosystems, over-zealous fire-fighting leads to unnatural fuel accumulation that can induce catastrophic fires in areas where more frequent but less intense fire regimes would normally control fuel accumulation.

Forest fires fall into these basic categories: 1) ground fires, which burn the organic matter and litter of the soil and have little impact on trees; 2) surface fires, which burn ground litter, herbaceous and shrubby vegetation, and small trees; and 3) crown fires, which burn virtually all vegetation from ground litter to tree tops. Crown fires are particularly destructive to timber, wildlife, and soil.

In the 1920s and 1930s intense fire-fighting operations began to prevent large-scale fires in commercial timberlands and to prevent the destruction of structures and loss of life. This evolved into a more comprehensive policy, practiced by most government agencies involved in forest management, to fight all naturally-started fires—a policy which permitted the build-up of huge amounts of fuel on the forest floor and resulted in more severe crown fires. By the middle of the 1970s, most agencies had adopted "let-burn" policies based on studies of regional fire history and which included letting naturally-set fires burn wherever practicable, fighting human-caused fires, and setting prescribed fires to mimic the natural fire regime. Nonetheless, decades of fire suppression allowed for significant fuel build-up. When combined with the extensive drought conditions in the Pacific Northwest throughout the 1980s, the scene was set for extensive and devastating fires. Fires in Southern Oregon in 1987 burned more than 200,000 acres; fires in October 1991 alone burned more than 250,000 acres, destroyed more than 100 homes and caused five deaths in Washington, Idaho, and Montana. One fire in the Columbia River Gorge nearly engulfed Oregon's historic Multnomah Lodge.

Though spilling over only marginally into the Pacific Northwest, the Yellowstone fire of 1988, which burned 45 percent of the Yellowstone National Park, has had a significant impact in the region. The Yellowstone conflagration focused public, scientific, and political attention on fire management on public lands. While scientists and agency representatives agree on the positive role of fire in forest ecosystems, to the public's mind a fire-scarred wilderness is an abomination. However, the rapid recovery of vegetation and wildlife in the Yellowstone area has vividly demonstrated that the ecosystem can recover from catastrophic fires, and that human enjoyment of the forest can continue undiminished.

Recognizing that fire plays an important and natural role in forest ecosystems, forest resource managers are developing new policies for fire management in Pacific Northwest forests. Reflecting the diverse uses of those areas, the policies differ for commercial forests, noncommercial timberland, and wilderness areas. In general, the programs call for the application of prescribed burns to supplement natural fires and to create as natural a spatial and temporal pattern of fire as can be obtained in a region heavily impacted by anthropogenic interference, especially by human-caused fire. Failure to mimic the natural fire regime will result in shifts in vegetation density, species composition, and wildlife habitat; yet allowing natural fires to burn uncon-

State	Douglas fir Tussock	Mountain Pine Beetle	Western Spruce Budworm	Dwarf Mistletoe	Root Disease
Idaho	14,200	57,445	2,399,378	3,224	1,929,000
Oregon	20	1,129,160	2,439,168	4,885	1,221,000
Washington	17,530	146,620	37,850	3,575	999,000
Western Montana*	0	1,492,074	2,545,326	2,416	1,400,000
Total	31,750	2,825,299	7,421,722	14,100	5,549,000

*Includes limited acreage in Eastern Montana

Table 4. Forest Areas with Insect or Disease Infestation.

trolled will endanger human lives and structures. The fire management policies in the wake of the Yellowstone fire will be based on our knowledge of the natural fire history of the Pacific Northwest, increased public understanding of the beneficial role of fire in forest ecosystems, and an awareness of the need to balance concerns and goals of timberland harvesters, other forest users, and wildlife.

Insect and Disease Management: To protect the valuable timber harvest, forest managers attempt to prevent and control damage from insects and disease. Major insects and diseases in the forests of the Pacific Northwest include western pine beetles, mountain pine beetles, western spruce budworm, root diseases, and dwarf mistletoe (Table 4).

Decades of fire suppression have made Pacific Northwest forests more susceptible to insects and disease. At the same time, widespread insect infestation makes the weakened trees more susceptible to fire damage, increasing the probability of catastrophic fires. In addition, the loss of forest overstory reduces thermal cover and grazing for wildlife. The problem has been aggravated by severe drought conditions in most non-coastal areas of the region. One area, the forests of the Blue Mountains in Eastern Oregon, has been ravaged by insects and disease after a decade of drought in the 1980s with more than 2 million forested acres dying from western spruce infestation. This ecological disaster was aggravated by destructive logging practices and extensive fire suppression. Repairing the damaged forests will require a decade-long program of tree thinning, prescribed burning, salvage logging, and reforestation, at an estimated cost of $5 million. The insect-plagued Blue Mountains are not unique in the Pacific Northwest, and the economic costs of such infestations have focused increasing attention on understanding and controlling the insects and diseases which infest the forests.

Insect infestation is a significant cause of tree mortality in the Pacific Northwest. The western spruce budworm infected more than 10 million acres during the 1980s, with wide-spread effects in central western Montana and Eastern Oregon. Although massive spray programs were planned for 1992, spray programs are seen as short-term fixes lasting only a few years. The mountain pine beetle has affected three million acres of lodgepole and ponderosa pine in the Pacific Northwest, especially east of the Cascades. The Douglas-fir bark beetle is making inroads into the region, and the European gypsy moth, which has defoliated 13 million acres in the East Coast, is now found in the Pacific Northwest. Though the gypsy moth prefers deciduous trees, forest managers throughout the Pacific Northwest are concerned because the moth will feed on conifers if preferred hardwoods are absent.

Root diseases, such as armillaria, and dwarf mistletoe produce chronic adverse effects and are common in the Pacific Northwest. Root diseases in particular affect the Douglas-fir and grand fir forests of western Montana. The dwarf mistletoe, common in Douglas fir and western larch, is a parasitic plant that absorbs nutrients and water from the host tree, particularly conifer hosts, affecting long-term health of forests. The net effect of insects and disease is the deterioration of the region's high mountain forests and loss of marketable timber. In 1991, the Umatilla Forest in Oregon and Washington, had an estimated 800-million-board-feet of insect-damaged trees in need of immediate salvaging. The normal annual harvest in this forest is 180 million-board-feet.

Various artificial methods for preventing and controlling insect and root damage exist, including treating diseased trees with antibiotics and applying insecticides and fungicides. Genetic engineering permits development of disease-resistant species. Clearcutting and burning of slash in affected areas can remove dead and infected trees, kill insects, and clear vegetation that serves as a host for disease organisms. Attempts to prevent importation of harmful insects into the Pacific Northwest have not been very successful in the face of parasites from abroad arriving on ships and in ballast water, and as overland pests such as the gypsy moth invade forested regions. Eradication programs can be expensive endeavors once pests arrive in the region. In 1991, a one-year, $2 million campaign to eradicate the Asian gypsy moth from Oregon and Washington port cities began. The program entailed trapping the moth and massive spraying with the biological control agent, *Bacillus thuringiensis*. Though the moth has apparently been controlled, the program must be continued indefinitely to ensure complete eradication and prevent new introductions.

Biological pest controls involving the use of natural predators, parasites, and pathogens (disease-causing microorganisms) can control a variety of pests, especially when used as part of a regulated and integrated pest management program designed for a particular forest. In the Pacific Northwest, such programs are being developed, particularly in heavily-infested forests in Eastern Oregon. However, the ultimate success of the programs will depend on long-term, comprehensive forest health projects that incorporate prescribed burns and ecologically sensitive logging practices.

Endangered Species: A major issue in forest management in the Pacific Northwest is the intense battle over old-growth forests, and the endangered species such as the northern spotted owl that inhabit those forests. The northern spotted owl (*Strix occidentalis lucida*) has become a symbol of the battle between environmental groups which seek to preserve old growth for its noncommercial values, and the timber industry, which seeks to harvest old growth for its high commercial value. The owl, which weighs about 22 oz., stands less than 2 feet tall, and ranges from Northern California through Washington, inhabits the ancient forests of the Pacific Northwest, preferring the mature old growth forests with old conifers (over 200 years old), snags, and fallen logs. Owl numbers have been declining due to heavy logging in these old-growth areas.

In 1990, the owl was listed as "threatened" on the Endangered Species List under the Endangered Species Act. This listing by the U.S. Fish and Wildlife Service means the owl's forest habitat must be protected to ensure its survival. More than 11 million acres, two-thirds on public land, have been identified by the U.S. Fish and Wildlife Service as important to the owl's survival. Very little is in designated wilderness, national parks or other protected designations; virtually all is in forest land designated for logging. Many definitions of old-growth have been proposed during the spotted owl debate. The Forest Service's definition of old growth is a forest with a deep, multilayered canopy, standing dead trees, and at least eight trees per acre older than 200 years or greater than 32 inches in diameter. As so defined, about 25 million acres of such forests once blanketed the Douglas-fir region of the Pacific northwest; today, two million acres remain. Old growth plays an integral role in regulating water levels and water quality, cleaning air, enhancing fish and wildlife productivity, and enriching and stabilizing soil. Old growth forests have all but vanished from private lands. Most remain on Forest Service and BLM lands, which are logged at the rate of 70,000 acres of old-growth forest annually. At current cut rates, the last ancient forest outside the limited wilderness areas will fall within 30 years, at which time mills designed or tooled for old-growth lumber will be forced to close or retool regardless of the spotted owl.

Environmentalists have long been concerned about the extensive logging of the virgin Pacific Northwest forests, especially in old-growth stands, and have argued for preservation of these ancient forests. Throughout the region small lumber towns feel particularly threatened by suggestions that logging be halted. Attention has been focused on the northern spotted owl, an indicator species for mature and healthy old-growth forests. There are several additional species listed as endangered or threatened in the forests of the Pacific Northwest. However, the real issue is forest management strategies. Pacific Northwest forests have been cut at rates far exceeding growth for decades; use of sustainable yield practices would protect owl habitat (as well as the habitat of less celebrated species) and would stabilize the regional timber supply although short-term economic impacts could be severe.

In addition to setting aside forest habitat for preservation, the effect of the spotted owl listing, depending on the preservation plan ultimately adopted, ranges from cuts in timber production by up to one-third, mill closings and cutbacks costing tens of thousands of jobs by the end of the century, falling real estate prices, community dysfunction, and reduced tax revenues at the federal, state, and county levels. Counties in particular would suffer lost revenue since they receive income from Forest Service and BLM timber sales.

Efforts to preserve the spotted owl habitat have already begun. Pending federal action, the Oregon State Department of Forestry banned logging near known nesting sites. To preserve the spotted owl's federal land habitat, it is estimated that the 1985–89 average of 4.4 billion-board-feet per year harvested on the national forests of Oregon and Washington must be reduced to 700 million-board-feet. In May of 1991 the U.S. District Court in Seattle banned timber sales on 66,000 acres of Forest Service land in Oregon and Washington pending adoption of a satisfactory plan for owl protection. In 1991 4,570 acres of Southern Oregon forest, representing 44 timber sales by the BLM, were protected from logging by court orders in law suits over the northern spotted owl. In December 1991, the federal panel directed to create a recovery plan for the northern spotted owl prepared a draft document that suggested timber sale reductions of about 2 billion-board-feet, significantly under the 1980s average but more than twice that advocated by many environmentalists. Court challenges to logging, as well as to owl habitat preservation, are expected to continue for decades, during which time additional species will be added to the Endangered Species List. For example in 1992, the marbled murrelet (*Brachyramphus marmoratus*), which also inhabits the old growth forests of the Pacific Northwest, was listed as a threatened species.

The stakes involved in balancing the economic and environmental interests go far beyond the Pacific Northwest. As America attempts to influence resource conservation abroad, particularly with respect to tropical rainforest preservation, the handling of the spotted owl issue will enhance or diminish U.S. credibility.

The owl is a focal point for the larger dispute over whether ancient forests should be logged or preserved. The issue has raised concern about protection of ancient forests as well as concerns about the availability of help for communities and workers affected by the downturn in logging. It represents an epic confrontation between different philosophies of the rights of nature and the place of humans within nature while raising fundamental questions of whether forests should be used and exploited or preserved and revered, of how much wilderness America needs, and how much commercial disruption can be tolerated in the name of conservation.

SPECIAL FOREST PRODUCTS: A SUSTAINABLE DEVELOPMENT STRATEGY

As projected revenues from Pacific Northwest timber harvests decline, other revenue-producing, forest-based activities are being explored. Of particular urgency is an economic activity which will sustain timber-dependent communities, where the specialization of the work force and physical isolation of the communities from major population centers makes economic diversification especially challenging. The collection and sale of Christmas greens, floral greens, edible mushrooms, medicinals, and other products, collectively called "special forest products," has been suggested as an alternative to or supplement for traditional timber harvesting. However, the basic parameters of the special forest products industry, such as sustainable harvest levels, potential markets, and investment and labor requirements, are just beginning to be understood. In addition, the fundamental question of whether special forest products harvesting is a viable sustainable development strategy is not yet definitively answered.

Special forest products can be divided into four categories: 1) floral greens, 2) Christmas greens (including cones), 3) edible mushrooms, and 4) medicinals.

Floral greens: Floral greens are used as complements to floral arrangements, and some of the products are used abundantly in funeral sprays. Important floral greens include bear grass (*Xerophyllum tenax*), salal (*Gaultheria shallon*), evergreen huckleberry (*Vaccinium ovatum*), baby's breath (*Gypsophila paniculata*), sword fern (*Polystichum munitum*), and various mosses.

Bear grass is found in forest stands of all ages in the Pacific Northwest but is most abundant after forest clearing and burning. It is dried and dyed and then sold primarily to floral retailers in the eastern United States and Europe. Salal and evergreen huckleberry, found in forested regions west of the Cascades, grow best under partial shade. Salal is used primarily in floral arrangements and appears in almost every funeral spray, but the berries can also be used for jam and wine. Evergreen huckleberry leaves are used for floral greenery, and the fruit is also marketed.

Baby's breath grows on the east side of the Cascades. The product is treated and dyed for use in floral arrangements and business decorations. Sword fern, also used in floral arrangements, is found west of the Cascades and grows best under a semi-full canopy of trees and is most productive under older stands of trees.

Mosses grow in a variety of environments but are most productive along creek beds where moisture and shade are abundant. Commercial uses of mosses include protecting and watering nursery plants during transport and for orchard plantings and floral displays.

Christmas greens: Christmas greens, found in forests throughout the Pacific Northwest, include coniferous boughs, cones, and holly. They are harvested almost exclusively during fall and early winter for use in Christmas decorations. Noble fir (*Abies procera*) is the preferred species for wreaths, swags and charms, although subalpine fir (*Abies lasiocarpa*) and silver fir (*Abies abilis*) are also popular. Western red cedar (*Thuja plicata*) is used for garland chains, and other fir, pine, and juniper species are used as accents. Douglas-fir (*Pseudotsuga menziesii*) is considered an inferior green because of its poor needle retention.

Medicinals: Medicinals include a wide variety of products. Today the most commercially significant products are dwarf Oregon grape (*Berberis nervosa*), Prince's pine (*Chimaphila umbellate*), and cascara (*Cascara segrada*). Other emerging medicinals include devil's club, various lichens and liverworts, bracken and lady's fern, wild ginger, vanilla leaf, and water parsnip. The bark of the yew tree (*Taxus brevifolia*) is the source of taxol, a potentially highly significant medicinal. Harvesting of yew bark is currently being conducted under special long-term contracts with a pharmaceutical company, pending laboratory synthesis of taxol, which is used to treat certain cancers.

Dwarf Oregon grape, which grows west of the Cascades, requires a semi-full canopy of trees to meet commercial grade. The rootstock is harvested for its alkaloids that are used to treat various digestive disorders and skin conditions. The berries, which are eaten by birds and other wildlife, also have medicinal qualities.

Prince's pine, or pipsissewa, is found in coniferous and hardwood forests above 2,500 feet. The entire plant is harvested and dried. Extracts of the leaves are used to treat indigestion and as a urinary antiseptic. It is also used as a flavoring agent and is the source of flavoring in cola products.

The cascara tree is associated with conifers in the Northwest, and its bark produces widely used cathartics. The bark is stripped from harvested trees; while the tree stump produces vigorous shoots and regenerates, stripping bark from a live tree kills the tree without regeneration.

Mushrooms: Edible mushrooms are found in forested areas throughout the Pacific Northwest, though commercial harvesting is concentrated in Oregon and Washington west of the Cascades. The conversion of mushroom gathering from a recreational activity to a growing multi-million dollar industry was largely sparked by European buyers looking for new sources of chanterelles in the face of reduced mushroom production in Europe in the 1980s. Growing domestic and international demand, continued poor mushroom production in Europe, and the development of a Japanese market for the American matsutake have contributed to the growth of the mushroom industry. In addition, mushrooms are collected for medicinal purposes, for floral arrangements, as sources of natural dyes, and for other uses. The scientific importance of mushrooms as indicators of forest health is increasingly being recognized.

The primary commercial species of wild forest mushrooms are the chanterelle (*Cantharellus cibarius*), the matsutake (*Tricholoma magnivelare*), the morel (*Morchella esculenta*) and the king boletus (*Boletus edulis*), though other species of edible mushrooms are collected commercially and recreationally. Chanterelles, matsutakes, and boletes are mycorrhizal fungi, meaning that they form a mutually beneficial relationship with the rootlets of trees. The rootlets provide the fungi with moisture and organic compounds and the fungi help the roots absorb minerals and may also play a role in disease prevention. Mycorrhizal fungi cannot grow without their hosts, and trees cannot compete successfully without their mycorrhizal partners. Morels are saprobes, living only on dead organic matter such as woody debris, dead heartwood or fallen leaves. Saprobes are important in decomposing the organic layers of the forest floor.

The western slopes of the Pacific Northwest mountains yield the most significant crops of chanterelles in the United States. Like other mycorrhizal mushrooms, chanterelles will reappear annually if weather conditions are favorable and the supporting forest system is intact. Harvested primarily in the fall, the chanterelle is in demand in the domestic market and for export, primarily to western Europe.

The primary market for matsutakes is Japan, where hunting and eating matsutakes is a fall ritual. Matsutakes generally have the highest value per pound of all Pacific Northwest commercial mushrooms, which has attracted increasing numbers of pickers. The most valuable matsutakes are young, unopened mushrooms. This has contributed to the potentially destructive practice of removing the moss cover with rakes and shovels to uncover the emerging matsutakes. Commercial harvest is concentrated in lodgepole pine (*Pinus contorta*) forests.

The king boletus is found in forests throughout the Pacific Northwest. The main fruiting is in the fall, but spring fruitings also occur, especially at higher elevations. The king boletus is most commonly associated with conifer species.

Morels are also widely distributed in temperate forests. Because morels are not mycorrhizal, they can be more easily cultivated but the technology is new and large-scale production is not yet significant. Morels are harvested in spring and are highly perishable. The Pacific Northwest supply varies widely from year to year and is influenced by the abundance of natural and controlled fires; most Pacific Northwest morels are harvested one to two seasons after a forest fire.

Potential Market for Special Forest Products

The special forest products industry has evolved without any direct large-scale investments in infrastructure, with limited access to capital markets, and often in spite of forest practices, which almost exclusively favor timber. The term "forest products" is generally synonymous with timber; other forest products are rarely mentioned in the forest management literature. Historically the Forest Service has used lease agreements designed for timber harvesting for special forest products, thereby often imposing conditions not appropriate for non-timber products. The development of the special forest products industry has also been hampered by the lack of standardization in regulations and in the permitting process. Permit requirements not only vary among forests, but also among districts and landowners. Identifying land ownership in large forested areas is often difficult, further compounding the confusion and information costs of special forest products harvesting.

Partly as a result of such confusion, management of the special forest products industry is beginning to change. In response to public demand and increased competition for special forest products, the Forest Service created the Western Oregon Special Forest Products Council in 1990. The initial impetus was to investigate the effects of increased harvesting on the occurrence and abundance of regional resources, but in 1992 a special forest products coordinator was hired to develop and manage a special forest products program.

Following suit, the BLM recently identified special forest product management as an important component of "total forest management," which is the BLM's emerging forest management philosophy. The report strongly recommended that special forest products be given a larger role in the BLM's management plans in terms of education, research, and management support.

Given the potential for increased institutional support, the growth potential for special forest products is large, though limits exist due to transportation costs (especially in the case of Christmas greens where product weight is a significant factor), trade barriers, and labor costs, particularly since special forest products harvesting and processing is labor intensive. Proximity to markets is also a significant factor in the industry, especially in the case of chanterelles where spoilage is rapid. However locational disadvantages can be compensated for by technology, such as cold storage facilities and investments in transportation infrastructure. With regards to mushrooms, the variability of local weather conditions will continue to limit any location's long-term dominance, and timber management practices will continue to influence special forest product potential.

Sustainable Harvest Levels of Special Forest Products

The increased harvesting of special forest products, particularly mushrooms, has created concern that the resource is being adversely affected. The Forest Service, the BLM, academics, environmental groups and mycological societies have made determining the sustainable level of harvest a top research priority. While there is considerable interest and urgency in assessing the sustainable harvest level, there is little information available on these products in the Pacific Northwest. Most studies have drawn on the long harvest history in Europe. Conclusions from preliminary research in the Pacific Northwest on the effect of harvesting on chanterelles suggests that picking may not adversely affect chanterelle productivity in the short term. However, long-term effects of picking on chanterelle productivity and forest health are not known

While overpicking may not harm mushroom populations, and may even cause disturbance-dependent species to increase production, air pollution and general forest decline may adversely affect harvest levels. In addition, indirect effects of harvesting also need to be evaluated when establishing harvest limits. Many mammals, including voles, mice, flying squirrels, deer, and bear, feed on mushrooms. Even if there is not direct competition for the resource, the presence of pickers and site disturbance can have adverse impacts on wildlife.

There is little specific information about the effect of harvesting on floral greens, but the general scientific literature indicates that removal of less than 25–30 percent of the foliage would have no significant effect on the plant. However, there is a great degree of variation in the amount of foliage removal tolerated. In addition, there is little information on regeneration rates, which is critical in determining whether regeneration can keep pace with a 25 percent foliage removal rate, or whether such a rate will lead to population decline in the long-term. A focus on foliage removal rates presumes that harvesters follow foliage removal guidelines rather than harvesting the entire plant. However, commercial harvesting has already caused depletion of Prince's pine in the eastern United States, and swordfern is scarce near some population centers. Harvesting can also disrupt the soil, leading to broader effects on ecosystem integrity. Many of the most popular greens are also widely used by wildlife for forage and habitat. Harvesting operations and the reduction in available biomass could adversely effect bird and mammal species.

Management of special forest products for sustainable development will have to incorporate consideration of the industry structure and regulatory environment affecting the products. The greens industry consists of five distinct segments: landowners, harvesters, producers, floral wholesalers or floral greens brokers, and floral retailers. Harvesters secure both long and short term leases from public and, increasingly, private landowners. Harvesters sell to producers, who process the product and in turn sell to wholesalers, brokers, or, on occasion, directly to retailers. Horizontal integration by producers is achieved by opening additional production facilities to increase the geographic scope of the business.

In the case of mushrooms, growth continues to be rapid with many new entrants into the industry. There is also a marked trend away from "mom and pop" operations to larger, better capitalized regional firms. This structural change is in response to customer demands for a wider inventory and a longer period of product availability. To be competitive, mushroom shipper/processors must be able to procure mushrooms throughout the Pacific Northwest, from British Columbia to Northern California, and be able to supplement these mushrooms with imports when local conditions are poor. Cultivated mushrooms, such as shitakes and oyster mushrooms, are often purchased to round out the product line and maintain inventory when wild mushrooms are out of season.

In addition to this horizontal integration, there has been a trend to vertical integration that has been the source of considerable conflict. Mushroom pickers have traditionally been loggers or other members of rural communities with loyalty to their own local mushroom spots. Mushroom buyers have traditionally been independent, small-scale operators. As shippers/processors have become larger and more regional, they have tended to send their own buyers and pickers into mushroom harvesting areas, competing very successfully with local buyers and pickers. Locals feeling their historical harvesting "territories" have been violated, contend that outside pickers do not have the same incentives for long-term stewardship of the forest products. The use of rakes by some pickers has also contributed to the fear that future resources are being adversely impacted.

In the past, special forest products harvesting has been unregulated or loosely regulated on public lands such as national forests. With the increased activity of harvesters, consistency is being demanded by users. In addition, both public and private land owners are looking at special forest products to make up some of the revenue lost from declining timber sales and are seeking compensation for products that they previously did not value. With the increased resource use, environmentalists are concerned about the direct and indirect environmental effects on the forest. Amateur mycologists, who are also increasing in number, are concerned that regulation of commercial harvesters will unnecessarily limit foraging for pleasure and education. Professional mycologists, also increasingly interested in assessing the resource, are being hampered in their research efforts by regulations aimed largely at commercial pickers. The public is concerned about the additional uses of these forest resources for commercial gain, as well as effects on recreation values and forest safety. Not surprisingly, while harvesters want consistency in the permitting system, they also want to continue to enjoy the historical low level of regulation.

The commercial harvesting of special forest products holds promise as a sustainable development strategy, particularly for timber dependent communities. The Pacific Northwest has a competitive advantage in producing high quality greens, mushrooms, and other products. At the same time, the trend toward larger, more capital intensive firms that employ non-local labor does not favor the objective of local community development. Nonetheless, public and private forest owners and users, rural communities, and parties interested in economic development and diversification, are intrigued by the potential of economic returns obtained on a sustained basis while maintaining essential forest ecosystem processes and structures (Table 5).

THE FUTURE OF THE FORESTS OF THE PACIFIC NORTHWEST

Though there are many conflicting interests in the forest management debate, all parties agree that forest resources are subject to increasing and competing demands. Private and public forests of the Pacific Northwest continue to be harvested beyond a sustainable yield while at the same time pressures are growing to allocate and protect greater forest areas for watershed protection and water supply, recreation, fishing, and hunting, as well as for habitat for many species of wildlife, endangered or not. The forest resources of the Pacific Northwest are not infinite, and balance between commercial and noncommercial uses of forests has yet to be defined and achieved. Nonetheless, the trend is to incorporate a broader range of interests and goals in the formulation of forest management policies. Hard choices will have to be made, and economic hardships may affect the entire region, especially communities dependent on an unlimited supply of timber. However, by developing better forest management strategies, based on sustainable yield and multiple use that draw on our increased understanding of forest ecosystems and forest stand dynamics, a truly balanced use of the forest resources of the Pacific Northwest may be achieved.

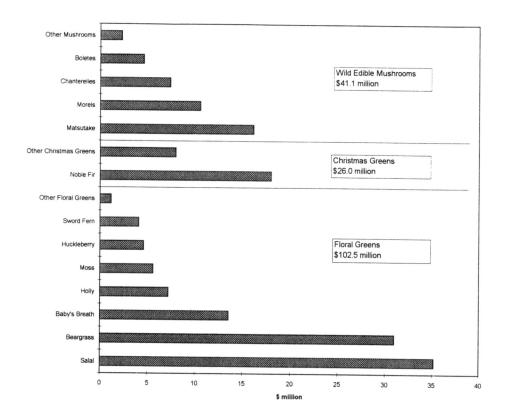

Table 5. Sales of selected Pacific Northwest non-timber products. Sales figures derived from surveyed product acquisition costs and estimated operating expenses. Floral and Christmas green survey based on producers in western Oregon, Washington and British Columbia; mushroom survey based on producers in Oregon, Washington and Idaho.

REFERENCES

Clary, David A. 1986 *Timber and the Forest Service.* Lawrence, KA: University Press of Kansas.

Deacon, Robert T., and M. Bruce Johnson. 1985. *Forestlands: Public and Private.* San Francisco: Pacific Institute for Public Policy Research.

Fairfax, Sally K., and Carolyn E. Yale. 1987. *Federal Lands.* Washington, D.C.: Island Press.

Franklin, Jerry F. and C. T. Dyrness. 1973. *Natural Vegetation of Oregon and Washington.* Corvallis, OR: Oregon State University Press.

Fuller, Margaret. 1991. *Forest Fires.* New York: John Wiley & Sons, Inc.

Kimerling, A. Jon, and Philip L. Jackson. 1985. *Atlas of the Pacific Northwest.* Corvallis, OR: Oregon State University Press.

Mather, Alexander S. 1990. *Global Forest Resources*. Portland, OR: Timber Press.

Molina, R. et al. 1993. Biology, Ecology and Social Aspects of Wild Edible Mushrooms in the Forests of the Pacific Northwest: A Preface to Managing Commercial Harvest. Portland, OR: USDA Forest Service General Technical Report PNW-GTR-309.

Norse, Elliott A. 1990. *Ancient Forests of the Pacific Northwest.* Washington, D.C.: Island Press.

Oliver, Chadwick D., and Bruce C. Larson. 1990. *Forest Stand Dynamics.* New York: McGraw-Hill, Inc.

Schlosser, W. and K. Blatner. 1990. *The Special Forest Products Industry 1989.* International Marketing Program for Agricultural Commodities & Trade (Information Series # 39). Pullman, Wa.: Washington State University.

Schlosser, W., K. Blatner, and B. Zamora. 1991. Economic and marketing implications of special forest products harvest in the coastal Pacific Northwest. *Western Journal of Applied Forestry*, 6 (3): 67–72.

Schlosser, W., K. Blatner, and B. Zamora. 1992. Pacific Northwest forest lands potential for floral greenery production. *Northwest Science* 66 (1): 44–55

Walstad, John D., Steven R. Radosevich, and David V. Sandberg (eds.). 1990. *Natural and Prescribed Fire in Pacific Northwest Forests.* Corvallis, OR: Oregon State University Press.

Williams, Michael. 1989. *Americans and Their Forests: A Historical Geography.* Cambridge: Cambridge University Press.

REGIONAL WATER RESOURCES

Chapter 8

Keith W. Muckleston
Oregon State University

INTRODUCTION

Water resources geography deals with the spatial distribution and characteristics of fresh water, the location and nature of water use, and the effects of its use on society and the environment. This chapter considers the where, what, why, and "so-what" of fresh water uses in the Pacific Northwest.

While this discussion focuses on water resources in the states of Washington, Oregon, Idaho, and part of western Montana, water related phenomena outside these regional boundaries may affect and in turn be affected by water use within the Pacific Northwest. These region-spanning interrelationships result from at least two conditions: First, surface waters flow across the borders of the Pacific Northwest. For example, a significant part of the Columbia River's discharge originates in Canada, as do smaller volumes of discharge from Nevada, Utah, and especially Wyoming. Second, water derived outputs move across regional borders. A case in point is the large volume of electrical energy generated at hydropower plants in the Pacific Northwest that is transferred to California's population centers over an electrical intertie system. The sale of electrical energy in California in turn affects river management of the Columbia system and has an impact on salmon in that system.

An understanding of water resources geography in the Pacific Northwest is enhanced when considered in the context of a system—the water use system (Figure 1). The water use system illustrates the concept that surface and groundwater function as raw material which the management subsystem transforms into a variety of outputs. In general, these outputs, or water related goods and services, are produced to enhance the social and economic well-being of society.

Water outputs result from the active physical utilization of water as well as policy and management decisions that do not directly alter water's attributes. An example of the first would be diverting water from a river and directing it to cropland. Irrigation increases the volume of food and/or fiber society derives from the land. An example of the other approach would be legislative and administrative actions preserving a reach of a surface stream as part of a wild and scenic river system. The

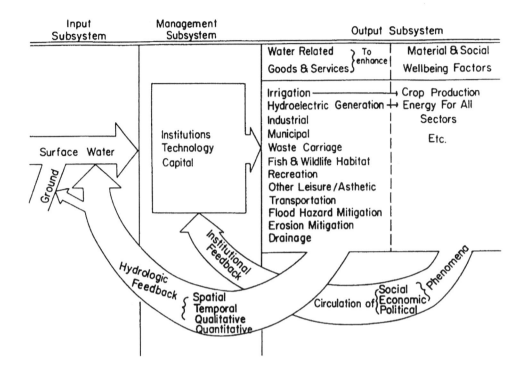

Figure 1. The Water Use System.

output would result in increased user enjoyment from outdoor recreational activities in a pristine environment that would not have occurred if traditional water development had taken place.

As with many other systems, feedback loops are present. The hydrologic feedback shown in Figure 1 refers to any of the four changes in surface and/or groundwater that can result from water use. Included in the hydrologic feedback category are changes in: 1) the location of the water; 2) its temporal characteristics; 3) its qualitative attributes; 4) and its quantitative characteristics. Hydrologic feedback may be instantaneous, or it takes place over an extended period of time. For example, downstream from an irrigation diversion point the stream will carry less water (a quantitative change). This reduction in flow may in turn adversely affect the quality of the stream by decreasing its capacity to carry and assimilate waste.

Hydrologic feedback may have a positive or negative effect on other water users. In the irrigation example above, adverse changes in quantitative and qualitative characteristics of surface waters below the point of diversion suggested potentially negative effects on downstream users. However, the same irrigation water may recharge groundwater on adjacent properties thereby providing a better source of supply to other water users.

Institutional feedback (Figure 1) refers to changes made within the management subsystem to adjust the manner in which a water related output is produced. This type of feedback is usually much less rapid than hydrologic feedback. Years or even decades may pass between the time it is generally recognized that changes should be made in a law or regulation and the date when that change is actually undertaken. This 'institutional lag' is illustrated by the decades it required to start water quality improvements in Oregon's Willamette River, which was notoriously polluted from the 1920's until the 1960's (Gleeson 1972, ii and 48).

Organization of the Chapter

This chapter is organized much the same as the water use system introduced above. First, attributes of the input subsystem in the Pacific Northwest are considered. Characteristics of surface waters in regional subbasins are presented, including selected spatial, quantitative, and temporal attributes. Groundwater characteristics are then noted but in less detail than surface waters.

Discussion of the water management subsystem in the Pacific Northwest is considered next. Most of this section is spent on the major characteristics of agencies that plan and manage water resources in the region. Included are the goals, general responsibilities, and areas of jurisdiction of the principal federal, state, and regional decision makers. The third part of the chapter considers the location and characteristics of the water related goods and services within the region. This section is divided between instream and offstream uses. A brief section containing a summary and conclusions rounds out the chapter.

THE INPUT SUBSYSTEM

Introduction

A popular myth in the United States is that the Pacific Northwest is awash with water. While the region as a whole does have considerable runoff, both in terms of volume and on a per capita basis, spatial and temporal supply characteristics are less favorable.

A brief examination of precipitation and runoff patterns reveals that extensive areas east of the Cascades are subhumid and semiarid, producing little runoff and relatively few perennial streams. In these drier parts of the Pacific Northwest, usually lying below 5,000 feet in elevation, rivers originating in distant mountains serve as important sources of supply. In these extensive areas east of the Cascades reliance on such rivers and/or ground-water more closely resembles water supply realities in the arid Pacific Southwest than in the more humid and relatively well watered areas west of the mountains. But even in these western parts of the region, the timing of

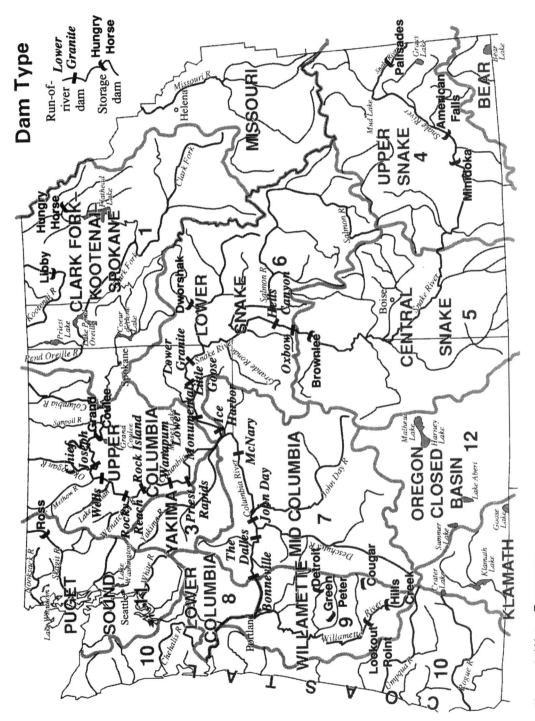

Figure 2. Water Resources.

precipitation is not well suited to water supply requirements, receiving little precipitation in the summer when crop demands and some other major water uses are high. Supply characteristics of surface and groundwaters are now considered.

Surface Waters

A major feature of surface waters of the Pacific Northwest is the sharp difference between areas east and west of the Cascades. The eastern portion (subbasins 1–7 and 12 in Figure 2) comprises roughly four-fifths of the region but only produces about one-third of the region's total discharge. Within the subregion east of the Cascades elevational differences account for major differences in precipitation and generation of surface runoff. While extensive areas below 5,000 feet elevation receive little precipitation (8–16 inches) and produce less than two inches of runoff, orographic precipitation produces substantial runoff along the mountains eastern and northern parts of the eastern subregion. The average water yield of the mountainous Subbasin 1, the Clark Fork-Kootenai—Spokane Subbasin, is about one-cubic-foot-per-second-per square mile (cfs/sq.mile), whereas subbasins 2, 4, & 5 yield less than one-quarter as much runoff per square mile (Table 1). It is noteworthy that the most productive subbasin east of the Cascades (Subbasin 1) produces only between 20 and 31 percent as much runoff per square mile as Subbasins 8–11.

When reviewing Table 1, it is important to recognize that limited production of runoff per square mile in a subbasin does not necessarily mean that surface supply will be limited accordingly. Supply may be derived from rivers fed by distant well-watered mountains that flow through drier subbasins. Such surface waters are termed "exotic rivers"; exotic rivers supply much of the water requirements in several subbasins in the Pacific Northwest. The extensive use of the Columbia River's mainstem in Subbasin 2 illustrates this point. In addition, groundwater may, in some cases, significantly augment water requirements in semiarid subbasins, as is the case in the Upper and Central Snake Subbasins.

When runoff by state is considered, Washington and Oregon are clearly dominant, contributing about three-quarters of the region's total (Table 2). Although Washington has considerably less territory than Oregon or Idaho, it is the leader in runoff contribution, reflecting a combination of physical and locational factors: high mountains relatively close to the coast that receive heavy orographic precipitation, and a larger area in the higher latitudes, which in general receive more precipitation. Idaho and western Montana together contribute a little over one-fifth of the region's total runoff, reflecting in large part their inland position on the leeward side of high mountain ranges.

Although Wyoming's contribution appears insignificant from the regional perspective, it contributes a large part of the runoff in the Upper Snake Subregion in Idaho. Table 2 includes flow contributed by the Canadian part of the Columbia River System. It is noteworthy that runoff originating in Canada and flowing into the Pacific Northwest is almost equal to the total contributions from Idaho, Montana, Nevada,

Subbasin Name	No.	Area (1000 Sq. Miles)	Average Discharge (1000 cfs)	Average Yield (cfs/sq. mile)
Clark Fork-Kootenal-Spokane	1	36.3	36.3	1.0
Upper Columbia	2	22.5	5.3	.24
Yakima	3	6.1	3.2	.53
Upper Snake	4	35.9	8.6	.24
Central Snake	5	36.8	7.7	.21
Lower Snake	6	35.1	29.9	.85
Mid Columbia	7	29.6	13.9	.47
Lower Columbia	8	5.1	25.0	4.9
Willamette	9	12.0	38.5	3.2
Coastal	10	23.8	87.6	3.7
Puget Sound region	11	13.4	52.1	3.9
Closed Basins	12	17.9	1.7	.109
U.S. Pacific Northwest Total		274.4	309.8	1.1
Canadian Part of Columbia Drainage		39.9	74.2	1.9
Total		314.3	384.0	1.2

Table 1. Selected Characteristics of Subbasins.
Source: Modified from Pacific Northwest River Basins Commission, *Comprehensive Framework Study*, Portland, Oregon, 1970, Appendix V, Vol. 1, Table 3, p. 34.

Subbasin Name	No.	ID	MT	NV	OR	UT	WA	WY	Total Subbasin
Clark Fork-Kootenal-Spokane	1	11.3	22.4	—	—	—	2.6	—	36.3
Upper Columbia	2	—	—	—	—	—	5.3	—	5.3
Yakima	3	—	—	—	—	—	3.2	—	3.2
Upper Snake	4	1.8	—	.14	—	.03	—	6.6	8.6
Central Snake	5	5.8	—	.77	1.2	—	—	—	7.7
Lower Snake	6	26.3	—	—	3.4	—	.24	—	29.9
Mid Columbia	7	—	—	—	9.3	—	4.6	—	13.9
Lower Columbia	8	—	—	—	—	—	25.0	—	25.0
Willamette	9	—	—	—	38.5	—	—	—	38.5
Coastal	10	—	—	—	52.9	—	34.7	—	87.6
Puget Sound region	11	—	—	—	—	—	52.1	—	52.1
Closed Basins	12	—	—	—	1.7	—	—	—	1.7
State Totals		45.2	22.4	.91	106.9	.03	127.7	6.6	309.8
% of U.S. Pacific Northwest		14.6	7.2	.29	34.5	<.01	41.2	2.1	
% of Region with Canadian Columbia Included		11.8	5.8	—	27.8	<.01	33.3	1.7	

Table 2. Runoff by State and Subbasin (average 1000 cfs).
Source: Modified from Pacific Northwest River Basins Commission, *Comprehensive Framework Study*, Portland, Oregon, 1970, Appendix V, Vol. 1, Table 3, p. 34.

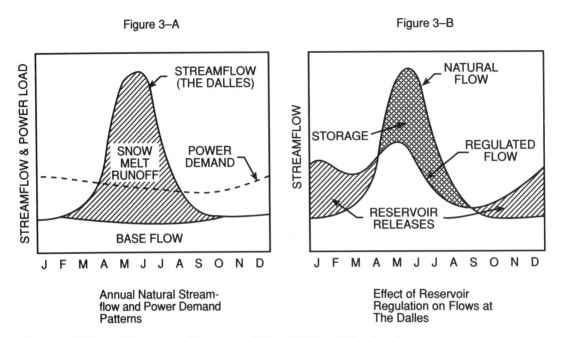

Figure 3. Effect of Upstream Storage on Columbia River Flow Regime.
Source: United States Department of Interior, Bonneville Power Administration, Draft Environmental Impact Statement, *Part 2, The Role of BPA* (part of a six part EIS entitled: *The Role of the Bonneville Power Administration in the Pacific Northwest Power Supply System Including its Participation in the Hydro-Thermal Power Program*), figs., VIII-2 and 3, p. VIII-7.

Utah, and Wyoming. The significance of runoff from Canada has led to an international treaty, selected aspects of which are discussed under sections on hydropower and flood hazard mitigation.

The hydrologic dichotomy between surface water supply east and west of the Cascades is also present when the seasonal runoff patterns are considered. The flow regimes of rivers west of the Cascades generally reflect the temporal patterns of precipitation (i.e., high in the late autumn and winter, decreasing in the spring, and low in the summer). For example, the Willamette River's regime mirrors precipitation patterns. While east of the Cascades, the timing of precipitation is similar to that west of the mountains, much of the precipitation is held during the late fall and winter in the form of snow and ice; therefore, the peak flow period is during the spring and early summer. The Columbia is a noteworthy example of a river with this type of regime (See Figure 3 A). The significance of river regimes on water use and management is discussed in later sections, particularly under hydropower.

Rivers and streams are the major source of water supply over much of the Pacific Northwest. Precipitation from extensive areas within drainage basins is concentrated

River	Rank of Discharge	Sub-basins Drained	Ave. Volume of 1000 cfs	Disharge MAF/Yr	Size of Drainage Area (1000 Sq. Miles)	Ave. Discharge cfs/sq. mile)
Columbia	1	1–9	248.7	180.1	259.0	1.04
Columbia [above the Dalles]	2	1–7	184.6	133.7	237.0	.78
Columbia [at the U.S. Canadian Border]	3	+	104.3	75.7	64.5	1.6
Snake*	4	4–6	50.9	36.8	109.0	.47
Willamette*	5	9	32.9	23.8	11.2	2.94
Kootenai	6	1	28.5	20.6	19.3	1.48
Pend Oreille/ Clark Fork*	7	1	25.8	18.7	25.2	1.02
Clearwater**	8	6	15.6	11.3	9.6	1.63
Rogue	8	10	15.6	11.3	5.1	3.06
Skagit	9	11	14.2	10.3	2.7	5.26
Salmon**	10	6	11.2	8.1	14.1	.79
Cowlitz*	11	8	9.6	7.0	2.5	3.8
Umpqua	12	10	7.4	5.4	3.7	2.0

* Part of Columbia Drainage
**Part of Snake Drainage

Table 3. Major Rivers of the Pacific Northwest.
Source: Modified from Pacific Northwest River Basins Commission, *Comprehensive Framework Study*, Portland, Oregon, 1970, Appendix I, figure 8A np., and Appendix II, p. 886.

in river channels, offering potential users large volumes of supply in confined locations. Knowing the location of rivers relative to population centers and water use characteristics helps in understanding water resources geography; in addition, data on amount of discharge, regime (timing), and reliability (probable range of flows) is helpful. Selected baseline data for the region's more significant rivers are included in Table 3.

Location of the region's major rivers shown in Table 3 may be established from Figure 2. The volume of discharge (Column 3) is a function of at least several factors upstream from the measurement point: area, average elevation, timing and amount of precipitation, rates of evaporation and transpiration, underlying geology, and such anthropogenic activities as dam construction, irrigation, municipal-industrial use, etc.

Runoff may be summarized as $R = P-ET \pm S$. Where R = runoff, P = precipitation, ET evapotranspiration, and S = storage, both natural and anthropogenic.

Table 3 illustrates that the Columbia River system is clearly dominant. Of the thirteen rivers and river segments entered, only three are not part of the Columbia System. The mainstem of the Columbia River is preeminent among surface waters of the region. Indeed, the Columbia carries more water to the Pacific Ocean than any other river in the Western Hemisphere.

Discharge of the Columbia at The Dalles (Subbasin 7) is included because most of the river's hydropower is above this point due to the steeper gradient. Between Grand Coulee Dam and The Dalles, the Columbia River drops over 1,200 feet in 405 miles, providing the potential for massive hydropower development. At The Dalles the average elevation of the Columbia is only 79 feet, but it is still over 190 miles from the ocean. Whereas the average descent of the river between Grand Coulee Dam and The Dalles is over three feet per mile, it is under five inches per mile below The Dalles—meaning there is relatively little energy production potential remaining in the river below this point. What little there is has been developed with the construction of Bonneville Dam, the tailrace water elevation of which is only eight feet above mean sea level.

As the Columbia River crosses into the Pacific Northwest from Canada, its average flow is over twice that of the Snake River, its largest tributary. It is noteworthy that a significant part of the flow that enters Washington State from British Columbia originated in western Montana and Idaho in Subbasin 1. This illustrates that some important Columbia tributaries flow in a northerly direction across the United States-Canadian border before joining the mainstem.

Extensive subhumid to arid areas in the Snake River drainage are reflected by the lowest yield entered in Table 3. Less than half a cfs/sq. mile is far less than most of the other river yields shown in Table 3, particularly those west of the Cascades (Subbasins 8–11).

Groundwater

Groundwater is the other part of region's input subsystem (Figure 1). Ground-water supplies about one-fourth of the total volume of water withdrawn for offstream uses in the Pacific Northwest (USGS 1992, Table 1). These subsurface waters are physically connected to surface waters although they are covered by separate water laws from those dealing with surface water. Much of the water that does not return to the atmosphere by evapotranspiration or travel as surface flow percolates downward and is (eventually) stored in aquifers, which in turn provide much of the runoff in the region's surface waters.

Conversely, some aquifers are recharged from rivers and streams. Thus, groundwater functions as storage between precipitation and runoff into the Pacific Ocean.

Where available, groundwater may be an important supplement to surface supply, especially where runoff is inadequate to meet demands. But not all of the Pacific Northwest has usable supplies of groundwater. About two-fifths of the region is underlain with aquifers that produce insufficient amounts of groundwater (PNRBC 1970(2), 42). These unproductive areas cover most of the following landform regions as delineated by Raisz: the Pacific Border, Cascade Mountains, and Rocky Mountains. It is fortunate that these unproductive groundwater areas are in sparsely populated portions of the region where little pressure for large scale water use exists, and where surface runoff is usually adequate.

The location of adequate groundwater supplies reflects the area's geology more than anything else, although landforms and amount of precipitation may also contribute. In the Pacific Northwest, areas with productive aquifers are often found in any one of three general types (USGS 1985, 193–6, 285–8, 355–8, 433–6). First are the alluvial and/or glacial deposits which are generally in topographic depressions and/or river valleys. Examples of productive aquifers with fluvial/glacial material are in the Puget Sound Lowlands (Subbasin 11), Spokane County, Washington, and adjacent parts of the Idaho Panhandle (Subbasin 1). Oregon's Willamette Valley (Subbasin 9) and valleys of the Clark Fork in western Montana (Subbasin 1) represent the valley-fill type of aquifer. The second general type is found in much of the Columbia Intermontane physiographic region. Major groundwater production areas of this type are located in the eastern half of the Snake River Plain (Subbasin 4), in Adams and Franklin Counties in eastern Washington (Subbasin 2), and in northern Wasco County, eastern Morrow, and western Umatilla Counties in north central Oregon (Subbasin 7). The third general type is the so-called volcanic rock-sedimentary-and basin fill alluvium. The most productive example of this occurs in the western half of the Snake River Plain (Subbasin 5). Productive aquifers in south and southeastern Oregon also fall into this third group, in Lake and Klamath Counties, for example (Subbasins 12 and Klamath, respectively).

Major areas of groundwater extraction in the Pacific Northwest include the following: The Snake River Plain, which is underlain by the region's most productive aquifer, the Snake Plain Aquifer and from the upper Snake, (Subbasin 4), over half of the region's groundwater is pumped; the contiguous Central Snake Subbasin, sharing the other half of the Snake River Plain, is second in groundwater pumped, producing just under 15 percent of the regional total. Thus the Snake River Plain accounts for about two-thirds of the region's total groundwater pumpage, over 90 percent of which is used in the agricultural sector. In total, Idaho accounts for approximately three-quarters of all groundwater pumpage in the region and is among the country's leading producers of groundwater.

In Washington State, irrigation is also the leading user of groundwater in the state—largely in the agricultural areas east of the Cascades. Large withdrawals of groundwater for public supply take place in Spokane County (Subbasin 1) and in Pierce and King Counties in the Puget Sound Lowland (Subbasin 11). Industrial use of groundwater is particularly significant from the terrace and valley fill aquifers near Vancouver (Subbasin 8) but also takes place in the Puget Sound Lowlands (Subbasin 11).

Although groundwater withdrawals in Oregon are about one-half of those in Washington, irrigation is again the leading single user of groundwater, but public supply and industrial uses also take place in the Willamette Valley and in northeastern and south central Oregon. Relatively heavy pumpage for irrigation is concentrated in Morrow and Umatilla Counties (Subbasin 7) and in Lake and Klamath Counties (Subbasin 12 and the Klamath River Drainage). In Montana's Columbia River drainage area, the only significant area of groundwater use lies in Missoula County, where industrial, public supply, and agricultural uses extract water from alluvial and basin fill aquifers.

THE MANAGEMENT SUBSYSTEM

Background

As described in the introduction, the management subsystem uses surface and groundwater as a raw material, transforming it into a variety of water-derived outputs (Figure 1). Institutions within this subsystem are comprised of actors and the laws and administrative procedures under which they operate. Actors in the management subsystem range from prominent federal agencies like the Corps of Engineers and Bureau of Reclamation, through the many and varied state agencies responsible for water rights, fish and wildlife, water quality, and outdoor recreation, to local entities that provide municipal water supply, waste treatment, permits for septic tank installation, and more.

Agencies use technology and capital to change water into outputs that society deems useful (e.g. hydroelectric power, inland water barge transport, irrigated croplands, etc. Figure 1). In this sense, technology includes both equipment (hardware) and processes. Dams, reversible pump turbines, and center pivot irrigation rigs are examples of hardware while methods of providing water for public consumption and managing waste water before returning it to rivers and streams exemplify processes. Of course, many processes require hardware to make them operable.

Capital, the third ingredient in the management subsystem, is necessary to purchase, operate, and maintain technology. It is also required to finance agency operations, e.g., to pay the salaries of thousands of highly trained planners, scientists, and engineers. Capital is derived by different levels of government from various sources, including special legislative appropriations, dedicated funding, property tax revenues, collection of user fees, and issuance of licenses. The remainder of this section focuses on the principal water management actors in the Pacific Northwest. They are considered under the following categories: federal, state, and regional.

Federal Actors in the Region

Federal agencies have been and continue to be responsible for much of the planning, operation, management, and regulation of water resources in the Pacific Northwest. Heavy federal involvement reflects several regional characteristics and historical developments: First, over half of the land in the region is owned by the federal government: 52 percent in Oregon, 28 percent in Washington, 65 percent in Idaho, and 68 percent in western Montana (PNRBC 1970 (1), 22). Second, the region lies in the western half of the conterminous United States and is therefore eligible for irrigation projects constructed by the U.S. Bureau of Reclamation. Third, the sheer size of the Columbia River requires such large capital investments for development that nonfederal interests often have been hesitant to make the necessary expenditures. Fourth, the relatively undeveloped state of resources in the Pacific Northwest coupled with the severity of the Great Depression in the region, encouraged the Roosevelt Administration to put into practice its contention that federal investment in natural resource development would stimulate regional growth (Muckleston 1990 (1), 25). Construction of the massive Grand Coulee Dam illustrates the heavy federal investments in natural resources development encouraged by New Deal philosophy. Lastly, national defense considerations prompted the federal government to develop much of the region's enormous hydroelectric resources. The drive to create the capacity to generate massive amounts of low cost hydroelectric energy for war related industries started in the late 1930's as hostilities in Europe and the Far East mounted, continued through the active participation of the United States during WWII, and extended well into the protracted Cold War period that followed.

The many and varied roles played by federal agencies in the water resources of the Pacific Northwest are summarized in Table 4. Agencies are divided into three groups according to their individual effect on water resources in the region. Whereas actors in the first group generally have multiple water related concerns and cause major water related impacts, those in the second group are usually concerned with less water related activities. With the exception of the Federal Emergency Management Agency (FEMA), actors in the third group are land managers whose resource use policies can indirectly affect water resources in other jurisdictions downstream from lands under their management. The three principal federal agencies in the region are discussed below.

The United States Army Corps of Engineers (Corps) is the preeminent water resources agency in the Pacific Northwest. While its primary responsibilities in the region are flood control, navigation (inland water transport), and hydropower, it has many secondary concerns including recreation, fish and wildlife, irrigation, and low flow augmentation (Table 4). In addition to the many large projects the Corps has constructed and operates through the Reservoir Control Center in Portland, Oregon,

Agency①	Significance②	WRCS of Primary Concern③	WRCS of Secondary Concern③	Functions④	Remarks
COE	1	N,FHM,H	WQ, REC, I,FW	P,C,Op	Coodinates Regional FHM measures & regulates use of wetlands.
BuRec	1	I,H	REC, F&W, FHM	P,C,Op	
BPA	1	H	F, F&W	P,O,C	Wholesales hydroelectricity produced by Feds. wheels electrical energy over its extensive high capacity grid. Since 1980, provides significant funds for F&W migration/restoration.
FERC	2	H		R	Licenses nonfederal hydroelectric facilities. May consider impact on F&W.
NMFS	2	F		P,R	Principal concern anadromous salmonids; plans and administrators programs under Endangered Species Act.
FWS	2	F&W		P,R	
EPA	2	WQ		R,O	Regulates waste discharge permits directly and/or through state agencies.
SCS	2	O,FHM	I, REC	P,C,Op	Principal concern for soil conservation/erosion control. Cooperative development w/ nonfederal entities.
FS	3	/	W&S	O	Administration of W&S corridors. Logging and grazing on FS lands affect timing and quality of runoff.
BLM	3	/	W&S	O	Administration of W&S grazing and logging on BLM lands affect timing and quantity and quality of runoff.
NPS	3	/	W&S	O	Administration of W&S
FEMA	3	FHM		R	Administers floodplain regulations through local governments participating in the National Flood Insurance Program.

① COE–Corps of Engineers; Bu Rec–Bureau of Reclamation; BPA–Bonnevile Power Administration; FERC–Federal Regulatory Energy Commision; NMFS–National Marine Fisheries Service; FWS–Fish and Wildlife Service; EPA–Environmental Protection Agency; SCS–Soil Conservation Service; FS–Forest Service; BLM–Bureau of Land Management; NPS–National Park Service; FEMA–Federal Emergency Management Administration.

② 1=Very significant; 2=Important and may be very significant for one; 3= Lesser role-generally land management that affects water indirectly.

③ 1 WRGS[s]= Water Related Goods and Services; H= Hydroelectric; N= Navigation; FHM= Flood Hazard Mitigation; I= Irrigation; F= Aundromous Solmonids; WQ= Water Quality; Rec= Recreation; F&W= Fish and Wildlife; W&S= Wild and Scenic Rivers.

④ P= Planning; C= Construction; Op.= Operation; R= Regulation; O= Other.

Table 4. Federal Agencies and Regional Water Resources.

it is responsible for planning, directing, and coordinating flood control operations at major reservoirs throughout the region, including three large reservoirs in Canada's part of the Columbia River basin. Therefore the agency exerts strong influence over flood control operations at reservoirs owned and operated by a variety of actors throughout the region and part of Canada. The Corps has also planned, directed, and underwritten most of the costs for thousands of miles of levees, extensive bank protection measures, and channel modifications in the name of flood control.

The Corps' responsibilities for navigation have resulted in profound modifications of the region's rivers, estuaries, and harbors. For example, eight Corps' dams have transformed 310 miles of the once free flowing reaches of the Columbia and Snake Rivers into a continuous chain of lakes. This transformation guarantees barge operators slackwater (currentless) navigation in waters no less than 14 feet deep between Bonneville Dam and Lewiston, Idaho. Although the Columbia's lower 146 miles are free flowing, the Corps has modified its channel through extensive dredging to enhance navigation by transoceanic ships.

The Corps is also the region's leading producer of hydroelectric energy, generating almost two-thirds of the total produced by federal agencies in the Pacific Northwest (CRWMG 1991, 71). In 1990, for example, the Corps generated enough energy to supply between six and seven metropolitan areas the size of Seattle. Most of the Corps' hydroelectric energy is produced at four multiple purpose dams on the mainstem of the Columbia (Subbasin 7) and four on the lower Snake River (Subbasin 6). Other noteworthy generating plants are at Chief Joseph, Libby, and Dworshak Dams in Subbasins 2, 1, and 6, respectively (Figure 2).

The Corps is also prominent when leisure time-environmental uses of water in the region are considered. On the positive side, several million recreation visitor days take place yearly at Corps' reservoirs. On the other hand, Corps' dams are responsible for significant losses of both adult salmon returning to spawn and juvenile fish migrating to the ocean. Water resources development and salmon are considered in more detail under instream water uses.

The Bureau of Reclamation (Bu Rec) has been very influential in the use of water and related lands in the Pacific Northwest. This agency is a product of the Reclamation Act of 1902 by which Congress sought to facilitate settlement by means of large scale irrigation projects in the 17 Western States. Unaccustomed aridity and chronic drought conditions had frustrated earlier attempts to establish viable agricultural operations in the relatively dry western half of the country. Climatic conditions in the Pacific Northwest east of the Cascade Mountains made many areas likely candidates for Bu Rec project water. Approximately half of the irrigated lands in the region receive water from Bu Rec projects.The gross value of crops irrigated by the project water from this agency approximated 2.46 billion dollars in 1990 (CRWMG 1992, 73).

Generation of hydroelectric energy is the Bu Rec's other major function in the Pacific Northwest. Proceeds from this energy allow this agency to subsidize its irrigation projects to an important degree. The Bu Rec generates approximately one-quarter of the total hydroelectric energy produced by federal installations in the Pacific Northwest. Grand Coulee Dam, the crown jewel of Bu Rec dams, has far more installed generating capacity than any other dam in the country. For many years the world's leader, Grand Coulee's approximate 7,000 megawatt capacity still makes it first in North America.

Storage capacity in Bu Rec reservoirs provide flood control for areas downstream and can be used for flow augmentation that in turn aids a number of other downstream users. Storage waters at reservoirs are also popular with recreationists, reportedly generating even more recreation visitor days than the Corps. On the other hand, projects built by the Bu Rec have not been compatible with salmon, the region's most important fish. This lack of compatibility is described in sections under water uses.

The Bonneville Power Administration (BPA) is the third major actor in the region. It qualifies for this top category of importance even though it builds no dams or powerhouses. The BPA handles all electrical energy generated at federal facilities in the region, wholesales much of this energy to public and private utilities in the region and in California, and directly supplies electrical energy to 22 regional industries that require large volumes of low cost electric energy. In addition, BPA transports (wheels) much of the total electrical energy produced in the region over its extensive grid of high-voltage transmission lines within the Pacific Northwest and between this region and California.

As the coordinator of operations by the Federal Columbia River Power System, the BPA makes numerous decisions that affect how many dams throughout the region are operated. This in turn affects the timing and volume of flows on many rivers and streams in the region. Revenues from BPA's energy sales are used to finance much of the operating costs of the Northwest Power Planning Council, which is an important regional actor in matters of power planning and fish and wildlife mitigation.

State Actors

Each of the region's four states is an active participant in the water management subsystem. Some of the state level functions are: 1) the issuance of water rights; 2) regulation of waste disposal into surface and groundwater; 3) management of fish and game, including the issuance of hunting and fishing licenses; 4) management of outdoor recreation, much of which is water oriented; and 5) coordination of the National Flood Insurance Program.

Similar to the organization of federal agencies, state agencies are specialized in regard to which function is managed. This means that each of the functions noted above is usually the responsibility of a separate agency in each state. This can result in problems of coordination because many of the water related functions are interrelated. For example, the issuance of a water right to irrigate 1,000 acres would probably affect some of the water's attributes (i.e. its quantity, quality, timing, and location Figure 1, see hydrologic feedback). These effects could in turn require adjustments by other state agencies responsible for fish and wildlife, water quality, or outdoor recreation. Despite these interrelationships, coordination at the intrastate level has not been characteristic of state level water management. In this respect, it is similar to the fractionated approach of federal agencies.

Among the many water related functions performed by state-level agencies, the issuance of water rights is the most important and has the longest tradition. Under the doctrine of prior appropriation, a water resources agency in each of the region's states is empowered to issue usufructuary rights for water to individuals, private firms and corporations, public entities, and, in some cases, to other state agencies. Even irrigation districts participating in federal Bureau of Reclamation projects must acquire water rights through their respective state agency. The most noteworthy characteristic of the prior appropriation doctrine is how it prioritizes rights to use water during periods of diminished supply. In the event of a water supply shortage, the continued right to use water depends on how long the right has been held. The more senior the right, the higher the priority. The phrase "first in time, first in right" is frequently used to describe this allocation system.

Water rights are specific as to the quantity, location, and timing of withdrawal, as well as the type and place of water use. Traditionally water rights were issued almost exclusively for offstream uses (see pp. 221–9 below). Irrigation continues to account for well over 80 percent of total withdrawals. As a result of increasing environmental concerns, some of the region's state agencies have begun to issue rights for instream uses. These rights are usually awarded to state fish and wildlife agencies to preserve adequate river flows for fish, but instream rights may also be granted to state agencies for preservation of flows in state scenic waterways or for recreation uses in all streams or for improved waste carriage capacity of selected streams (Root 1993).

Interrelationships between water uses exist across state lines. States of the Pacific Northwest have made limited use of interstate compacts to address interstate water problems.These compacts must be approved by Congress but do not have a federal representative. Washington and Oregon have used this approach to agree on salmon harvest in the Columbia River where it forms a border between the states. Idaho and Wyoming have a compact on the Snake River that allocates designated percents of flow for use in the two states. Oregon and California have a similar arrangement on the Klamath River which rises in Oregon and flows into California on its way to the Pacific.

No comprehensive regional compact exists on allocating the quantity of water in the Columbia-Snake River basin between states with territory in that drainage area. Agreement to do so has not been achieved despite decades of sporadic effort. Therefore, in theory at least, Idaho could legally consume (use up) all of the Snake River flow before it reached Washington, while Montana could do the same to the Clark Fork before it flowed into Idaho.

It must be concluded that traditional interstate compacts in the region do not provide the means to comprehensively manage water in the Pacific Northwest. As described in the following section under regional management, since the early 1980's a new type of interstate compact provides comprehensive water resources planning and management for two water-related outputs.

Regional Actors

Actors that operate at the regional level have had some effect on water resource use in the Pacific Northwest although much less than that exercised by federal and state agencies. Those favoring regional organizations over conventional federal and state agencies emphasize that placing the Columbia River drainage area under one planning/management unit would improve water resources management by allowing more comprehensive approaches to problems common in large river systems that cross more than one state.

Two examples of regional water issues illustrate these problems. First, if irrigation in Idaho's Snake River Plain were doubled, as some agricultural interest groups have called for, water quantity and quality would decline appreciably downstream in Washington and Oregon. One of the potential negative spillover effects would be a decrease in hydropower output in both states. This decrease would in turn require development of more expensive sources of electrical energy to replace the amount of generation reduced by increased consumptive use of water upstream in Idaho and to meet new demands for energy in the future. Since most of the consumption of electrical energy in the region is in downstream states, the negative effects of increased irrigation in Idaho would fall disproportionately on consumers in Washington and Oregon. A second example of interstate water problems in the region is over salmon harvest in the Columbia Snake System. Heavy fishing pressure by downstream residents (in Washington and Oregon) on salmon runs returning to Idaho adversely affect fishing interests in that upstream state. Over the past decades fishing interests in Idaho have pressed unsuccessfully for an official voice in the decision process by which Washington and Oregon determine salmon harvest in the Columbia. If a regional organization were adequately empowered, it could help resolve these and other regional issues. Unfortunately, experience in the Pacific Northwest has shown that when regional entities did exist they were not adequately empowered.

An evolution of regional-level actors has taken place in the Pacific Northwest over the last half century, with the states playing an increasingly important role in the process (Muckleston 1990 (1), 36–7). From the 1940's until the late 1960's, the Columbia Basin Interagency Committee (CBIAC) functioned as a coordinating group for federal agencies managing waters and related lands in the region. Although CBIAC consulted with various state agencies about projected actions by the federal agencies, states had very little meaningful input into plans that were often voluminous, complex, and already too far advanced to readily modify. From 1968 until 1981, the Pacific Northwest River Basins Commission (PNWRBC) formulated plans for water developments in the region that were intended to be comprehensive, coordinated, and jointly prepared by federal and state agencies. Under the authority of the Water Resources Planning Act of 1965, the states of Washington, Oregon, Idaho, Montana, and Wyoming participated as members of the PNWRBC, providing a vice chairperson and some financial support to the PNWRBC's operating costs. The plans crafted by the PNWRBC were intended to provide the U.S. Congress with guidelines and priorities of proposed water projects in the region. Although budget constraints had stopped most federal water resources developments by the second half of the 1970's, it is problematic that the PNWRBC's plans would have been followed. The PNWRBC was abolished by the Reagan Administration in 1981, but a new regional actor had taken life one year earlier from passage of the Pacific Northwest Power Planning and Conservation Act (the Act).

The Act enabled the states of Washington, Oregon, Idaho, and Montana to form the Northwest Power Planning Council (NPPC) that was charged with designing two detailed plans: one that will ensure the region has an adequate supply of electrical energy, and the other offers wide-ranging protection and improvements for fish and wildlife populations, with particular emphasis on salmonids in the Columbia Basin (Muckleston 1990 (2), 12).

The NPPC is a unique and innovative regional institution comprised of eight gubernatorial appointees, two from each state, whose plans for electrical energy and fish and wildlife must in many cases be substantially complied with by federal agencies. This aspect makes the NPPC unique in that federal agencies must comply with at least some of the plans of an organization that does not have a federal representative. The Act also guarantees the NPPC a substantial funding base from the BPA's energy based revenues—funding that is crucial to successful operations and was not available to CBIAC or the PNWRBC. A weakness of the NPPC is that it may plan comprehensively for only fish and wildlife and electrical energy; it has no authority over other water-related outputs in the region even though their production can seriously affect both categories under its jurisdiction.

During the 1980s and early 1990s, the NPPC was widely perceived to have had a positive effect on salmon survival, a role which may continue into the next century. On the other hand, the NPPC's impact on electrical energy planning was still largely untested by the early 1990s: due to a large surplus of electrical generating capacity in the 1980s there was relatively little reason to implement plans that had been formulated.

References to NPPC activities in the Pacific Northwest appear in the section on water resources outputs of this chapter under the subsection on fish and wildlife.

THE OUTPUT SUBSYSTEM

Introduction

The outputs derived from water are one of the principal concerns of water resources geography. This section discusses the location, characteristics, and selected consequences of water resources development in the Pacific Northwest. A few of the outputs are water and may be measured by conventional units of water: millions of gallons per day (mgd) of municipal supply or waste water treatment. But most outputs are not water per se. Examples of water derived outputs measured by non-water indicators include: kilowater hours of hydroelectric energy, increased production of food and fiber from irrigation, ton miles of commodities transported on inland waterways, and recreation user days. In this section the water outputs are considered under two major source types: instream and offstream uses. In the case of the former the use takes place within the banks of rivers or canals while the latter requires removal of water from its channel or the ground before utilization.

Instream Uses

Instream uses include: hydroelectric generation; waste carriage and assimilation; inland water transport; flood damage mitigation; fish and wildlife habitat (for both recreational and commercial purposes); recreation, on both slack water reservoirs behind dams and on free flowing streams; and preservation of free flowing streams in their natural state.

Multiple uses of instream water are common. For example, the same 50–mile stretch of river may be used to generate hydroelectric energy at one or more dam sites, carry barge traffic, provide habitat for fish and wildlife, support recreational use, and carry wastes resulting from anthropogenic activities within the drainage basin. Obviously, there are limits to multiple use: tradeoffs and conflicts among user groups are common, which will become evident in the following pages.

Hydropower

The Pacific Northwest is world renowned for producing vast amounts of inexpensive hydroelectric energy. The region produces almost half of the country's total hydroelectric output (USGS 1992, Table 27). The generation of such large amounts of inexpensive hydroelectric energy in a region with only about three percent of the nation's population has had far reaching effects. First, it has allowed electrical energy to be substituted for some energy needs usually met by fossil fuels. This is important because the region produces very little fossil fuels and would have to import them if hydroelectricity was not used as a substitute. For example, in the Pacific Northwest electrical energy meets much more of the requirements for heating homes and buildings and for hot water heating and cooking than it does in the United States as a whole. Thus it substitutes to a significant degree for natural gas, petroleum, and coal. Second, types of manufacturing oriented towards large blocks of inexpensive electrical energy have been attracted to the region: The Pacific Northwest accounts for an important part of the United States aluminum production. And third, large-scale production of hydroelectric power allows the region to export significant volumes of electrical energy to California. During the first 20 years of export operations (1968–1988) a daily average energy equivalent of 37,000 barrels of oil moved over the extra-high-voltage intertie from the Pacific Northwest to California (USDE 1990, 14). This not only reduced the cost of electricity bills to users in the Pacific Northwest but also saved California utilities millions by allowing them to forego purchase of petroleum for thermoelectric generation. In addition, air quality in California was improved because burning 37,000 barrels of oil daily for this period would have produced 2.7 million tons of CO_2 that would have been disposed of in the state's airsheds (USDE 1990, 14).

Physical characteristics of the Pacific Northwest provide the ingredients for a very large hydroelectric potential: numerous mountains that are relatively close to an abundant source of moisture (the Pacific Ocean) and which lie in the path of prevailing westerly winds result in heavy precipitation and large runoff from the mountains. Since much of the runoff originates in elevated areas, a significantly steep gradient (difference in elevation) adds another prerequisite for hydroelectric potential. Geology favorable for dam construction and numerous entrenched rivers round out the physical conditions facilitating hydro power development.

Tremendous volumes of water are used to generate hydroelectric energy in the Pacific Northwest. In 1990, for example 1,250 *billion* gallons per day (bgd) were passed through hydroturbines (USGS 1992,Table 27). This volume of water weighs approximately 5,600,000,000 tons. This is about four times the total average runoff originating in the region and is almost equal to the total average runoff from the conterminous United States. The remarkable volume of water used for hydroelectric production in the Pacific Northwest results from the continuous chain of dams and hydroelectric plants on some of the region's principal rivers. This allows the same gallon of water to be used repeatedly as it passes through consecutive power houses on its way from uplands to the ocean. For example water flowing across the United State-Canadian border within the banks of the Columbia River can be utilized at 11 powerhouses between Grand Coulee Dam and Bonneville Dam (Subbasins 2 and 7). If the path of water originating behind Hungry Horse Dam (on the Middle Fork of the Flathead in Subbasin 1) in western Montana is traced, it can be utilized 19 times for hydropower production before reaching the ocean.

Most of the region's production of hydroelectric energy takes place in the Columbia River basin; and much of the Columbia system's energy output comes from about 20 large plants, particularly on the mainstem of the Columbia River (Subbasins 2 and 7). Over a 725 mile stretch of river, 11 dams with a total installed generating capacity of about 19,000 MW account for an important part of the hydroelectric energy generated in the region.

State (a)	Water Use (BGD)	% of Water Used for Hydro in Region (b)	Hydroelectric Generation (GWh)	% of Regional Hydroelectric Generation (b)	National Rank in Hydroelectric Generation
WA	654	52.3	82,100	57.8	1
OR	455	36.4	41,500	29.2	2
ID	107	8.6	12,500	8.8	6
WMT	34	2.7	5,800	4.1	—
Total	1250(c)		142,000(c)		

a) Individual state amounts calculated for % of 1985 water use and generation.
b) May not total 100% due to rounding
c) 1990 data.

Table 5. Selected Characteristics of Hydroelectric Generation in the Pacific Northwest. *Sources:* For 1990 data, U.S. Geological Survey, *Preliminary Estimates of Water Use in the United States,* Open-file Report 92-63, Table 27. For 1985 data, U.S. Geological Survey, *Estimated Use of Water in the United States in 1985,* GPO, 1988, Table 20, p. 47.

Other noteworthy hydropower developments in the Columbia System take place in Subbasin 1 on the Kootenay and Clark Fork/Pend Oreille Rivers that are owned by both federal agencies and private utilities; and in the Snake River drainage at five dams owned by the Corps of Engineers (Subbasin 6), and in three owned by the Idaho Power Company in Subbasins 5 and 6. The regions most important hydroelectric development outside the Columbia System is on the Skagit River (Subbasin 11) where dams built by the City of Seattle account for most of the generating capacity.

Differences in the intraregional hydropower production between the states may be explained in large part by the location of the Columbia System vis-à-vis state boundaries. Washington State, which has the mainstem of the Columbia and lower Snake flowing through its territory, is the nation's leading producer of hydroelectric energy (Table 5). Oregon, which shares with Washington most of the Columbia below the Columbia-Snake confluence, is the second largest producer in the Unites States, but generates just slightly more than half as much as its neighbor to the north (Table 5). Idaho has mountainous terrain favorable for hydroelectric generation but relative to Oregon and Washington considerably less runoff; it ranks sixth among United States states in hydroelectric generation.

A significant part of the electrical energy generated by the powerhouses on the mainstem of the Columbia results from cooperative management of the river with Canada. Under natural conditions much of the Columbia's discharge occurs in June and July from snowmelt (See Figure 3A). This type of flow regime is unfavorable for hydropower generation, particularly because the demand for electrical energy is high in the fall and winter when the natural flow of the river is low.

To meet this problem utilities called for the construction of storage reservoirs in the mountainous headwaters of the system. The function of such upstream storage reservoirs is to impound spring and summer snowmelt and then release it later in the autumn and winter in order to increase the generation of electrical energy during the periods of greatest demand (See Figure 3B). The existence of multiple dams and powerhouses on the middle reaches of the system meant that every unit of water stored upstream could be used many times after its release.

During the 1940s and 1950s interests favoring upstream storage were unable to provide as much as they felt was necessary in Idaho and Montana where most of the region's potential upstream storage sites are located. After lengthy negotiations, Canada agreed to provide a significant volume of upstream storage in its part of the Columbia Basin. It did this in return for half of the hydropower and flood control benefits that would result in the Pacific Northwest from Canadian storage operations. In 1964 electric utilities in the Pacific Northwest paid Canada $254 million to cover one-half the value of additional electric energy that would be produced at the region's hydroelectric facilities on the Columbia because of Canadian storage operations. This sum covered the first 30 years of the 60 year life of the treaty; negotiations over the next 30 years of sharing downstream hydroelectric benefits were underway in the 1990s.

In sum, an international treaty on sharing downstream benefits was necessary before storage in Canada could be provided. As a result of the treaty, 20 of the 44 million acre feet of useable storage in the Columbia System are in Canada.

Inland Water Transport

Commercial navigation is an important instream use of water in some parts of the Pacific Northwest. Where the correct channel conditions are present, bulk products—such as grain, sand and gravel, petroleum, etc.—are transported at relatively low cost by water carriers; savings to shippers result from much lower energy requirements than those of rail or truck.

Two types of carriers use the region's inland waters. First, vessels on coastal or transoceanic voyages use the Lower Columbia as far as Portland (Subbasin 9), approximately 110 miles from the Pacific. The large volume of freight handled yearly by Portland, Vancouver, and other Lower Columbia ports contributes to the economic well-being of northwestern Oregon and southwestern Washington. Access by large maritime vessels as far as Portland is made possible by the Corps of Engineers, through dredging and maintenance of a navigation channel in the river. In the Seattle area maritime shipping is provided access from Puget Sound to Lakes Union and Washington via locks and a canal. Although heavily used, this facility plays a relatively small role in the economy of metropolitan Seattle but it does afford recreational boaters and commercial fishers access between the area's fresh and salt waters.

The second type of commercial navigation is that which takes place entirely on inland waters. The principal inland waterway in the region stretches from near Astoria, Oregon, (Subbasin 10) 465 miles up the Columbia-Snake System to Lewiston, Idaho (Subbasin 6). A free flowing reach of 146 miles lies between Bonneville Dam and the Pacific; it is maintained at a depth of 15 feet from Bonneville Dam to Portland and at 42 feet between Portland and the mouth of the Columbia. From Bonneville Dam to Lewiston, Idaho's eight large multipurpose dams completed by the Corps of Engineers between the late 1930s and 1975 provide a continuous chain of reservoirs with a guaranteed depth of 14 feet. Adequate water depth and little current in these impoundments markedly improved shipping conditions over those of the natural river system, which descends from 736 feet elevation at Lewiston to eight feet above mean sea level at the tailwater of Bonneville Dam. Shallows, rapids, and sand bars that markedly impeded commercial navigation have disappeared beneath the waters of these reservoirs.

Spacious lock dimensions (86 feet wide by 675 feet long) allow use of large barges that reduce shipping costs through economies of scale. These barges have a cargo capacity of 3,200 tons, which is more than twice the capacity of barges used on the Mississippi River. It is noteworthy, however, that the total volume of freight carried on the Columbia is only a small fraction of that on the Mississippi, which has access to much larger centers of commodity production and consumption.

The volume of commercial cargo carried on the inland waterway system above Portland has grown unevenly since completion of the Bonneville Lock in the late 1930's. Between 1940 and 1970, the increase was relatively modest, growing from approximately one million to three million tons. Growth accelerated after completion of the Snake River portion of the waterway in the 1970s, which gave water carriers better access to grain producing areas. By 1990 more than 11 million tons of cargo passed through the locks at Bonneville Dam (CRWMG 1992, 77).

Downstream movement of goods is considerably heavier than upstream transport, reflecting the commanding position of grain shipments destined for export after transshipment at lower Columbia River ports. The dominance of downriver movement is more pronounced on the Snake River portion of the system, where ratios of downstream to upstream movement are in some years well over ten to one (CRWMG 1992, 76). On the Columbia reach of the inland navigation system, from the Tri Cities downward, ratios are less skewed in favor of downriver movements but still range between four and six to one. Less skew results from relatively heavy upriver shipments of fuels, chemicals, and fertilizers, much of which is offloaded in the Tri Cities area.

Although the early navigation plans designated Wenatchee, Washington, (Subbasin 2) as the head of inland water transportation on the Columbia, it is doubtful that the navigable waterway on the Columbia mainstem will be extended past the Tri Cities because of both economic and environmental reasons. Construction of costly navigation locks would be necessary at three existing hydroelectric dams between the Tri Cities and Wenatchee, and with the era of large federal water projects over, it is doubtful that Congress would agree to invest public funds to expand the systems. In addition, navigation expansion would require dredging the last free flowing section (the Hanford Reach) of the Columbia River between Bonneville Dam and the Canadian border. Dredging this reach would jeopardize valuable spawning habitat of wild salmon, an action which has been and would continue to be vehemently opposed by influential groups representing fishing interests, environmental values, and Indian tribes.

Fish and Wildlife Habitat

Another noteworthy instream use of water is the provision of habitat for fish and wildlife. This aspect of water resources is very important in the Pacific Northwest for several reasons. First, the viability of salmon runs is widely interpreted as an indicator of the region's environmental well-being. Second, a variety of interests continue to compete vigorously for declining fish and wildlife resources. And third, innovative institutional arrangements have been introduced to rebuild fish and wildlife species—with special emphasis on anadromous salmonids—while maintaining the output of traditional water related goods and services at an acceptable level. This is a formidable challenge because influential interest groups associated with various water related outputs, such as e.g., hydroelectrical energy, irrigation, and inland water transport, may not be expected to willingly reduce the benefits they receive from water use in favor of improving fish and wildlife habitat.

The potential for conflict between traditional utilitarian and environmental values associated with water resources is shown by the Water Use System in Figure 1, which illustrates that hydrologic feedback resulting from the production of various water related goods and services can adversely affect fish and wildlife by altering one or more of the characteristics of its water based habitat. As indicated in Figure 1, water's quantity, quality, timing, and/or spatial distribution may be changed when, for example, society uses a river to produce hydropower or irrigation water or most of the other water related outputs. Any one of the four attributes above may be of crucial significance to biota depending on those waters as habitat.

In other less frequent cases, the development of water resources enhances habitat for fish and/or wildlife. For example, reservoirs created to store water for irrigation or flood control may also be stocked with game fish that were previously absent from that particular locality or area. The provision of improved fish and wildlife habitat is often a planned part of water resources development but may also be unintended.

Perhaps the most striking example of unintended fish and wildlife benefits has taken place in the Columbia Basin (Subbasin 2, Figure 2) as a result of the Bu Rec's extensive irrigation project there. As noted in the section on irrigation, massive amounts of water are pumped out of the Columbia River behind Grand Coulee Dam and conveyed between 50 and 85 miles south to irrigate lands. Transfer to and application of this much water in an arid area, which receives on average only 8 to 12 inches of precipitation yearly, markedly raised its groundwater levels and changed its surface hydrography. For example, the number of lakes increased from 35 to 140, many of which have since been stocked with game fish and attract large numbers of anglers (USDI 1981, 384). Increases in the population of migratory waterfowl in and around the irrigation project are also noteworthy. A major waterfowl migration route, the Pacific Flyway, passes through the area; the numerous lakes and marshes created by the irrigation project have attracted heavy concentrations of ducks and geese. After the United States Fish and Wildlife Service included it in its Columbia National Wildlife Refuge, part of the area functions as a game preserve while other portions are open to hunters. Grant County, Washington, in the heart of the irrigation project, is the undisputed leader in duck harvest among regional counties and ranks among the top three in the harvest of geese (Matzke 1985, 133–4).

On the other side of the coin, the Columbia Basin Irrigation Project had large negative effects on stocks of anadromous salmonids above Grand Coulee Dam. Construction of the dam in the 1930s cut off more than 1,100 miles of salmon habitat (Muckleston 1992, 386) causing the permanent loss of extensive reaches of salmon spawning and rearing habitat. Water storage capacity provided by Grand Coulee Dam also brought with it river regulation that made feasible the construction of additional hydroelectric projects downstream, which in turn further degraded salmon habitat and migration conditions.

During the three decades following completion of Grand Coulee Dam, eight additional dams were built below it on the mainstream of the Columbia River. Although seven of the eight allow passage of migratory fish, losses do occur to both adults moving upriver to spawn and to juvenile salmonids (smolts) heading downriver toward the Pacific Ocean. Mortality levels of smolts have been particularly high; significant cumulative losses of salmonid stocks that must pass more than three or four dams between spawning areas and the Pacific Ocean continue into the 1990s.

During the post World War II period construction of several dams on the Snake, Cowlitz, and Deschutes Rivers contributed significantly to the decline of salmon in the Columbia Basin. Dam construction was accompanied by increased logging, irrigation, mining, and urban-industrial development while fishing pressure increased in coastal waters and even on the high seas. Whereas the number of adult salmonids returning to the Columbia River had ranged from 10 to 16 million annually in the 19th century, by the early 1980s only about 2.5 million were returning to the Columbia, with hatchery fish comprising approximately 70 percent of that number (Muckleston 1990 (2), 11–12).

Growing concern over disappearing stocks of salmon was one of the principal forces behind the 1980 passage of the Pacific Northwest Electric Power Planning and Conservation Act (The Act). The Act has set in motion a number of actions that have altered the region's water resources management. The new management calls for reducing hydroelectric output in favor of fish. These actions are planned by a unique, somewhat controversial, and innovative regional institution whose existence was made possible by the Act. As noted in the section on the Management Subsystem, the fish and wildlife plans formulated by the Northwest Power Planning Council (NPPC) must be substantially complied with by Federal agencies.

The NPPC's restoration efforts have focused on improving passage of salmonids along the main stems of the Columbia and Snake Rivers. These efforts include releasing guaranteed volumes of water from upstream storage during the spring and early summer to speed downstream movement of smolts. This practice departs markedly from the original major purpose of the storage dams which was to store water until the release period of autumn and winter in order to increase the hydroelectric output during the seasons of higher energy demand. Thus, considerable energy revenues are foregone in order to decrease smolt mortality.

NPPC plans also call for construction of bypass facilities at major dams on the Columbia system. The bypass systems would allow smolts to move past the dams without going through the turbines or over the spillways. Until all of the expensive bypass systems are installed by the middle-late 1990s, some water will be diverted away from the hydroturbines and directed over the spillways. This allows smolts to use a less dangerous route to get past the dams. Again, this approach reduces revenues from the hydroelectric system. In addition the NPPC has designated more than 40,000 miles of undeveloped rivers and streams in the region as protected areas to preserve good salmon habitat (NPPC 1988, 7–8). Such an action prohibits development of hydroelectric projects in the waters so designated and limits future hydroelectric generating capacity in the region. The NPPC's designation of streams as protected areas prohibited the construction of 202 of the 327 hydroelectric projects that had been proposed in the region as of mid-1988. The 202 potential hydroelectric projects could have produced 688 average megawatts (aMW) of energy. Although this is enough to supply greater Portland, considerable hydroelectric potential remains. The 125 proposed plants, which were unaffected by NPPC's designation in 1988, would produce 800 aMW if developed (NPPC 1988, 8).

In the early 1990s some of the region's stocks of salmonids were listed under the Endangered Species Act. The listing means that fish recovery programs might call for significant tradeoffs from a number of water user groups in the Pacific Northwest. Considerably more water might be released from upstream storage to enhance downstream migration of smolts. This would result in further losses of hydroelectric output. In addition, other water uses such as irrigation, recreation, and inland water transport would probably be required to make significant sacrifices in order to enhance the salmon restoration program.

Recreation

Recreation has become an important use of water. Opportunities are high for recreational use of water in the Pacific Northwest for several reasons: The region has a bountiful supply of high quality surface water relative to the size of the resident population, many of these surface waters are set in attractive natural landscapes, and most of these waters are available to recreationists for little or no charge.

Leisure-time use of water encompasses a broad range of activities, including outdoor recreational activities of all types that may take place in, on, or by the water; sports fishing; hunting waterfowl; and enjoying water's aesthetic qualities in both developed and natural settings. Recreational waters may be classed as: slackwater (without current), which are conditions found on natural and artificial lakes; placid, slowly flowing water, for example on the lower 146 miles of the Columbia River below Bonneville Dam; and white water, which refers to rapidly moving water usually in the upper reaches of the region's rivers and streams where steep gradients produce such conditions.

The Pacific Northwest has an abundance of each type. Since the 1970s approximately 2.5 million acres of surface waters are available for leisure-time uses, (PNRBC 1971, 1). However, much of this activity takes place on larger bodies of water—lakes and reservoirs with a surface area greater than 1,000 acres. Table 6 illustrates the amount of surface waters in these large bodies of water by subbasin. Data on recreational use are available for many reservoirs but largely absent for lakes where little organized recreational management is present. In 1991 recreational use of Corps' and Bu Rec reservoirs was 5.6 and 9.2 million 12-hour visitor days, respectively (CRWMG 1992, 78-9). While recreational use of these agencies' impoundments has become large enough to sometimes influence reservoir operations—timing, and extent of drawdown, or filling schedules—it is not usually allowed to significantly reduce the traditional, utilitarian outputs for which the reservoirs were originally created. That is to say, longer established uses like irrigation, hydropower production, navigation, and water supply usually continue to have a higher priority than leisure-time uses of water.

The preservation of free flowing rivers is another significant aspect of instream water management and leisure-time uses of water in the Pacific Northwest. In the previous section, streams preserved as salmon habitat served as an example. Preservation is a radical departure from the traditional approach to water management which generally requires the modification of one or more of water's spatial, temporal, quantitative, or qualitative attributes in order to produce water derived outputs (see Figure 1). Preservation of free flowing rivers and their adjacent lands is also accomplished through the National Wild and Scenic Rivers Act, but Oregon, Washington, and Idaho also have state administered systems. Rivers maintained in their free flowing state attract a small fraction of recreationists relative to slackwater settings, but the quality of a whitewater experience in natural surroundings is generally held to be superior to mass participation in an artificial environment.

The Pacific Northwest has a relatively large number of rivers in the National Wild and Scenic Rivers System. This reflects the region's abundant supply of relatively unaltered environments and the large proportion of the region under federal owner-

Subbasin	Large Reservoirs[1]		Lakes and Other Slack Water[2]	
	(No.)	(1,000 Acres)	(No.)	(1,000 Acres)
1	30	335.1	63	58.3
2	17	223.3	28	19.6
3	6	16.0	4	1.2
4	25	196.2	6	22.1
5	36	121.1	3	0.7
6	8	59.8	11	6.0
7	17	149.2	15	11.9
8	5	14.6	14	60.5
9	19	35.7	12	53.4
10	9	5.0	22	32.1
11	20	45.3	69	47.2
12	1	2.9	13	152.2
Total	193	1,204.2	260	465.2

1) Reservoirs with over 5000 ac-ft capacity and regulated lakes
2) Includes natural lakes over 100 surface acres and water surfaces of the Lower Columbia and Willamette Rivers.

Table 6. Area of Fresh Surface Waters in Large Lakes and Impoundments.
Source: Modified from Pacific Northwest River Basins Commission, Comprehensive Framework Study, Portland, Oregon, 1971, Appendix XIII, Table 1, p. 22.

ship. The latter is critical because the Act is difficult to implement when private property is the predominant type of ownership along the banks of streams and rivers under consideration for inclusion in the system (Root 1989, 91–4). When the National Wild and Scenic Rivers Act was first passed in 1968, Congress designated only eight rivers in the entire country for immediate inclusion in the system, making provisions to add more rivers after they were studied and judged suitable. Three of the original "instant" rivers are in the Pacific Northwest; and the length of the region's designated segments was approximately half of the United States total in 1968. These three original river segments in the region included 84 miles of the Rogue in southwestern

Oregon (Subbasin 10) and two in central Idaho (Subbasin 6)—185 miles of the Middle Fork of the Clearwater and 104 miles of the Middle Fork of the Salmon (Figure 2).

Since 1968 many additions to the system have kept the Pacific Northwest disproportionately well represented in the United States total. By 1987, 20 years after passage of the Wild and Scenic Rivers Act, 13 river segments were in the region with a total length of more than 1,200 miles. Much of this additional length was added in the mountainous Subbasins 1 and 6 although more river segments were placed in the system in several other subbasins within the Pacific Northwest. In 1988 there was a remarkably large addition to the National Wild and Scenic Rivers System in the Pacific Northwest. Sponsored by Oregon's influential Senator Mark Hatfield, the Oregon Omnibus Rivers Bill added more than 40 rivers or river segments to the system (Root 1989, 94). The length of these additions was greater than the total length of then existing wild and scenic rivers in the Pacific Northwest and placed Oregon first in the United States in the number of rivers within the national system.

Flood Hazard Mitigation

Flood hazards result when people and/or their property occupy floodplains. Floodplains are lands which under normal conditions are occasionally reclaimed by adjacent streams and rivers. The frequency, magnitude, and duration of inundations vary widely. Flood hazards reflect the following: the types and intensities of floodplain use; the existing degree of flood related adjustments (flood hazard mitigation measures); and the hydrologic characteristics of the streams and rivers adjacent to the floodplains. Thus, flood hazards result from a combination of natural and anthropogenic factors.

Floods have been costly in the Pacific Northwest. In the early 1990s it was estimated that yearly costs from stream bank erosion were $23.6 million (TFIFMT 1992, 19).The value of insurance claims distributed by the National Flood Insurance Program is another indicator of flood hazards in the region: between 1978 and 1987 the Program paid out about $13.2, $2.4, and $.5 million in claims for flood damage in Washington, Oregon, and Idaho, respectively (TFIFMT 1992, 41). Even during the 1991 water year, when flows in most of the Columbia Basin were below average, flood damage in the basin above Vancouver, Washington, was more than $10.2 million (CRWMG 1992, 65).

Measures designed to mitigate flood hazards may be divided into two general types. First, the structural approach includes dams, levees, flood walls, stream channelization, etc., which are designed to lower or contain flood crests. This approach attempts to keep water away from people and their property. Second is the nonstructural approach which includes floodplain regulations, flood insurance, and warning systems. This approach attempts to keep people and their flood-vulnerable properties away from flood hazardous areas. Most of the expenditures on flood hazard mitigation in the region have been for structures.

Flood hazard mitigation is included under instream uses although it is not a water use per se. It is included there because flood hazards and protective works are often within stream banks and flood hazard mitigation is one of the water-related outputs shown in Figure 1. In addition, "flood control" was often one of the principal goals when multipurpose projects were planned, particularly when dams with storage capacity were included. Regulation of river flows downstream from such dams can markedly affect other instream water uses, including hydropower production, inland water transport, fish and wildlife habitat, recreation, and cost of waste treatment.

Construction of storage dams, levees, and channelization was widely employed in the Pacific Northwest between the mid-1930s and about 1970. Since then building additional flood control structures has been severely limited due to a growing concern for environmental values and increasingly severe federal budget constraints on this type of activity; instead more attention is paid to flood plain regulation.

Tables 7 and 8 illustrate the extent to which structural measures were employed near the end of the construction period. Most of these structures are still in place and may be expected to remain in use for the foreseeable future. In a few subbasins where rapid suburban growth is taking place on floodplain lands, levee construction may be expected to continue.

Most of the approximate 1,800 miles of protective levees in the region are concentrated in several subbasins (Table 7). Extensive lowlands in the Puget Sound area, Subbasin 11, are afforded various degrees of protection by almost 600 miles of levees. These include productive agricultural lands and more recent urban-residential developments sprawling outward from the established urban areas in the Nooksack, Skagit, Snohomish, Green, and Puyallup drainage basins. Rapid urban growth makes Subbasin 11 the most likely candidate for continued levee construction.

In Subbasin 8, the lower Columbia, levees protect very large areas of agricultural land, but the most important protective works are found in the Portland-Vancouver reach of the Columbia River lowlands. That levees differ in the degree of protection they provide is illustrated by three contiguous drainage districts along the Columbia River in north Portland. The first, Peninsula Drainage District #1, includes the site of former Vanport City, an urban area completely destroyed in the floods of May-June, 1948. The city has not been rebuilt and the drainage district remains in open space uses—a golf course and radio transmitter antennas—because its dikes are rated not sound enough to withstand a regulatory flood. A flood of this size has a probable return frequency of one percent in any one year and is often referred to as the "100 year flood." Because of the flood hazard, the City of Portland will not issue building permits to would-be developers, thereby maintaining open space land uses within the district that are harmonious with potential flooding.

Immediately to the east lie the lands of Peninsula Drainage District #2. It was also flooded in 1948, but since its levees were improved in 1972 this district has been the site of rapid urban, commercial development. While its protective system is rated to

Subbasin	Levees miles	Channel Improvement & Stabilization miles	Bank Protection miles
1	161	1,168	488
2	20	203	205
3	64	128	—
4	171	119	131
5	48	192	1,042
6	29	436	407
7	72	366	106
8	449	100	176
9	—	368	115
10	202	383	129
11	578	—	203
12	—	—	20
Total	1,794	3,463	3,022

Table 7. Local Flood Protection Projects.
Source: Pacific Northwest River Basins Commission, *Comprehensive Framework Study*, Portland, Oregon, 1971, Appendix VII, Table 2, p. 28.

withstand a regulatory flood, it is not rated to protect the district if the larger but less frequent standard project flood occurs.

On the eastern border of District #2 lies the Multnomah Drainage District, site of Portland's international airport and many the urban commercial activities attracted to large airports. The district's lands were also flooded during the 1948 flood but since then have been protected by levees rated to withstand the standard project flood. Thus, along a relatively short reach of the Columbia River in north Portland, lie three drainage districts where three different levels of levee protection are provided. Intensive urban, commercial developments are allowed to take place in the middle and eastern districts while in the western one such land uses are illegal due to a greater flood hazard.

In addition to levees and channel improvements, floodplain uses in the Portland-Vancouver area benefit from flood crest reduction created by the construction of

Subbasin	Allocated Primary	Flood Control Joint Use	Incidental Major Res.	Incidental Farm Pond and Small Res.	Totals	Usable for Control of Columbia River Floods
Canada						20,900
1		10,319	335	24.0	10,678	10,552
2		5,232	24	21.0	5,276	5,232
3		—	1,071	4.5	1,075	—
4	10	1,716	3,108	51.0	4,883	1,600
5	97	2,270	2,120	97.0	4,584	2,783
6	—	2,000	—	11.0	2,011	2,000
7	8	577	471	38.9	1,094	500
8	100	260	1,850	3.3	2,213	—
9	—	1,703	—	17.7	1,720	—
10	—	65	—	9.7	74	
11	106	226	1,719	7.0	2,058	
12	—	—	—	155.3	155	

Table 8. Flood Control Storage. Storage in 1,000 Acre-Feet.
Source: Pacific Northwest River Basins Commission, *Comprehensive Framework Study*, Portland, Oregon, 1971, Appendix VII, Table 1, p. 27.

storage dams far upstream. Provision of upstream storage in the distant and mountainous headwaters of the Columbia system has lowered the height and duration of flood crests along many hundreds of miles of rivers downstream where more intensive types of land use often takes place. Table 8 indicates the approximate volume of storage available by subbasin. The two left columns include flood control storage that was intended to be an integral part of the water development project.

Flood control is the overriding purpose of storage space designated as "allocated primary." Most of the active flood control storage, however, is under joint use. Flood control joint use means that other planned water uses share the reservoirs' storage space because they are complementary. For example, the storage in Roosevelt Lake, the reservoir formed by Grand Coulee Dam (Subbasin 2), is used jointly for flood control, hydropower, and irrigation. These uses are complementary given the Columbia River's flow regime (Figure 3A) and differing seasonal requirements for the three water related goods and services in question. Roosevelt Lake is lowered prior to the spring snow melt upstream, providing storage space for flood reduction downstream. The water stored from spring and early summer snow melt is then used later in the summer and early autumn for irrigation; and in the autumn and winter for hydropower production, the season of peak demand for electrical energy. Hydropower from this storage is not only produced at Grand Coulee but also at the ten dams downstream that have very little storage of their own (Figure 3B).

By far the largest volume of storage available for flood control in the Columbia System is located in Canada (Table 8). As noted under the section on hydropower, the procurement of this storage required lengthy negotiations between the United States and Canada, and resulted in a treaty which incorporates the principle of sharing downstream benefits. When the treaty was finalized in 1964, Canada—the upstream country—received $64.4 million from the United States—the downstream country (Fernald 1967, 210–11). This sum represents one-half of the value of property damage that will not occur in the United States portion of the Columbia system due to upstream storage in Canada over the 60 years of the treaty life.

The considerable addition of upstream storage capacity that was created in the 1960s in the upper reaches of the Columbia River system proved to be very beneficial. In 1972 and again in 1974, rapid melting of heavy snow packs across the Columbia's mountainous drainage areas would have caused flood flows in the Portland-Vancouver area roughly equal to those of the devastating 1948 flood had extensive upstream storage not been provided. Major storage projects in the Columbia System are credited with preventing $474 and $546 million of damage during the high water years of 1972 and 1974, respectively (CRWMG 1992, 67). Between 1981 and 1990, a total of about $538 million damage was prevented by those structures, reflecting an absence of potentially high flows experienced in 1972 and 1974 (CRWMG 1992, 67).

West of the Cascades, flood control and hydropower are also produced from the same water resources developments although the seasonal complementarity noted in the foregoing paragraphs is generally lacking. For example, in western Washington, Subbasins 8 and 11, flood control is provided as an incidental benefit of hydropower development (Table 8). This is illustrated by Seattle City Light's High Ross Dam that provides significant flood crest reduction along the floodplains of the lower Skagit Valley but whose primary purpose is hydroelectric production.

Flood control and irrigation are complementary uses of storage reservoirs in all of the region's subbasins, although in some subbasins (e.g., 8 and 11) little irrigation is practiced. Noteworthy volumes of incidental flood control storage exist in the Yakima, Upper Snake, and Central Snake Subbasins as a result of building storage for irrigation projects (Table 8).

The provision of flood control can also create a number of additional water related goods and services. In the Willamette River system, Subbasin 9, provision of flood control was the primary goal behind the construction of a system of reservoirs.

Authorized by the Flood Control Act of 1938, the Corps of Engineers constructed 13 dams over the following three decades that provide approximately 1.7 MAF of storage (Table 8). Because the flood season in Subbasin 9 runs from November through February—in contrast to the flood period in the Canadian Columbia and Subbasins 1–7—the reservoirs are lowered during the autumn to provide storage capacity for high inflows. After the flood season passes, these reservoirs are filled in the spring and are then used for a number of purposes including outdoor recreation oriented towards slack water activities.

Of greater significance, however, are the releases of water stored in these reservoirs which approximately double the normal flow of the Willamette River during the months of July, August, and September. This additional water markedly increases the Willamette's capacity to carry and assimilate waste. Thus, provision of flood control storage in the headwaters of the system has been instrumental in greatly improving water quality in Oregon's most populous river basin. Without this additional water for waste dilution, the large expenditures for waste treatment plants made by municipalities and industries in the decades following World War II would not have "cleaned up" this once grossly polluted river in Oregon's heartland (Gleeson 1972, 62). Both increased flows and better treatment were needed. Because water quality in most of the Willamette system has improved markedly, a number of water related goods and services can be provided at relatively low cost, including high quality fish and wildlife habitat, and all forms of outdoor recreation in, on, and by the river.

OFFSTREAM USES

Unlike instream use, the other major category of water utilization is spatially removed from its water source. Among the uses commonly considered in this category are: irrigation, public supply, industrial, commercial, domestic, and thermoelectric power. Several sequential steps are common in this category of use: water is first withdrawn from surface or ground sources, then conveyed to its point of use (which can vary in distance from a few feet to many miles), utilized, and then the unconsumed portion is returned to the nearest feasible disposal site, usually a stream or river. For most uses the unconsumed portion of the water is treated and then conveyed via pipe or open channel to the point of disposal. The nature of irrigation, however, precludes such conveyance and the unconsumed water is allowed to find its own way back to surface waters. This often takes place via groundwater aquifers. Treatment of unconsumed return flow by irrigators is not feasible.

As noted above, offstream uses consume part of the water that was withdrawn. The term "consumptive use" refers to the water that does not return to its stream or aquifer of origin, having been either incorporated into the product, irretrievably lost during conveyance, or changed from its liquid to gaseous state through evaporation or transpiration. Evaporation takes place from land, water, or other surfaces; transpiration refers to water used by plants and trees during photosynthesis. The percent of water used by consumptive use ranges from as high as 90 in some irrigation operations to as little as one or two for thermoelectric cooling. It is noteworthy that consumptive use refers only to quantitative changes. The production of some types of water related outputs do not consume much water, but the quality of the water returned to the river of origin may be very poor, decreasing the utility of the water for users downstream.

As shown in Figure 1, changes in the characteristics of return waters (the hydrologic feedback) can include their quantitative, qualitative, timing, and/or locational attributes. Any of these changes may adversely affect users downstream. The severity of the effect depends on the nature of the return flow in question, the waste assimilating capacity of the receiving stream, and the nature of use downstream from the point or area of deposition. Less commonly, return flows may have positive effects on subsequent users. Some of the effects of return flows and consumptive use are discussed below under specific offstream uses.

When compared to other major water resources regions in the United States, the Pacific Northwest ranks among the leaders in the quantity of fresh water withdrawn in 1990, averaging more than 36,800 million gallons per day (mgd) (USGS 1992, Table 1). Table 9 shows the approximate volume and the origin of water withdrawn by various categories of offstream uses. The pattern of uses is similar to that of most western states: irrigation with other agricultural uses are dominant. Thermoelectric cooling, which is the preeminent use in many states east of the Mississippi, accounts for about one percent of the total withdrawals in the Pacific Northwest.

Irrigation

Irrigation may be defined as the purposeful application of moisture to the root zone of crops and other desirable vegetation. This water use accounts for more than 85 percent of the total fresh water withdrawals from surface sources in the Pacific Northwest. It is also the preeminent consumptive use of water in the region, accounting for over 95 percent of the regional total. In addition, irrigation is responsible for practically all the conveyance losses in the region that leads the nation in this regard. It is noteworthy that some of the water lost in conveyance recharges groundwater aquifers and is subsequently withdrawn by other water users.

The large volume of water extracted from surface and ground sources for irrigation reflects two major factors; 1) a significant part of the region is subhumid and semiarid, lying east of the Cascades where for many crops irrigation is necessary; 2) the bulk of precipitation both east and west of the Cascades falls during the autumn and winter when plants are dormant, so even in the relatively well watered areas west of the Cascades, irrigation is necessary for some crops and desirable as a moisture supplement for many others.

Consumptive use by irrigation can seriously degrade riverine environments and aquatic habitat. This has been particularly damaging in small river systems in the semiarid parts of the region where salmon runs and resident fish habitat have been destroyed by dewatering streams. Dewatering results from heavy irrigation withdraw-

State		Public Supply	Domestic	Commercial	Irrigation	Livestock	Industrial	Mining	Thermo-electric
ID	Surface Water	28	0	0	12100	0	26	7.8	0
	Ground Water	173	48	16	6620	560	170	.6	6.1
	Total	201	48	16	18720	560	196	8	6
OR	Surface Water	365	7.6	704	6300	18	254	.6	15
	Ground Water	105	57	7.3	563	3.2	31	1	0
	Total	470	65	711	6863	21	284	2	15
WA	Surface Water	441	0	.4	5280	8.4	397	.6	330
	Ground Water	434	104	27	754	22	104	2.4	4
	Total	875	104	27	6034	30	501	3	334

Table 9. Offstream Water Uses (mgd).
Source: U.S. Geological Survey, *Preliminary Estimates of Water Use in the United States,* 1990 Open-file Report 92-63, Table 3, p. 9.

als and consumptive use. Even when water remains in these small systems, quality may be dangerously degraded by irrigation return flows that often contain high levels of silt, salts, and various agricultural chemicals.

Idaho is the undisputed leader in irrigation among the states of the Pacific Northwest (Table 10). The Upper and Central Snake Subbasins (4 & 5) have the largest irrigated areas and withdraw the most water for irrigation, accounting for about 60 percent of the total surface water withdrawals for irrigation in the region. A national comparison of total fresh water withdrawals for irrigation reveals that Idaho is second only to California. When a comparison of per capita withdrawals for irrigation is made, Idaho's approximate 18,500 gallons per day per person makes it the undisputed national leader in this category (USGS 1992, Table 2).

Although not oriented towards irrigation to the same extent as Idaho, Washington and Oregon also rank among the country's leading irrigation states. In the Upper Columbia Subbasin, irrigation is dominated by the federal Columbia Basin Irrigation Project. This is the Bu Rec's single largest reclamation project in the country, irrigating more than 500,000 acres. Yearly, an average of 2.9 MAF of water is lifted

State		Withdrawals (mgd)	% of State Total	% of Three State Total
ID	surface	12,100	65	51
	ground	6,620	35	83
	total	18,720	100	59
OR	surface	6,300	92	27
	ground	563	8	7
	total	6,863	100	22
WA	surface	5,280	88	22
	ground	754	12	9
	total	6,034	100	19

Table 10. Irrigation Withdrawals.
Source: Modified from U.S. Geological Survey, *Preliminary Estimates of Water Use in the United States,* 1990, Open-file Report 92-63, Tables 3, 4, 5, & 7.

about 280 feet from Roosevelt Lake, formed behind Grand Coulee Dam, and then conveyed many miles south before being distributed to the fields (Muckleston 1992, 384–5). When the project was authorized in the 1930s, Congress designated that more than one million acres be brought under irrigation. Thus, the country's largest irrigation project is only half completed. Although the second phase of the project is technically feasible, significant concerns over environmental and economic questions make project completion problematic. Environmental concerns are chiefly related to the survival of anadromous salmonids, while economic questions reflect the fact that about 90 percent of the Columbia Basin Project's irrigation costs are presently paid from the sale of hydroelectricity generated by Grand Coulee Dam and other federal water projects in the region. The public has become concerned over such large subsidies for irrigation, which could be markedly increased in the area if the second phase of the project were built.

An additional concern is the projected loss of energy output from the Columbia River hydropower system if the second phase of the Columbia Basin Irrigation Project is completed. Adding another 500,000 acres to the irrigation project would require a diversion of about two-million acre-feet of water. Therefore, significant losses of hydropower would result at the 11 dams with 180 generating units located downstream from the diversion point on Roosevelt Lake. Consumptive use of water at the project is at least double the regional average of 40 percent, therefore relatively little water would find its way back to the Columbia where it could be used for hydropower production and for other water related purposes. The Corps of Engineers estimated that completing the second phase of the irrigation project would cause a loss of 966 aMW in the hydropower system (Muckleston 1982, 387). This is the approximate amount of electrical energy used by Seattle, the region's most populous city.

Moreover, considerable additional energy would be required to operate the second phase of the Columbia Basin Irrigation Project: millions of additional acre feet of water would be lifted 280 feet from Roosevelt Lake to supply the new irrigation lands. Energy to lift this much water is considerable, as one acre foot of water weights approximately 1,357 tons. Considerable additional energy would also be required to pressurize and operate the sprinkler systems used to irrigate fields.

In Subbasin 3 the Yakima Irrigation Project has made Yakima County the region's leader in agricultural production. Irrigation is absolutely necessary here because the Yakima Valley receives only about eight inches of precipitation yearly, far too little for rainfed agricultural requirements. This project is one of the Bu Rec's earliest irrigation developments and also one of its most successful in terms of value of crop production per dollar of federal expenditure.

In the Mid Columbia Subbasin much of the irrigation development by the Bu Rec is in Oregon's Deschutes and Umatilla drainage basins. In the 1970s considerable irrigation development by private interests took place in the eastern part of Subbasin 7. The development of irrigation in this rolling, sandy land formerly classified as unirrigable became possible through the use of center pivot technology (Muckleston and Highsmith 1978, 1124).

West of the Cascades, in Subbasins 8–11, irrigation is much less common; therefore, much less water is withdrawn and consumed than in Subbasins 2–7 discussed above. Due to more humid conditions in Subbasins 8–11, when irrigation is practiced it often is used to supplement rainfall although the dry summers preclude the growth of many crops without at least provision for supplying some additional moisture. In the Willamette Valley (Subbasin 9) frequent dry summers and production of some water demanding crops result in considerably more irrigation than in the Puget Sound lowlands (Subbasin 11).

Two relatively small but noteworthy areas west of the Cascades do depend almost entirely on irrigation. Both the Middle Rogue Valley in Oregon and the northeastern corner of the Olympic Peninsula in Washington are on the leeward sides of the Siskiyou and Olympic Mountains, respectively. Because these mountain ranges are significantly higher than other coastal ranges in Oregon and southwestern Washington, the dry shadow effect is more strongly developed there, which results in precipitation amounts well below the minimum required by most crops.

In Oregon's part of the Coastal Subbasins considerably more irrigation takes place than in Subbasin 10 in Washington. This is due to a greater area of irrigated pasture in Oregon and also because large agricultural developments in the Rogue Valley are included in this subbasin.

Public Water Supply

Public Water Supply is defined by the United States Geological Survey as ". . . water withdrawn by public and private suppliers and delivered to groups of users" (USGS 1988, vi). This use category accounts for the second greatest amount of water withdrawn from surface and underground sources in the region. Although the volume of water withdrawn is only about five percent of that extracted by irrigation, public supply is of great significance because it provides domestic water to most of the region's population. In 1990 approximately 7.1 million residents of the region received their domestic water from public supplies. In a few of the region's subbasins, where little irrigation takes place, public supply is the dominant offstream use. Public supply withdrawals exceed two-thirds of the total in the Puget Sound Subbasin and in Washington's part of the Coastal Subbasin.

Public water suppliers must treat the water they withdraw so that it becomes potable, or fit for human consumption, before distributing it to residential units, commercial establishments, industrial customers with modest requirements for water, and for other public uses such as street washing and fire suppression. Obviously not all these end users require potable water, but public suppliers must meet drinking water standards for all water distributed.

Relative to irrigation, public supply consumes a small part of the water it withdraws, generally less than 20 percent. But the proportion consumed increases markedly in many communities during the summer when heavy lawn watering is practiced. An indirect result of public supply is water quality degradation downstream from population concentrations as return flows from residential, commercial, and industrial uses are disposed in surface waters. Since the 1970s, implementation of state and especially federal pollution laws have mandated improved treatment of waste before discharge; as a result, water quality problems resulting from point sources have become less serious in many cases.

Table 11 summarizes public water supply in the region by state and source of supply. In some parts of the region groundwater is the dominant source of public water supply. Groundwater is in general the preferred source of supply if unpolluted, adequate, and dependable amounts are close to the area of demand because relative to surface water costs of development, treatment, and conveyance are lower. In 1990 groundwater accounted for 46 percent of the total public supply withdrawals. Frequently, however, conditions favoring groundwater use are not present, requiring the development of surface waters. The predominant use of surface waters for supply of the region's populous Puget Sound Lowlands and Willamette Valley is encouraged by the proximity of normally well watered mountains. The relative high quality of the source water reduces treatment costs, while delivery by gravity allows inexpensive conveyance to points of demand.

Notable patterns of reliance on surface or groundwater are present in the region. Differences exist between states, within states, and between communities based on size. When comparing states, public suppliers in Idaho are most dependent on ground-

State	Sources and Totals	Withdrawals (mgd)	% of the State Total	% of three State Total
ID	surface water	28	14	3
	ground water	173	86	24
	total	201	100	13
OR	surface water	365	78	44
	ground water	105	22	15
	total	470	100	30
WA	surface water	441	50	53
	ground water	434	50	61
	total	875	100	57

Table 11. Public Supply Withdrawals.
Source: Modified from U.S. Geological Survey, *Preliminary Estimates of Water Use in the United States,* 1990, Open-file Report 92-63, Tables 2, 3, & 4.

water, relying on subsurface sources to supply 86 percent of the total. In Oregon, by contrast, groundwater constitutes only about 22 percent of the supply; in Washington it supplies about half. Groundwater is a more important source of supply east of the Cascades and for small public supply systems throughout the region. A few larger urban areas do depend on groundwater, however, with Spokane and Tacoma, Washington, and Missoula, Montana, serving as notable examples.

Public water supply systems also provide water for industrial use. Industries that obtain water from public suppliers generally use relatively small amounts of water by industrial standards. Nevertheless, about 15 percent of the total industrial use comes from public suppliers. The proportion used by end consumers varies between and within states. In Idaho, the region's least populous and most rural state, almost all of the public supply is delivered to domestic/commercial customers; whereas in Oregon and Washington, the proportion of public supply delivered to domestic/commercial users declines roughly in proportion with increasing size of population and degree of industrialization.

Self-Supplied Industrial Use

Industry withdraws the third largest amount of water in the region. This section focuses on self-supplied industrial use where individual plants often withdraw relatively large volumes of water. Principal industrial water uses include processing, in-plant transport of materials, washing, and cooling. Water for cooling of thermoelectric plants is considered in the next section. Relative to agriculture, consumptive use by industry is low, but industrial waste water can degrade the utility of surface waters for users downstream from plant discharge points. The effect of the degradation depends on a number of things: the nature of the discharge water (amount and type of pollutants); the assimilative capacity of the receiving stream; the distance between a discharger and subsequent intake points; and the water quality requirements of the downstream users.

Industrial water use in the Pacific Northwest reflects the region's economic orientation toward forest products although some individually noteworthy withdrawals take place by producers of primary metals, chemicals, and refined petroleum products. Withdrawals by industrial users have declined slightly from those of the previous decade, reflecting the national trend of more efficient inplant water use and a decline in the region's forest products industry. Unlike industrial water use in some of the other United States water resources regions, in the Pacific Northwest there is little reliance on saline water and/or reclaimed sewage by some types of industry.

The industrial water use sector in Idaho relies heavily on groundwater. Most of this subsurface source supplies food processing plants in the Snake River Plain (Subbasins 4 and 5) where large volumes of groundwater are available. In the Idaho panhandle (Subbasin 1) surface water is used to supply the needs of industries processing forest products. In western Montana the processing of forest products is also the principal industrial user of water. Although most of these plants in Montana use surface water, the largest single user relies exclusively on groundwater.

In Oregon more than half of the industrial water use is accounted for by 11 pulp and paper mills located in the Coastal and the Willamette Subbasins. Much of the water used by food processors is found in the widely scattered agricultural areas of the state while chemical plants are concentrated in the Portland Metropolitan Area.

In Washington almost two-thirds of the state's industrial water use takes place in only three counties. The first two counties, Clark and Cowlitz, are located in Subbasin 8 along the lower Columbia River. This subbasin is the clear leader in industrial water withdrawals, accounting for about one-third of the regional total for withdrawals from surface waters. In these counties the manufacture of forest products accounts for most of the industrial water use. Although in Clark County these industries are close to the Columbia River, groundwater from the terrace and valley fill aquifer supplies approximately half of the industrial water withdrawals there. In Benton County (Subbasins 2 & 3), the manufacture of chemicals accounts for most of the water used by industry. The presence of the Hanford Nuclear Energy complex and agricultural chemical production would account for most of this use.

Thermoelectric Energy

Water use by thermoelectric power plants is the last category considered under offstream uses. Thermoelectric power refers to the production of electricity using either fossil fuels or nuclear energy. Large amounts of water are required to cool this type of equipment while smaller volumes of water are converted to steam that drives turbines and generators.

Although at the national level thermoelectric power rivals irrigation for total freshwater withdrawals, in the Pacific Northwest relatively little water is used for this purpose. This reflects the region's heavy dependence on hydropower. During dry years, when lower than average runoff reduces hydropower output, thermoelectric generation becomes more important in the region although even in these circumstances it accounts for a relatively small part of the region's total production of electricity.

Water use by thermoelectric plants is concentrated in a few places within the region. In Oregon, the Trojan nuclear plant is sited on the lower reaches of the Columbia River (Subbasin 8) about 40 miles north-northwest of Portland. Plant owners originally planned to operate Trojan until 2011, but in 1993 it was closed permanently for early decommissioning in 1996. Oregon's other thermoelectric facility is the Boardman coal-fired plant in the north-central part of the state (Subbasin 7).

In Washington State thermonuclear generation of electricity takes place in the Hanford area (Subbasin 2). West of the Cascades a large coal-fired plant near the cities of Chehalis and Centralia (Subbasin 10) accounts for the remainder of water used by thermopower plants in the state. In Idaho relatively small amounts of water are withdrawn for this purpose while no withdrawals are recorded in western Montana.

SUMMARY AND CONCLUSIONS

The geography of water resources in the Pacific Northwest has been presented within the context of the water use system. Sources of supply for the input subsystem are distributed unevenly in the region. The major surface water supply areas are separated by the Cascade Mountains, which divide the region into a relatively well watered western portion and much larger eastern part, much of which is subhumid and semiarid. In the eastern part of the Pacific Northwest productive aquifers and exotic rivers provide water for the input subsystem. Throughout the region the temporal distribution of precipitation, which provides little during the warmer months of greater water need, increases the challenges faced by the water management subsystem.

Several federal agencies dominate the management subsystem, but the issuance of water rights by state level agencies is also noteworthy. Extensive modification of the temporal patterns of river discharge by providing storage has been widely used in the region as a management tool. Although water management actions can affect a wide variety of physical, biological, and sociocultural phenomena, coordination between agencies remains relatively undeveloped. The Northwest Power Planning Council has shown some progress in this direction since the 1980s but has responsibility for only two of the many water related outputs.

Instream water uses in the region are very important, particularly the generation of hydroelectric energy. Hydropower development has been a major output for much of the 20th century, but since the 1980s a growing concern over the viability of anadromous salmonids has brought the interest groups representing these two instream uses into increasing conflict.

The development of a high-capacity inland water transportation route has contributed to significant modifications of the Columbia and Snake Rivers, but heavy use of the waterways by commercial bargelines still remains a potential rather than a reality. Attempts to mitigate flood hazards through significant modification of channels, banks, and river regimes has been pursued both as a single purpose approach and as part of multipurpose schemes. The latter appears more significant in the overall scheme of water resources development in the region; and it includes an important part of an international treaty over the provision of upstream storage with Canada. Benefits in the United States from flood control and hydroelectric generation that result from storage in Canada are shared by both countries. Outdoor recreational use of water, although increasing in the region, remains a relatively insignificant output.

Irrigation dominates the offstream uses, reflecting the large area within the region that has subhumid to semiarid conditions. Heavy consumptive use by irrigation can adversely affect other users downstream. This is one of the reasons that the second half of the Columbia Basin irrigation project, the region's largest, remains uncompleted decades after its authorization. Patterns of public supply generally correspond to the size and distribution of urban concentrations in the region. Groundwater is a much more important source of public supply in Idaho than in Oregon and Washington. Characteristics of self-supplied industrial use reflect the region's noteworthy output of manufactured products originating from forestry and agriculture. Withdrawal of water for thermoelectric cooling is conspicuously small, reflecting the region's highly developed hydroelectric generating capacity.

Over the last decades it has become increasingly evident that water resources in the region, particularly the Columbia River system, are not an inexhaustible source of water derived outputs that can be continually developed by traditional management techniques. Achieving sustained development in the coming decades will require new management techniques, which among other things will necessitate acceptance of tradeoffs and even changes in patterns of consumption by individuals and interest groups who are benefiting from water resource use and development in the Pacific Northwest.

REFERENCES

CRWMG (Columbia River Water Management Group) 1992. *Columbia River Water Management Report for Water Year 1991*, Portland, Oregon, CRWMG.

Fernald, G.H. 1967, "Columbia River Treaty Implementation," *International Conference on Water for Peace*, Washington, D.C., USGPO.

Gleeson, G.W. 1972. *The Return of the River*, Corvallis, Oregon, Oregon State University's Water Resources Research Institute.

Matzke, G.E. 1985. "Hunting and Fishing," Kimerling, A.J. & Jackson, P.L. (eds.), *Atlas of the Pacific Northwest*, 7th Edition, Covallis, Oregon, Oregon Sates University Press.

Muckleston, K.W. 1982. "Attempts to Reconcile Conflicting Demands over the Columbia River," Lowing, M.J. (ed.), *Optimal Allocation of Water Resources* (IAHS Pub. No. 135) Wallingford, UK, International Association of Hydrological Services.

_____ 1990 (1). "Integrated Water Management in the United States," Mitchell B. (ed.), *Integrated Water Management*, London and New York, Belhaven Press.

_____ 1990 (2). "Salmon vs. Hydropower: Striking a Balance in the Pacific Northwest," *Environment* V. 32, No. 1.

_____ 1992. "Grand Coulee Dam," Janelle, D. G., (ed.), *Geographic Snapshots of North America*, New York and London, The Guilford Press.

_____ & Highsmith, R. M. Jr., 1978. "Center Pivot Irrigation in the Columbia Basin of Washington and Oregon: Dynamics and Implications," *Water Resources Bulletin*, Vol. 14, No. 5.

NPPC (Northwest Power Planning Council) 1988. "Protected Area Issues Open for Comment," *Northwest Energy News*, Vol. 7, No. 3.

PNRBC (Pacific Northwest River Basin Commission) 1970 (1). CNPCFS (Columbia North Pacific Comprehensive Framework Study), *Land & Mineral Resources*, Appendix IV, Vol. 1, Portland, Oregon, PNRBC.

_____ 1970 (2). CNPCFS, *Water Resources*, Appendix V, Vol. 1, Portland, Oregon, PNRBC.

_____ 1971. CNPCFS, *Recreation*, Appendix XIII, Portland Oregon, PNRBC.

Root, A. L. 1989. *The Wild and Scenic Rivers Act: Problems of Implementation in Oregon*, Unpublished M.S. Thesis, Department of Geosciences, Oregon State University, Corvallis, Oregon.

_____ 1993. *Improving Instream Flow Protection in the West: An Evaluation of Strategies with an Analysis of Oregon's Program*, Unpublished Ph.D. Dissertation, Department of Geosciences, Oregon State University, Corvallis, Oregon.

TFIFMT (The Federal Interagency Floodplain Management Taskforce) 1992. *Floodplain Management in the United States: An Assessment Report*, Vol. I (FIA-17), Washington, D.C., Federal Emergency Management Agency.

USDOE (US Department of Energy) 1990. *1989 Bonneville Power Administration Annual Report*, Portland, Oregon.

USDI (US Department of Interior) 1981. *Bureau of Reclamation Project Data*, Water Resources Technical Publication, Washington, D.C. USGPO.

USGS (US Geological Survey) 1985. *National Water Summary 1984*, (Water Supply Paper 2775), Washington, D.C., USGPO.

_____ 1988. *Estimated Use of Water in the United States in 1985*, (USGS Circular 1004) Washington, D.C., USGPO, USGS and USDI.

_____ 1992. *Preliminary Estimates of Water Use in the United States in 1990*, Open File Report 92–63, Reston, Virginia, USGS.

ENERGY RESOURCES OF THE PACIFIC NORTHWEST

Chapter 9

Alex Sifford
Sifford Energy Services

INTRODUCTION/BACKGROUND

The Pacific Northwest contains abundant energy sources. Society uses these energy sources in a variety of ways. Houses and offices use energy for space heating and cooling; industry uses energy to supply process heating or cooling. Energy can be converted into electricity to light our surroundings. And we all use energy to travel from place to place.

Historically, energy use began with simple wood burning for cooking by native Americans. This was followed by pioneer water-power-driven uses such as in grain milling. Early timber harvesting required wood-fired steam "donkeys" to move logs. More recently, many electric utilities in the region trace their origins to dams built to supply electricity to early lumber mills and settlements.

The Pacific Northwest is fortunate to have sizable renewable energy sources. Chief among these is converting falling water into electricity or hydroelectric power. Wood wastes (biomass) are used by many industrial firms to supply their own steam and electricity. Wood stoves provide heat to as many as a third of the region's inhabitants. Solar hot water heaters are found on thousands of homes throughout the region. Homes in Boise, Idaho and Klamath Falls, Oregon use geothermal water to provide heating. Two wind farm sites on the Columbia Plateau may be developed soon, and, until very recently, a wind farm operated on the Oregon coast.

Conventional fuels in the Pacific Northwest include fossil and nuclear energies. Fossil fuels found in the region include coal and natural gas, although quantities are limited. One sizable coal mine operates in the region and coal supplies a few power plants and large users in the Northwest. While gas is a significant fuel for homes, industry and new power plants, nearly all gas comes from outside the region. Oil—all of which is imported into the region—is a minor fuel for homes, and a backup fuel for industry and new power plants. Three nuclear plants exist in the region, using uranium fuel from mines in other areas.

Interregional transfers of both fuels and electricity are common today. No region is isolated from external energy sources. Our global dependence on oil illustrates this point well. Areas adjacent to the Northwest are critical: utilities buy supplies from them, make sales to them, via transmission lines connecting us. Our integrated energy system is not limited to electricity. Nearly all fossil fuels come into the region from other areas. Long-distance gas transmission pipelines lead to the Pacific Northwest from Alberta, British Columbia, Colorado and Wyoming.

Transportation is the single largest use of energy, both in our region and the country. Virtually all transportation energy comes from oil, most of which is now imported. Reducing our automobile driving is the single largest energy conserving act we can make. This chapter focuses on energy used in residential, commercial, industrial and utility applications.

We first examine uses of energy. Every facet of human activity involves some energy use. Energy supplies in the region are then discussed, and options to meet future energy needs are highlighted.

CONSUMPTION/USES

We all use energy every day. We do so directly through turning on light switches or heaters or driving a car. Indirect energy use is easy to miss but occurs every time we eat, drink or wash. How humans consume energy varies according to activity or application. Energy use is typically broken down according to sector, that is, according to residential, commercial, industrial, and transportation sector use.

Table 1 shows a matrix of applications and energy sources of those uses. While only uses that are commercially available are shown, some, such as solar cars, are close enough to commercial use to be included.

Electricity, while not a fuel, shows up in both supplies and uses because it serves both functions. This brings up an important first point: distinguishing fuels from energy. Fuels, like gas, oil and coal, are commodities that follow the laws of supply and demand. Fuels are able to supply energy in the form of heat and electricity. In contrast to fuels, *energy* is a physical concept (measured in British thermal units, calories or kilowatts for example) that follows the laws of thermodynamics. Curiously, what's driving current electric utility industry changes is the evolution of electricity from energy into a commodity.

Residential Use

The energy we use in our residences is perhaps easiest to see. In fact, to see at all in any building requires energy. Houses and apartments need light, which typically requires electricity. Although lighting is critical, the biggest residential energy uses are space and water heating, refrigeration, cooking and drying laundry.

APPLICATION

FUEL-TYPE	RESIDENTIAL	COMMERCIAL	INDUSTRIAL	TRANSPOR-TATION	ELECTRICITY*
Biomass	X	X	X	X	X
Coal			X		X
Electricity	X	X	X	X	X
Geothermal	X	X	X		X
Hydro					X
N. Gas	X	X	X	X	X
Nuclear					X
Oil	X	X	X	X	X
Solar	X	X		X	X
Wind	X				X

* Functions as both supply and use.

Table 1. Energy Supplies and Uses.

Heating building spaces is done by circulating hot air or water. Common space heating devices in the Northwest include baseboard heaters, furnaces and woodstoves Fireplaces, with few exceptions, are so inefficient that they add only aesthetic value. Woodstoves (as spelled by the modern hearth products industry) are radiant heat sources and heat only the room they are in. Fans and air ducts distribute heat from woodstoves and furnaces respectively. Ducts connect heat sources to individual rooms. Modern residences have thermostats and duct vents to control which rooms get heat. Most furnaces burn fuel when the thermostat dictates and blow the warmed air through the ducts to each room. Electric furnaces use coils to heat the air before entering the ducts. Geothermally heated homes in Boise and Klamath Falls use hot water to heat the air prior to blowing it throughout the house. Many fuels provide space heat in the Pacific Northwest, with gas and electricity the most common.

(Air conditioning is less important than heating in the Pacific Northwest, but is common in the warmer climates. Room size (radiant) or complete home (ducted) air conditioners are available. The conditioner itself is a heat pump or refrigerator-like device that runs on electricity.)

Domestic water heating is commonly done via a tank that is heated by either gas or electricity. Tank location (in an enclosed space versus outside) and insulation are important factors. Gas and electricity are the most common water heating fuels. Water heating is an application that lends itself well to solar energy, discussed further below.

Appliances to cook, refrigerate and freeze food use energy. Ranges, ovens and microwaves cook food using gas and electricity. Refrigerators and freezers have compressor-expander equipment which transfers heat from the enclosed box to the room. Electricity supplies nearly all refrigerators and freezers in the Northwest. Propane refrigeration and lighting is common in remote applications.

Washing and cleaning, whether dishes, clothes or ourselves, are residential activities that use energy. It takes energy to collect and clean water before it gets to your door. Once there, energy to heat domestic water and power washers is then needed. Electricity powers virtually all washing appliances. Foreign manufacturers now market dishwashers and clothes washers that use dramatically less water and energy.

Commercial Use

Commercial energy use varies widely. Stores and other businesses use energy to heat and cool space and water, to provide lighting, sound, cooking, refrigeration and other services. The variety of commercial businesses is in itself responsible for wide-ranging energy use: retail stores, restaurants, lodging, offices, etc. are all in this group. While commercial buildings heat space and water much the same as residences, the larger spaces of commercial buildings dictate that heating and cooling systems be optimally sized. Much energy is expended simply moving energy from one place to another.

A distinction for many commercial businesses is the increased heat energy given off by two factors: 1) intensive lighting (to boost worker productivity or consumer sales) and 2) high numbers of people. Large office buildings require cooling, even on winter days. Reducing excessive lighting saves energy two ways: using less light energy and cooling energy. Lighting research continues to spur advances in technology that now give us lower energy lights which illuminate better. Homes are now taking advantage of these same lights.

Appliance energy use is another application that crosses over both residential and commercial technologies. The pumps, motors and compressors in commercial appliances are simply bigger or run more often. Again, larger commercial appliance applications initially spurred research into ways to cut those energy costs. Current research into improving pump, motor and compressor efficiency is yielding appliances that perform as well or better with reduced energy inputs.

Industrial Use

Industrial firms use energy to transform raw and unfinished materials into products. Raw material or process firms are the major energy users. They include chemical, food, lumber, and minerals. Most industrial energy use occurs at just a few sites.

Fuel use accounts for most industrial energy consumption. Fuel is burned in boilers to make steam for cooking, drying, and other processing. Direct fire process heat, in which hot gases transfer heat, is another major fuel use.

Energy is used to power machinery, process materials and heat or cool buildings. Electricity powers motors to compress, convey, cut, grind, freeze and pump. Lighting and space heating claim a tiny share of industrial energy use.

Table 2 lists the largest industrial customers in the region. It also includes non-aluminum industries typical of other similar plants throughout the region. Plants in Table 2 are all direct service customers of the Bonneville Power Administration. What separates them from other industries, however, are the aluminum plants.

One group of industries unique to the Pacific Northwest is aluminum smelters. No other region of the United States has these energy intensive plants. In fact, quipsters note that these smelters really convert electricity into aluminum, given their energy-intensive product. Smelters usually are located just outside cities. One exception is the smelter next to the Columbia River near Goldendale, Washington, where the availability of both raw material (via rail and barges) and electricity from nearby John Day Dam dictated this location.

SPATIAL PATTERNS

Humans use energy where they live, work and play. A map of energy use therefore coincides with the population map, only with different units of measure. Large circles indicating concentrated industrial and commercial energy use are found in all major metropolitan areas. Ringing large cities is a more diffuse residential energy pattern consistent with suburban growth. Farther out is agricultural energy use, typically for irrigation.

Energy use in the Pacific Northwest is concentrated first along the north-south Interstate 5 freeway corridor. East-west interstates like I-90 and I-84 show secondary levels of inhabitants and subsequent energy use. Next, highways link smaller cities and towns, with their attendant tertiary level of energy use for homes and businesses. Agricultural energy use is concentrated on dry-land farming sites east of the Cascades where irrigation is common.

Transportation energy use nearly exactly duplicates traffic patterns. High commuter freeway use in major cities and suburbs is most energy intensive. The category of secondary highways, with much less use, is followed by the rural road network. Airline and rail energy use again follows large cities.

SUPPLIES

Hydroelectricity

Though much of the Pacific Northwest has a semi-arid or arid climate, mountain ranges trap moisture from wet Pacific air and passing weather systems. This orographic phenomenon results in significant precipitation along the Pacific Coast and at higher elevations throughout the region. Precipitation appears as both snow

	Company Name	Plant Location	Primary Products	Contract Demand (MW)	Electricity Capacity (MW)
ALUMINUM	Alcoa	Wenatchee, Washington	Primary Aluminum	360	216
	Columbia Aluminum	Goldendale, Washington	Primary Aluminum	296	279
	Columbia Falls Aluminum Co.	Columbia Falls, Montana	Primary Aluminum	427	343
	Intalco	Ferndale, Washington	Primary Aluminum	468	459
	Kaiser	Mead, Washington	Primary Aluminum	738	411
	Kaiser	Tacoma, Washington	Primary Aluminum	1	149
	Northwest Aluminum	The Dalles, Oregon	Primary Aluminum	174	161
	Reynolds	Troutdale, Oregon	Primary Aluminum	701	253
	Reynolds	Longview, Washington	Primary Aluminum	2	419
	Vanalco	Vancouver, Washington	Primary Aluminum	235	230
	Total Aluminum			3,399	2,920
NON-ALUMINUM	Alcoa, Northwest Alloys	Addy, Washington	Ferro-Silicon and Magnesium	3	67
	ACPC	Vancover, Washington	Aluminum Cable	5	3
	Carborundum	Vancover, Washington	Silicon Carbide	34	0
	Georgia Pacific	Bellingham, Washington	Chlorine and Caustic Soda	34	34
	Gilmore Steel	Portland, Oregon	Ferro-Alloys & Calcium Carbide	30	0
	Kaiser	Trentwood, Washington	Aluminum Plate	1	67
	Nickel Joint Venture	Riddle, Oregon	Ferro-Silicon, Nickel	120	103
	Oremet	Albany, Oregon	Titanium	18	11
	Pacific Carbide	Portland, Oregon	Calcium Carbide	9	0
	Atochen	Portland, Oregon	Chlorine and Caustic Soda	84	78
	Port Townsend Paper Corp.	Port Townsend, Washington	Paper Products	17	13
	Alcoa, Vanexco	Vancouver, Washington	Aluminum Extrusions	3	4
	Total Non-Aluminum			351	380
	TOTAL			3,750	3,300

Notes: 1) Included with Kaiser Mead, 2) Included with Reynolds Troutdale, 3) Included with Alcoa Wenatchee, 4) Plant has been dismantled, 5) Currently served by PGE, 6) Contract demands not reduced for expected conservation/modernization program savings.

Table 2. Bonneville Power Administration's Direct Service Industries.

and rain, providing for great hydroelectric potential: large volumes of snowmelt and rain flowing from high elevations to sea level.

One major river system, the Columbia, provides the bulk of the Pacific Northwest's hydropower. The Columbia system drains the western slopes of the Rocky Mountains in British Columbia, Idaho, Montana and Wyoming. It also drains slopes of the Cascades in Washington and Oregon. Prominent Columbia tributaries include the Snake, Willamette and Yakima rivers. Figure 1 shows the Columbia River drainage basin and power plants in and adjacent to it. The Columbia River's discharge, about 165 million acre-feet per year on average, is second only to that of the Mississippi system in the U.S. This enormous flow, coupled with significant elevation drop, accounts for the great level of hydroelectric power in the Pacific Northwest.

In addition to the Columbia River system, hydroelectric plants are found on shorter streams draining western slopes of both the Cascade and Coast ranges. For example, the Umpqua and Rogue rivers in Oregon and Klamath River in Northern California provide sizable amounts of electricity to regional customers.

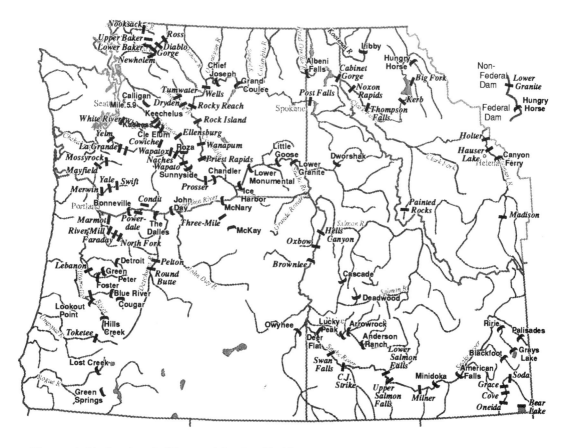

Figure 1. Hydroelectric Plants of the Pacific Northwest.

Outside the region, other rivers supply our energy needs. The McKenzie system drains the eastern slopes of the Rocky Mountains in northeastern British Columbia and in Alberta. Individual rivers, including the Athabaska, Peace, and Laird drain to the McKenzie which flows northwest into the Arctic Ocean. Developments on the Peace River provide much of British Columbia's hydropower resources. B.C.'s hydropower in turn supplies our region via transmission lines.

Development of dams for energy plants began in the Pacific Northwest in the late 19th century. Early developments used the force of falling water to push huge grinding stones and make flour from wheat. The Butte Creek Grist Mill in Eagle Point, Oregon is one such pioneer operation. It continues to operate today as it has since 1872. Shortly after the Butte Creek Grist Mill opened, small dams with Pelton wheels and, later, turbines, were built for the sole purpose of producing electric power. They were located at sites close to users because of limited transmission capability. Many continue in use today, often with the original generating equipment. Figure 2 shows the relatively early hydroelectric facility at Willamette Falls, Oregon.

The federal public works program of the Great Depression prompted massive dam development on the Columbia. Perhaps best known are the Bonneville and Grand Coulee dams on the mainstem Columbia. Dam building was justified under a range

Figure 2.

of objectives. Power generation budgets paid for part of the cost along with irrigation, navigation, flood control, and recreation funds. When federal dam building slowed, utilities, both public and private, stepped in to continue large dam projects in the region. Dam projects on major rivers of the entire U.S. Pacific Northwest continued until the mid-1970s, by which time acceptable sites were used up and conflicting uses began to be recognized.

Table 3 lists large federal system hydroelectric projects which provide the foundation for Pacific Northwest energy supplies. Nine projects operated by the U.S. Bureau of Reclamation have a combined average capacity of over 2,000 megawatts (MW). Twenty-one projects operated by the U.S. Army Corps of Engineers have a combined average capacity of over 5,100 MW. Bonneville Power Administration (BPA) markets the electricity generated at these dams.

Table 3 also points out the difference between the maximum power a plant can produce (peak capacity) and what it typically generates in a year (average energy). The Northwest's ample hydropower base has tremendous capacity, but because it can't be sustained without draining reservoirs, average energy is a more accurate indicator of true power output. Hydroelectric plants average about 35 percent of rated capacity. This is referred to as capacity factor.

Table 4 lists all sources of primary Northwest hydroelectric supplies. Private electric utilities have an additional 3,100 average MW of large and small hydro projects (shown on Figure 1) in the Pacific Northwest. Public or consumer-owned electric utilities have an additional 15 large and small hydro projects, totalling about 2,000 average MW.

Most recent large-scale hydroelectric power comes from adding generation capacity at previously developed sites. Small scale hydroelectric plants account for nearly all new dams since the late 1970s. Rising oil prices, cost increases for thermal power plants, and wise national policies promoting development of renewable energy created the environment for small hydro plants. Smaller, more remote and otherwise previously overlooked sites were developed by both independent developers and utilities. Many such sites were not developed due to fish concerns, declining utility power prices and a subsequent surplus of energy.

The regional hydro system generates an average of about 12,400 MW of firm power under critical water conditions. Projects range in size from less than a megawatt to the Grand Coulee project of nearly 6,700 megawatts capacity. All hydroelectricity is limited to the amount of water flowing down the river. So while a dam like Grand Coulee can have a huge capacity, energy generation is limited to how much water passes downstream at any given time. Thus, regions heavily dependent upon hydroelectricity, like the Pacific Northwest, have abundant generating capacity and limited energy. One way to optimize hydroelectric supplies is coordinated operation of all projects on a river.

Where topography was suitable, projects were designed to store large amounts of water. These storage projects permit water flows to be regulated to improve power

Project	Initial Year of Service	Number of Units	Generating Capacity (peak MW)	Firm Energy (aMW)
U.S. Bureau of Reclamation				
Grand Coulee	1941	33	6,998	1,875
Hungry Horse	1952	4	428	98
Palisades	1957	4	164	61
Anderson Ranch	1950	2	30	11
Minidoka	1909	7	16	9
Roza	1958	1	13	4
Chandler	1956	2	13	7
Black Canyon	1925	2	10	7
Subtotal		55	7,672	2,072
U.S. Army Corps of Engineers				
Chief Joseph	1955	27	2,614	1,156
John Day	1968	16	2,484	911
The Dalles	1957	22	2,074	723
Bonneville	1938	18	1,186	477
McNary	1953	14	1,127	662
Lower Granite	1975	6	932	217
Lower Monumental	1969	6	930	218
Little Goose	1970	6	932	188
Ice Harbor	1961	6	693	177
Libby	1975	5	600	205
Dworshak	1974	3	460	164
Lookout Point	1954	3	138	24
Detroit	1953	2	115	34
Green Peter	1967	2	92	22
Lost Creek	1975	2	56	23

Table 3. Federal System Hydroelectric Projects.
Source: Bonneville Power Administration, 1994.

Project	Initial Year of Service	Number of Units	Generating Capacity (peak MW)	Firm Energy (aMW)
Lost Creek	1975	2	56	23
Albeni Falls	1955	3	49	26
Hills Creek	1962	2	35	14
Cougar	1964	2	29	12
Foster	1968	2	23	10
Big Cliff	1954	1	21	10
Dexter	1955	1	17	8
Subtotal		149	14,607	5,306
Total Federal Projects		204	22,279	7,378

Table 3. Continued.

production. Major storage projects are located at upstream sites on the Columbia River. Other sites were developed as run-of-river projects. These have relatively limited water storage capacity and generate power strictly in response to stream flow. Operation of downstream run-of-river projects is generally coordinated with that of upstream storage projects. Major dams with substantial water storage on the upper Columbia River expanded the hydropower base of British Columbia and improved the generating capability of downstream U.S. projects. Canadian and U.S. Columbia River dam operation is coordinated by treaty.

Source	January Peak	Annual Average
Federal System	22,279	7,378
Public Utilities	4,461	2,029
Private Utilities	7,026	3,107
TOTAL	33,766	12,514

Table 4. Primary Pacific Northwest Hydroelectric Supplies.
Source: Bonneville Power Administration, 1994

Though hydroelectric systems are often expensive to construct, hydropower operating costs are generally the lowest of any major generating resource. This is due to the "free" cost of the fuel. However, all energy sources have social costs. The direct social costs of hydroelectricity that were formerly ignored are best shown by the controversy surrounding two dams on the Elwha River in Washington: the Glines and Elwha dams built on that river have no fish passage facilities. As a result, all anadromous fish runs there died off and federal agencies are seriously considering removing the dam in order to restore fish runs. On a much larger scale, many dams on the Columbia system with fish passage facilities may still be responsible for endangering salmon runs. This issue is far from settled, but utilities and irrigators now accept responsibility and are taking steps to restore fisheries.

The latest chapter in the electricity industry is unfolding at the national and international levels: deregulation. Electricity has since the 1930s been a heavily regulated monopoly industry. Major changes worldwide are underway. Foreign countries are selling state-run electric utilities. Nationally, mergers between utilities of all kinds are occurring. Regionally, more open competitive markets have lead Puget Sound and Portland area electric utilities to seek mergers or buyouts. All electric utilities and many large industrial buyers now have access to multiple suppliers and new kinds of services. Whether the same benefits will accrue to residential customers remains to be seen. Each state is deregulating the industry in slightly different ways.

Natural Gas and Oil

Petroleum and natural gas are found in porous rock formed from sedimentary accumulations in geologic basins. Finding oil is difficult, as many times it is covered with a "cap rock" of impervious strata. Gas found in these formations was derived from organic material accumulated in the basins along with the inorganic sediments that eventually formed the present rocks.

The Pacific Northwest has limited petroleum and natural gas resources. The Mist gas field in northwestern Oregon is the only commercial field in the region. (Eastern Montana produces oil, gas and coal.) Instead we get most of our natural gas from British Columbia and the Rocky Mountains. Long-distance gas transmission pipelines lead to the Pacific Northwest from Alberta, British Columbia and Colorado. One major oil pipeline starts at refineries in the Puget Sound and send product down the I-5 corridor as far as Eugene, Oregon. Other large oil pipelines connect Montana with eastern Washington and along the I-84 corridor between Walla Walla and northern Utah.

Gas is extensively used as a fuel for homes, industry and new electric power plants. Where available, natural gas supplies space and water heating in most new homes in the Northwest. Many industries have "dual fuel" boilers capable of burning either oil or gas, depending on price. Although historically, natural gas markets have been heavily regulated, major changes made in the late 1980s promote more open, competitive markets. Buyers now have access to more suppliers and new kinds of services. As a result, while plenty of gas will be available to Northwest buyers they will have to compete for it.

Gas may be used to generate electricity using several generating technologies, including steam-electric plants, combustion turbines and cogeneration. Steam-electric technology involves burning gas in a boiler to first make steam, which drives a turbine and generates electricity. Most older U.S. plants use this technology. Combustion turbines are modified jet engines that burn gas or oil and, instead of boiling water into steam, use the combustion gases to drive turbines and generate electricity. Virtually all new utility-scale power plants use commercially available combustion turbines with heat recovery steam turbines. A typical application, such as found next to Intersate 84 at Hermiston Oregon, has a combustion turbine generating 170 MW of power in series with a steam turbine generating an additional 80 MW from the exhaust gas heat, and finally steam sales to an adjacent food processing plant. Total plant efficiency approaches 45 percent, compared to 33 percent for steam turbine plants. Cogeneration refers to a simultaneous use of steam for process and electricity. Older applications involved steam for a turbine plant and low pressure steam for an adjacent lumber kiln dryer. New cogeneration plants are much like the Hermiston project described above, with combustion turbines and a food processor.

Natural gas plants are relatively simple in design, inexpensive to construct and readily adapted to use either gas or oil. This gives plant owners flexibility to switch fuels as prices dictate. A new combustion turbine should be available 85 to 90 percent of the year. Because of high variable costs, combustion turbines have historically operated only about one-third of any given year. Because of their relatively low efficiency (and subsequently high variable costs), combustion turbines are generally designed to operate during peak demand times. If natural gas prices stay low, these plants will operate on a regular basis.

Plants using natural gas required little air pollution control equipment until recently when nitrogen oxide emissions became regulated. Carbon dioxide contributions still result from gas plants, affecting global climate change.

There are now 11 combustion turbine plants in the Pacific Northwest, with a total capacity of 2,433 MW. Average plant size is about 220 MW. Average plant output is 72 MW, indicating an approximate 33 percent availability for these plants. Plant sizes range from 1 MW up to the 534 MW Beaver plant in Oregon. The 20 MW cogeneration unit at Great Western Malting in Vancouver, Washington is a good example of recent appropriate-sized gas cogeneration plants.

Biomass

The Pacific Northwest has an abundance of biomass resources which provide fiber, fuel and food to the region. Primary biomass resources are wood, agricultural and solid waste and sewage. Other potential sources include grass straw and logging residues.

Wood energy content varies according to how much moisture is present. Completely dry wood contains about 8,500 British Thermal Units per pound ((Btu/pound). Green wood has about 50 percent moisture content. Dry cordwood typically still has about 20 percent moisture, and hence contains about 6,800 Btus/pound.

Perhaps the best known bioenergy use is wood stoves. New wood stoves pollute less and are more energy efficient than old stoves. About one-third of the region's households use wood heat on a regular basis. Another 15 percent have fireplaces, which provide aesthetic value more than heat.

Wood wastes have long been used to supply steam and power in the lumber and pulp and paper industries. These wastes are a byproduct of forest products mill operations and include bark, sawdust and shavings. Low-value wastes (not otherwise useful for fiber) are ground up in a "hog" and called hog fuel. Typical energy content is low due to average 50 percent moisture content. Table 5 shows estimates of industrial wood energy use by state. The internal operations of forest products firms are accounted for here. These substantial bioenergy users used to consider their current fuel waste for many years, burning it in "wigwam" style incinerators.

In recent years several independent power plants and one utility plant fueled by wood wastes were built in the Pacific Northwest. Plants range in size from 3 to 50 MW. All but the 47 MW Kettle Falls plant in Washington are related to the operation of forest product firms and cogenerate both steam and electricity.

With greater constraints developing on landfill disposal of municipal solid waste, more and more of this waste is being burned to generate electricity. The municipal solid waste burner built in 1986 in Brooks, Oregon supplies 11 MW of power to Portland General Electric customers. The plant operates 365 days a year, reducing garbage to 15 percent of its previous volume. Tacoma City Light renovated an old 50 MW power plant to burn a combination of wood, coal and municipal solid waste. Regional municipal solid waste burners that recover energy are listed in Table 6.

State	Industrial Plants	10^6 Btus per Year
Oregon	137	142,993
Washington	59	64,608
Idaho	34	25,900
Montana	31	11,288
TOTAL	251	244,789

Table 5. Primary Pacific Northwest Bioenergy Use.
Sources: Idaho, Montana, Oregon & Washington State Energy Offices, 1987–1990

Site	Tons Per Day	MW capacity
Spokane, WA	680	22
Brooks, OR	550	11
Tacoma, WA	270	50
Skagit, WA	137	2
Bellingham, WA	100	1

Table 6. Pacific Northwest MSW-to-Energy Plants.
Source: Washington State Energy Office, 1992

Growing quantities of assorted biomass fuels such as methane from sewage treatment and landfills are being used to generate electric power. Methane or biogas results from the sewage digestion process. Biogas is lower in energy content than natural gas, but still quite usable. Most large sewage treatment plants use methane to heat the digesters and optimize the process. One innovative arrangement has a roofing company buying gas from a neighboring sewage plant in Portland.

Landfills generate methane as a result of decomposition. Capturing that methane and putting it to use takes an environmental hazard and creates energy from it. The Emerald Peoples Utility District in Oregon started a 3.2 MW landfill gas power plant in 1991. The utility invited developers to build this project using the Short Mountain landfill in Eugene. Wells and collection pipe lead to modified tractor engines which generate electricity.

Biomass development costs vary widely, depending on use. Equipment to store, prepare, convey and combust fuel is needed, with air pollution control equipment, chiefly for particulates or small particles, necessary for large uses.

Geothermal

Geothermal energy is the heat of the earth. Geothermal resources take many forms: hot springs, steam geysers, fumaroles and boiling mudpots. Even normal groundwater, at about 50° Fahrenheit, contains useful geothermal energy.

Geothermal energy is useful to heat buildings and domestic hot water, and to generate electricity. Geothermal energy heats buildings by passing hot spring water through furnace coils, room radiators, or pipes under floors. More common is using geothermal water to heat a closed loop of makeup water for those applications. Geothermal energy meets the same needs as are met by fossil fuels such as oil and natural gas.

Geothermal heating systems are more efficient than combustion units, so customers use less energy. Because user buildings neither store nor burn fuel, geothermal systems are safer than fossil fuel systems. Geothermal energy is relatively clean, with controllable air pollution impacts. Air quality improves in areas where fossil fuel and wood burners convert to geothermal heating.

Geothermal energy is widespread in the Pacific Northwest. Hot springs exist from the western Cascades of Oregon and Washington to the high desert valleys of Idaho and Montana. Figure 3 shows geothermal sites in the Northwest. Geothermal resources were used by native Americans long before settlers arrived. Place names such as Warm Springs Avenue in Boise, Idaho and Hot Springs Addition in Klamath Falls, Oregon indicate how long those cities have used geothermal energy. The entire Idaho Capital Mall complex of state buildings uses 165°F water for heating, saving over $100,000 of fossil fuel costs.

About 1,165 billion Btu of geothermal energy now supplies homes, schools, commercial and institutional buildings throughout the region. Table 7 shows estimates of geothermal energy direct use by state. These estimates do not include water- and ground-source heat pumps. These devices use 50°F ground temperatures to supplying heating and cooling.

How renewable is geothermal energy? A look at a local situation best answers this question. The 500 plus geothermal users in Klamath Falls Oregon recognized in 1985 that they were depleting the local geothermal resource. Under an ordinance passed that year which took effect in 1990, all users must reinject if they withdraw geothermal water. That is, no more geothermal water is pumped to the surface and then wasted: if pumped, it is put back into the ground. Significant reinjection is already working, with a reported average rise in water level of almost 10 feet in 10 years.

Electricity can be generated using geothermal steam to drive turbine generators. In power generation, the earth simply acts as the boiler. Geothermal power plants get steam from the ground. Thermal generating equipment—steam or hot water wells—is unique to geothermal power. An analogy would be the case of a gas power plant located next to the gas wells. Instead of gas (and its requisite combustion), geothermal steam comes out of the ground ready to drive a turbine and generate electricity. Figure 4 shows a geothermal well being tested at Newberry Volcano in Oregon. Pollution control equipment is only slightly different from that at fossil fuel plants. The cost of geothermal plant pollution controls is similar to that of fossil fuel plants, and is included in cost estimates.

Generating power from geothermal resources is typically done in several major stages. Environmental permitting occurs before each stage and activities overlap between stages. Typical stages include: Exploring for the resource; Developing the wellfield and gathering system; Constructing the power plant and related facilities; Operating and maintaining resource supplies; and Operating and maintaining plant facilities. New developments in California, Nevada and Utah all use state-of-the-art air pollution controls. Design and management at newer plants also stress conserving resources as renewable and not exploiting them as was done, for instance, with natural gas.

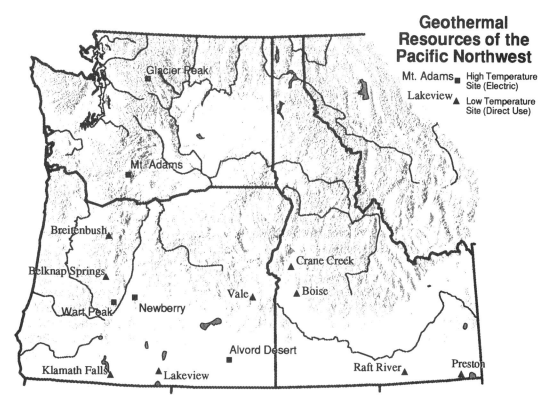

Figure 3. Geothermal Resources of the Pacific Northwest.

State	Applications	10^6 Btus per Year
Idaho	45	629,100
Oregon	39	409,100
Montana	19	125,600
Washington	2	1,200
TOTAL	105	1,165,000

Table 7. Pacific Northwest Geothermal Energy Use
Sources: Lienau, Culver & Rafferty, 1990

Geothermal resources suitable for electric power development likely occur in the Cascade mountains of Oregon and Washington. Recent plants in similar environs (transition zone between High Cascade and High Desert provinces) offer exciting opportunities for environmentally acceptable development in the Northwest. Two recently completed power plants are near Mammoth Lakes, California. Combined with the first plant there, almost 30 megawatts of power are generated representing a truly renewable geothermal electric project. Pacific Energy Company operates three plants that use air cooling, and reinject 100 percent of the 330°F water back into the ground. The project is in a popular high elevation recreation area, akin to sites in central Oregon and Washington.

Lower temperature steam and hot water at about 300°F, still useful for electric energy production, is scattered throughout the Basin and Range province of Oregon, Nevada and Utah.

Both developers and electric utilities in the region investigated geothermal energy as a supplemental source to their existing power plants. Geologically speaking, regional geothermal power potential at the 20 most promising sites could supply as much as 4,000 average MW. Economically speaking, all that potential is expensive at a low cost of about 8 cents per kilowatt-hour (kWh). Geothermal energy costs should remain level or drop in the future, at which time it should be more cost-effective (competitive with other sources). A 30 MW project at Newberry Volcano ended in 1996 with high temperatures but lack of sufficient resources, at least based on two deep wells.

Solar

The clear skies of the inland Pacific Northwest provide significant solar energy. Areas such as Bend, Oregon and Spokane, Washington receive 80 percent of the solar radiation measured in southwest states such as Arizona. Until recently, BPA funded a solar radiation monitoring network with stations throughout the northwest.

Solar hot water heaters are found on 14,500 homes in Oregon alone. Throughout the region, perhaps three to four times that amount or over 50,000 homes use solar domestic hot water heating. Probably the simplest way to use solar energy is to orient houses facing south. Direct solar gain provides significant heating for residential and also commercial buildings. The Lee Metcalf state office building in Helena, Montana is solar-assisted with a conventional facade.

Utility use of solar power for central station generation of electricity is growing as evidenced by projects in California. Central station solar receiver and parabolic dish plants now total 450 MW there. Capital costs have dropped from $15,000 per kW in 1984 to $3,500 per kW in 1988. Target costs are $1,500 per kilowatt (kW) by 1998. At the same time, capacity factors have risen from 25 percent to 35 percent. Photovoltaic technologies likely hold the greatest promise for widespread use.

Photovoltaic technology is commercial for a growing number of uses. Commercial photovoltaic applications today include aircraft and navigation beacons, area and

Figure 4. Geothermal well

security lighting, livestock water pumping, and transmission line switches. Technology costs for photovoltaic solar cells have dropped from $500,000/kW for the first satellites in the 1960s, to about $50,000/kW in 1980, to about $3,500/kW today. The industry target is $1,000 in the late 1990s. It should be noted that the last decade's advances were made with little support from the federal government.

Many remote homes in rural areas of the region use photovoltaic power simply due to the high cost of connecting to the local utility. One forecast predicts 40,000 units will be sold each year over the next 5 years. Pacific Gas & Electric Co. in northern California has 840 photovoltaic applications now. Figure 5 shows a commercial-scale photovoltaic installation at a BPA substation in Redmond Oregon.

Solar energy costs will continue dropping in the future, as photovoltaic technology continues efficiency gains. Northwest solar energy is cost-effective using new PV technology at remote sites. Individual applications in more populated areas of the region will grow in proportion to natural gas costs.

Four Northwest utilities are participating in a demonstration photovoltaic water pumping project. Remote sites with irrigation and cattle water needs are to be fitted with micro PV (<1 kW) systems. Costs are expected to range from $2,000 to $15,000 per site. BPA and three rural electric cooperatives, Glacier and Vigilante in Montana and Oregon Trail in Oregon, are participating.

Wind

Wind resources suitable for electric power production are found in many locations throughout the Pacific Northwest (Figure 6). Ideal wind resources are not necessarily found in frequently mentioned areas, i.e., the coast. In fact, too much wind is not useful, and corrosive environments add significant costs. Yet one advantage to coast sites is the winter-peaking nature of the storms which produce wind. Consistent, moderate winds are most desirable. The eastern Columbia Gorge area is one such area. The Gorge's new sizable sail board industry indicates that wind resources in the area are already world-renowned. Northwestern Montana holds the region's greatest potential, specifically Blackfeet Tribe Indian Reservation lands.

Knowing wind strength is crucial to the cost and quantity of power. The amount of energy in the wind rises with the cube of the speed. Because wind turbines can capture only a portion of the energy, annual energy rises with the square of annual average wind speed. Even so, small differences in wind speed cause large differences in power output. Because wind turbines are capital intensive, those same small differences in wind speed cause large differences in the cost of power. As a result, wind developers consider good sites to be those with annual average wind speeds of 14 to 16 miles per hour, measured 40 feet above the ground.

Wind power is an old technology. The Dutch have been using wind power for well over a century. Pioneers in the Northwest used wind to pump groundwater for irrigation. Commercial development in California began with about 7 MW of power on line in 1981. Today over 17,000 turbines exceeding 1,500 MW of capacity operate

Figure 5. Solar Application

at California wind farms, primarily at Altamont, Tehachapi and San Gorgonio passes. California wind farms now use second generation machines and will have third generation commercial machines of even greater efficiency on line soon.

Three prototype 2.5 MW Boeing MOD-2 turbines were erected at Goodnoe Hills in southcentral Washington. Sponsors included the US Department of Energy, BPA, NASA, and others. The project field tested machines, materials and noise. Site development began in May 1980 and experiments ended in 1986. MOD-2 research confirmed the inherent viability of small scale (up to about 400 kW) wind turbines.

The Whiskey Run wind project was built in 1983 on the southern Oregon Coast and is the only commercial wind farm in the region to date. The project originally started with 25 wind turbines with a total rated capacity of 1.25 MW.

Regional wind energy potential is something over 4,500 average MW. There are sixteen sites that could supply about 3,200 average MW at less than 10 cents per kWH. Wind energy costs will continue dropping in the future, as new generation machines come on line. Northwest wind energy could be cost-effective with new machines at identified sites.

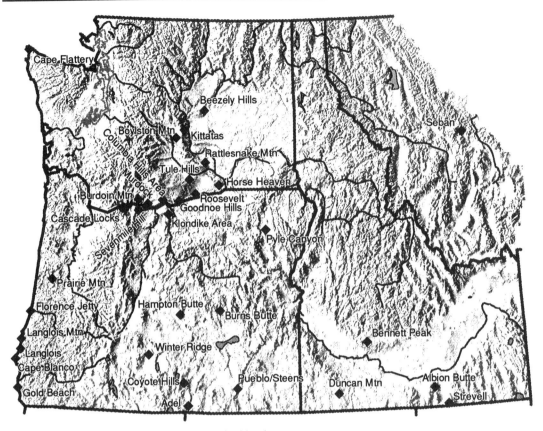

Figure 6. Wind Resources of the Pacific Northwest.

Coal

While the Pacific Northwest contains isolated deposits of coal, only Washington has commercial coal deposits. (Eastern Montana and Wyoming have significant coal mines, which in turn provide electricity to the Northwest.) The two coal mines in Washington are Bucoda in Thurston County and the Centralia field in Lewis County. Pacific Northwest coal deposits are relatively low in energy value. This partly explains the region's lack of widespread coal use, either directly to heat homes and run factories or to generate electricity.

The amount of heat coal contains is critical in determining its value for supplying energy. Heat content varies by coal type: bituminous, subituminous and lignite. Bituminous coals have heating values in the range of 10,500 to 14,000 British Thermal Units (Btu) per pound. These coals are found near the east slopes of the Rocky Mountains. Subituminous coals have heating values ranging from 8,300 to 11,500 Btu/lb and are found farther to the east of the Rocky Mountains and in the Centralia

Figure 7.

field of Washington. Lignites (heating values less than 8,300 Btu/lb.) are found in northeastern Oregon and Washington, and along the mid-Oregon coast.

Western coal is well-suited for electricity production. The seams of the Great Plains, Rocky Mountains and Colorado Plateau are often thick, flat and near the surface allowing the coal to be easily mined. Major deposits there supply electricity for our region via transmission lines. Coal and power plants in the Powder River and Green River Basins can be considered "part" of our region.

Sulfur, which naturally occurs in coal, forms oxides of sulfur when burned. Sulfur oxide released to the atmosphere may form sulfuric acids, suspected of contributing to "acid rain" and its attendant effects on the environment. Fortunately, western coals generally have very low levels of sulfur, and require the minimum level of sulfur removal (70%) when burned in new plants complying with the Clean Air Act. Western coals are also characterized by low ash and sodium—desirable properties for combustion.

Unlike in the East, coal was not widely used for electricity generation in the Northwest prior to the 1960s. With development of most of the cost-effective hydropower completed by the 1960s, utilities began to turn to coal to meet their baseload needs. This trend accelerated during the 1970s with the rapid increases in petroleum and natural gas costs and enactment of the Fuel Use Act of 1978, which forbade most uses of petroleum and natural gas fuels for baseload applications. There are now two coal-fire power plants with a peak generating capacity of nearly 2,000 megawatts currently supplying the Pacific Northwest. Unit sizes are 560 MW at Boardman and 1,460 MW at Centralia.

The Boardman coal plant is a good example of cooperative regional energy supply. It was built and operated by private Portland General Electric. Other partners include Idaho Power Co. and Pacific Northwest Generating Co. (PNGC). PNGC is a group of Idaho, Oregon and Washington public utilities. Ownership is thus shared by both public and private utilities in three states. The 530 MW plant is in Northeast Oregon and burns Wyoming coal.

Due to declining load growth and increasing construction costs, orders for new coal units declined rapidly in the late 1970s. Like their nuclear counterparts, a number of coal-fired plants have been canceled in recent years. The last large coal plant proposal affecting the Pacific Northwest ended in 1991. Still, some schemes call for resurrecting an old coal plant proposal in eastern Washington.

Fuel transportation costs are a significant component of total electricity costs for a coal plant. Coal plant siting must therefore balance the cost of coal transportation with the cost of electricity transmission to load centers. Other important factors include air quality and the availability of water for waste heat removal. Load centers are avoided because of existing air quality problems in these areas. For these reasons, coal plants tend to be sited farther from load centers than nuclear plants (Figure 1).

Nuclear

Nuclear power plant construction trends in the Pacific Northwest follow nationwide trends. Following construction of a few midsize commercial designs in the 1960s, unit sizes were rapidly scaled up in pursuit of economies of scale. Nuclear plant orders were placed in response to rapidly growing loads, on the basis of erroneous cost advantages, and without regard to waste disposal. Now nuclear plants under construction are either cancelled or mothballed or converted to gas. No new plants are being ordered.

In the late 1970s and early 1980s, forecasted rates of load growth failed to materialize. Massive increases in construction costs and longer than anticipated construction schedules eroded nuclear's cost advantage. Any remaining units were only completed years behind their originally scheduled completion dates.

The Pacific Northwest has a place in history with the Washington Public Power Supply (WPPSS) nuclear plant bond default. WPPSS started building five nuclear plants at two locations in Washington. Capital to build the plants came from municipal bond sales. As in the case of the Boardman coal plant, WPPSS sold interests in these plants to other utilities and gained financial backing from BPA. Fortunately for WPPSS, BPA backing of plants 1, 2, and 3 allows it to avoid defaulting on bonds for those plants. Pacific Northwest ratepayers will pay for those bonds even though only one plant, WPPSS 2, was ever completed. The remaining two plants were terminated. Only BPA, i.e., ratepayer, backing of WPPSS 1 and 3 has avoided a default on those plant bonds. Construction of two of the plants, WPPSS 4 and 5, was terminated. WPPSS did not have guarantees behind plants 4 and 5 and defaulted on the bonds, causing the largest municipal bond default in history.

Trojan nuclear plant in western Oregon was shut down permanently in November 1992. WPPSS-2 in eastern Washington produces about 660 average MW of capacity when it operates. This represents about three percent of the region's generating capacity. No additional nuclear plants are planned or under construction in the Pacific Northwest. Although these plants were designed to be available for operation 80 percent or more of the time, actual operating experience has been disappointing, and availabilities of 55 to 65 percent have been typical. Also, the issue of disposal of the radioactive wastes generated by these plants is still not resolved. Trojan is the first commercial nuclear plant in the region to be decommissioned.

Imports

Interconnecting transmission lines with neighboring systems make power transfers between the Northwest and other regions possible. Figure 8 shows only the main BPA transmission grid, giving the reader some idea of how many lines traverse the region. Total firm resources available to this region account for the net effect of these transfers. These transmission interconnections also support sales of nonfirm (secondary) power to other regions. Nonfirm power sales, however, have no effect on the need for long-term, firm energy.

Interregional transactions involve the transfer of two basic commodities, energy and capacity. Energy transactions generally involve either outright sales of energy or exchanges of energy. Capacity contracts, on the other hand, are less straightforward, because they do not necessarily result in the use of the commodity (capacity). Capacity is the maximum power output for which a generating plant is designed to operate continuously. A utility may purchase the rights to all or part of the capacity of an out-of-region resource in order to ensure that it will have adequate generation to meet its daily peak demands. The purchasing utility may never call upon that resource for power, but it pays a fee for the right to the generation, even if no energy is ever delivered. If energy is delivered during peak hours, an equivalent amount of energy is then returned to the selling utility during off-peak hours. This type of transaction is more predominant among utilities whose firm resource mix is made up primarily of thermal resources.

Typically, transactions fit into five basic categories:

Capacity Sales. Payment is made in dollars for capacity provided during the peak demand hours of the day. If energy is delivered, an equivalent amount of energy is returned to the sending utility over the lightly loaded hours of the night and on weekends. No net energy is transferred between regions over the specified time period, usually a week.

Capacity/Energy Exchange: This transaction is similar to a capacity sale, but payment for capacity provided is made in energy instead of dollars. As in a capacity sale, capacity is provided during the peak demand hours of the day. If energy is delivered, an equivalent amount of energy is returned to the sending utility. Payment for the capacity provided is made in additional energy returned to the sending utility. This additional energy may be returned during the same week or during a different part of the year. This type of transaction represents a net energy import for the region.

Seasonal Exchanges. Capacity and/or energy are provided to a utility during a specified part of the year. An equivalent amount of capacity and/or energy is later made available to the sending utility during a different part of the year. Usually in these arrangements no money is exchanged. This type of transaction is most beneficial when used by two regions whose system loads peak in different seasons.

Firm Energy Sales. Energy is purchased on a guaranteed basis. Firm energy sales can be either long-term or short-term. Typically, transactions which span periods of time greater than 18 months are referred to as long-term sales. Energy may be delivered 24 hours a day or during the peak demand hours only. Sometimes energy is delivered only during a specified season of the year. Often these types of transactions also specify a maximum amount of capacity to be provided along with the equivalent amount of energy.

Economy (Nonfirm) Sales. Energy is delivered on an hour-by-hour as-available basis, usually scheduled one day in advance. These transactions take advantage of the diversity that exists in short-term operating costs due to different fuel sources in the different regions and the short-term variability in water supply in the Pacific Northwest.

Technically speaking, the region imports more firm power than it exports. This is due primarily to imported energy from Pacific Power and Light's thermal resources outside the region which are used to met regional loads. The sum of all the power exchanges represented a firm net energy import to the region of about 830 MW in

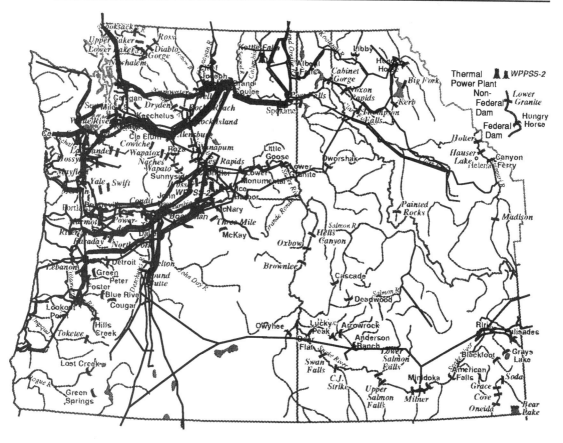

Figure 8. BPA Transmission Grid.

1996. This amount will change through time as old contracts expire and new contracts are put in place.

As the reader now knows, Pacific Northwest electricity comes from many resources. Table 8 summarizes those resources and the power they provide the region. The dominance of hydroelectricity is what separates our region from the rest of the country. This renewable energy foundation also provides an example of future options, particularly for wind, geothermal and solar resources. But, like the rest of the U.S., we need to first be more efficient with the energy we already use.

Conservation/Efficiency

It is only recently that conservation was thought of as a "supply." Conservation is in fact a tremendously significant source of new energy. Efficiency is another way of viewing how to conserve energy from existing uses. The incentive to conserve—or be more efficient—varies indirectly with energy prices.

SOURCE	Firm MW	%
Hydro	12,312	59
Coal	3,862	19
Imports	1,734	8
Cogeneration and NUG*	1,012	5
Nuclear	819	4
Natural Gas/Oil CT**	544	3
Misc.	363	2
TOTAL	20,646	

*NUG = non utility generators, e.g., lumber mills & small hydro plants
**CT = combustion turbines

Table 8. Pacific Northwest Generating Capacity (average capacity as of January 1996)
Source: Bonneville Power Administration, 1994

Historically, low electricity prices discouraged a high level of electrical conservation in the Northwest. Yet, beginning in the 1970s, heating bills were high enough to cause residential, commercial and industrial consumers to get an audit of energy use from local utilities. Based on the audit, steps could be taken to reduce heating costs.

Experiments in large scale energy conservation began in the mid-to-late 1980s. Utility conservation research measures range from weatherizing all the homes in a town (Hood River, Oregon) to jointly investing in industrial process equipment and sharing energy savings. Utility conservation programs have saved an estimated 350 MW to date. The timing was good as we entered the 1990s. Electricity supplies were scarce enough to eliminate reserves for many utilities. Conservation was bid along with other supplies in power supply solicitations from utilities. More will be coming: other areas with higher energy costs continue to provide examples of future options available to us.

State energy office programs saved an additional 200 MW. Aluminum and other direct service industries have conserved much as well. That brings a total of over 550 MW of conservation capacity online. This is as much as one new coal plant or several renewable energy plants. Such conservation is significant, and more potential follows.

Estimates of future electricity conservation supplies in the Northwest range from about 1,400 MW up to 2,900 MW. The range depends on which growth rate occurs, but appears to be on the low end for now. Further energy savings can be had in the transportation sector. Although gasoline prices are currently low, congestion more than gasoline costs may force us to conserve car use.

RESIDENTIAL CONSERVATION

Conserving home energy use does *not* mean doing with less. Residential conservation demonstrates this best. Residences use energy to heat and cool the living space, to heat domestic water, and for lighting and other miscellaneous electrical needs. Residential conservation does not mean lower thermostats. In fact, during a particularly cold spell in February 1989, the most efficiently constructed homes in the region (built to "model conservation standards") saved on average 2 kilowatts each day. More important, these larger than average homes had indoor daily temperatures that were kept higher than homes that used more energy. Cumulatively, these efficient homes cut regional demand for power by 200 megawatts. Energy savings also meant almost $7 million in that period. It is unknown how much additional gas, oil and even wood was saved by such well-insulated homes.

For existing homes, much can usually be done to reduce energy use. First and foremost is having an audit performed by the primary energy supplier or utility. From this inspection, a look at past heating bills, if available, and subsequent calculations, the auditor recommends specific measures. Conservation measures can be divided into no-cost, low cost and higher cost. Examples of no-cost measures include the following:

- cleaning furnace filters and vent registers;
- turning the thermostat down at night and when no one is home;
- covering crawlspace vents in the winter.

Examples of low-cost measures include the following:

- caulking around window and door exteriors;
- weatherstripping doors and windows;
- installing foam gaskets in electric outlet-switch covers;
- covering water pipes with foam sleeves; and
- putting an insulating wrap around the water heater.

Examples of higher cost measures include the following:

- adding insulation to the attic;
- adding insulation to the floor;
- insulating walls, and
- padding storm windows.

Each house is different and will only need some of these measures. Yet all existing residences can conserve some energy by adopting the most basic measures. Higher cost measures may or may not save energy quickly enough for the homeowner to afford them.

The Northwest Power Planning Council estimates 495 MW of residential conservation will be available by the year 2000.

COMMERCIAL CONSERVATION

Commercial sector users can conserve energy by reducing lighting and space conditioning energy. Swapping fluorescent bulbs and ballasts for indoor incandescents saves energy. Outdoor or other users of mercury vapor lights can switch to high pressure sodium or metal halide lamps to save energy. Installing economizers on air conditioning systems is the primary way of reducing that energy load.

The Northwest Power Planning Council estimates 435 MW of commercial conservation will be available by the year 2000.

INDUSTRIAL CONSERVATION

Industrial energy users conserve through reducing energy needed to manufacture products. Seven areas of industrial conservation include motors, refrigeration, pumping, compressed air, process heating, air conditioning and lighting.

The primary motor conservation measure is replacing standard efficiency motors with high efficiency motors. Refrigeration measures include reducing condensing pressures, increasing suction pressures and using automatic controls. The main pumping efficiency improvement is replacing existing units with variable speed drives. Reducing leaks and pressures of compressed air systems will save industrial energy. Insulating steam piping will reduce process energy use. Economizers on air conditioning systems apply here as well as the commercial sector, as does the swapping of fluorescent lights for incandescents and mercury vapor for high pressure sodium or metal halides.

The Northwest Power Planning Council estimates 235 MW of industrial conservation will be available by the year 2000.

FUTURE SUPPLY OPTIONS

The Northwest experienced a period of electric energy surplus in the 1980s. Electric utilities in the region entered the decade expecting a shortage that soon vanished. We are now in a period of load/resource balance, and even seasonal surpluses. Those same utilities now operate in the 1990s in a time of both growing demand and expanding supplies. Barring another recession, energy demand will rise at a modest rate. Growth is the driving force behind need for energy in this area.

Areas of high growth within the region need new power supplies first. The Puget Sound area of Washington is experiencing such growth now. It is therefore no surprise that Puget Sound Power and Light was the first utility to need power. Puget adopted the national trend and solicited bids to meet that need. Developers submitted bids to Puget in 1989. It selected several power suppliers in 1991, for resources that came online in 1993. Other utilities are now repeating the process to acquire power later in the decade.

Many supply options face Northwest utilities. There are nine supply options available to all utilities. Table 9 lists those options. Many investor-owned utilities intend to solicit bids like Puget and B.C. Hydro to the north have. For example, Puget also contracted with a 125 MW gas cogeneration plant. In contrast, most public utilities will continue to purchase power from BPA, which faces the same options as private utilities (Table 9).

Natural gas pipeline expansions occurred in 1993–1995 with a commensurate increase in regional gas supplies. Northwest Canadian gas supplies both our region and other western markets. Gas utilities and cogenerators lead new demand for this fuel.

Using conservation, natural gas and renewable energy the Pacific Northwest should be able to meet future energy needs without sacrificing our high quality of life.

- Efficiency/Conservation
- Transmission & distribution system improvements
- Contract options, i.e., recalling power sales
- Power purchases from BPA
- Purchases from other utilities
- Purchases from independents: gas
- Purchases from independents: renewable energy
- Purchases of other utilities
- Combining nonfirm hydro power with gas turbines
- Power plant capacity expansions

Table 9. Utility Power Supply Options

REFERENCES

Bonneville Power Administration, *1994 Pacific Northwest Loads & Resources Study*, DOE/BP-2522, Portland, OR, 1995.

Northwest Power Planning Council, *1991 Northwest Conservation & Electric Power Plan, Volumes 1 and 2,* Portland, OR 1991.

Home Power Inc., *Home Power—The Hands-On Journal of Home-Made Power*, Ashland, OR, various issues.

Idaho Department of Water Resources Energy Division, Montana Department of Natural Resources & Conservation Energy Bureau, Oregon Department of Energy & Washington State Energy Office, *Miscellaneous Biomass Energy Publications*, 1986–1996.

Lienau, P., Culver, G., & K. Rafferty, *Direct R&D Assistance Final Report September 1990*, Geo-Heat Center, Oregon Institute of Technology, Klamath Falls, OR, 1990.

Solar Energy Research Institute, *Photovoltaics: New Opportunities for Utilities*, prepared for U.S. Dept. of Energy, DE91002168, Golden, CO, 1991.

National Renewable Energy Laboratory, *Photovoltaics for Municipal Planners*, prepared for U.S. Dept. of Energy, DE93000099, Golden, CO, 1991.

Sparling, S., *Public Power Essentials*, Public Power Council, Portland, OR, 1987.

MINING Chapter 10

Robert Whalen

GEOLOGIC SETTING

Access to infrastructure and customers influences the location of mines. Some mines were developed in the Pacific Northwest because of the availability of inexpensive hydroelectric power. The main determinant of mine locations, however, is geology. Successful mines are developed in places where there are economic mineral deposits. Mining activity in the Pacific Northwest clusters around areas which are defined by their favorable geologic settings.

All rock contains metallic elements. Geological processes can separate these elements from one another and form concentrated zones of individual minerals. If these can be mined economically, these zones are called "ore deposits."

Metallic ores are found in nearly all rock types, but most are associated with a heat source. This is why many metallic ore deposits in the Pacific Northwest are found in or near igneous rock intrusions. Igneous rocks are created by the solidification of molten rock. If the molten material reaches the surface, it is called "lava" Underground, molten rock is called "magma."

Minerals precipitate out of solution from magma and other mineral-rich fluids. This usually is a gradual process prompted by subtle changes in the temperature, pressure, and chemistry of the fluid. To create a large mineral deposit, extensive amounts of fluids must be exposed to the appropriate conditions for deposition over very long periods of time. Typically, this happens at great depths. For such large deposits to be found and exploited by miners, they need to be brought closer to the surface by extensive erosion. This situation is what characterizes the geological setting in some of the major mining districts of Idaho. Washington and much of Oregon, however, are covered by thick layers of young, mineral-poor volcanic rock and, as a consequence, have had relatively few large metallic ore mines.

Water acts as carrier of minerals in the earth. Groundwater contains small amounts of dissolved metals and, under the right circumstances, can create large ore deposits. Magma itself often has a high water content. As magma solidifies, fluids rich in water and minerals tend to migrate away from the igneous intrusion. Sometimes valuable minerals form on the margins of these intrusions. The magmatic fluids can flow long distances away. As these fluids encounter changes in temperature, pressure, and chemistry, they deposit various minerals. They may form veins by filling voids and fractures in local rock. These veins can be unusually rich sources of minerals, but they are often small.

Mineral-rich solutions cool as they near the surface. Compounds of copper, lead, and zinc form early in the cooling process. Closer to the surface and at lower temperatures, gold drops out of the solutions. This is followed by arsenic. Mercury, which is among the most volatile elements in these solutions, is typically the last to be deposited. Some springs in the Pacific Northwest contain dangerously high levels of natural arsenic and mercury.

The deposition pattern of mineral solutions is made use of by prospectors. Geologists looking for gold often use surface occurrences of mercury as an indicator of underground gold deposits. Historically, most of the major copper, lead, and zinc districts of the Pacific Northwest started out as areas where gold was discovered.

The rapid solidification and exposure to surface conditions experienced by volcanic rock are favorable for the formation of large concentrations of certain nonmetallic minerals. These can be mined as sources of industrial minerals. Industrial minerals and construction aggregates account for 60 percent of the value of mine output in the United States.* Industrial minerals are usually mined in bulk and are used for their physical or chemical properties. Examples of these include clays, phosphates, and pumice.

Metallic ore deposits can form as a consequence of erosion. Placer gold is the best known example of this. Gold is eight to ten times denser than most rocks and minerals. Erosion from fast-moving water washes away lighter rocks and minerals, leaving dense gold particles behind. It is a highly effective and natural way of concentrating gold. Extensive but uneconomic occurrences of gold can erode into rich, small pockets of placer gold. Other placer minerals found in the Northwest include tantalum, platinum, titanium, and tin.

Gradual weathering is another type of erosion that yields mineral deposits. Soil that is exposed to long periods of wet tropical and subtropical weather breaks down leaving behind only the most resistant minerals. These deposits are called laterites. Nickel, iron, and aluminum laterite ores are found near the Pacific coast of the Northwest.

Extensive erosion can expose deep igneous intrusions. The metallic minerals these intrusions contain become more accessible to exploration and production. Many of the metallic ore mines in northeastern Washington and Idaho and western Montana work such deposits.

Because coal, oil, and natural gas are created from organic matter, deposits of these substances are almost exclusively found in sedimentary rock formations. For

---

* Based on data from the *Minerals Yearbook, Area Reports: Domestic 1993–94, Volume 2*, U.S. Bureau of Mines. In 1994, non-fuel mineral output from U.S. mines totaled $28.5 billion. Of that $11.4 billion was metallic mineral ore, $10.3 billion construction aggregate, and $6.8 billion industrial minerals.

commercial deposits, the sedimentary rock must be porous so that oil and gas can be economically extracted. Other important minerals found in sedimentary formations are vanadium, phosphate, potash, uranium, and salt.

Some mineral products occur only in metamorphic rocks. Metamorphic rocks are created when sedimentary or igneous rocks are subjected to extreme pressures and temperatures that are so harsh that the structure of the rock changes and new minerals are formed. Among the materials mined from metamorphic rock formations are marble, talc, garnet, asbestos, and slate.

Construction aggregates, a high-volume, low-dollar-value, inert material usually used to provide bulk to construction projects, can be mined from a wide range of rocks. Aggregate is the main ingredient in concrete and asphalt pavement and is also used to establish a base upon which buildings and roads are constructed.

The most common forms of construction aggregate are crushed rock, sand, and gravel. While these materials can be found everywhere, the poor quality of many deposits makes them of little or no commercial value. A usable deposit must contain large amounts of relatively clean and uniform aggregate that is inexpensive to mine and transport to customers. Good quality sand and gravel deposits are often found in rivers. This has led to conflicts between mining interests and those who are concerned about preserving fish habitats, clean water, and the recreational uses of rivers.

Although the Pacific Northwest has a strong history of mining, the political climate has turned unfavorable for development. Because of the high-paying jobs they provide, mines have strong impacts on rural economies. There used to be strong support for mine development.

Now, the political balance in the Pacific Northwest is shifting as the region becomes more urbanized. Residents of cities and suburbs are generally more prosperous and less concerned about economic development through natural resource extraction than rural residents, and are unlikely to support rural development projects if they believe the environment may be compromised. Even in some rural communities, attitudes are changing and in some cases prohibitions have been placed on mining. Contributing to this shift is the large inward migration of affluent urbanites and retirees into rural areas of the Pacific Northwest.

While modern techniques can ensure that mines are designed and run in an environmentally safe way, popular perceptions of mining are more often colored by the past. The Pacific Northwest has many dangerous old smelter and mining sites which operated years ago when little attention was given to environmental concerns. Distrust of the mining industry is, therefore, widespread, making it increasingly difficult and expensive to get new mines opened.

The growing urban-rural conflict over mining has resulted in the abandonment of many exploration projects. The costs and uncertainties of permitting new mines have risen so high that many mining companies are no longer pursuing such projects in the Northwest. There is still interest in old, inactive mines whose sites were permitted years ago. However, these often have marginal reserves. This inhospitable political environment for siting new mines has had a severe impact on the economies of some rural communities.

OVERVIEW OF MINING IN THE PACIFIC NORTHWEST

About 60 different mineral products are produced in the Pacific Northwest. Total output in 1993 was valued at $1,392.7 million.

By far, construction aggregates accounted for the largest share of production. Nearly 164 million tons of aggregate worth $640.6 million were extracted from mines. As a high-tonnage, low-value product, aggregate needs to be mined close to where it is used. Aggregate is consumed and produced in every county in the Pacific Northwest.

Ranking second in production value at $212.4 million was gold. Over 85 percent of the gold output came from ten mines in eastern Washington, western Montana, and Idaho.

An estimated $135 million worth of copper was mined in 1993, nearly all from western Montana. The value of coal production, at $103.5 million, ranked fourth. All of the coal came from two mines in Washington. The fifth leading mineral was phosphate, with $78.4 million worth of phosphate rock mined in Idaho.

Mineral Commodity	Value* of 1993 Production ($ Mn.)	% of Total Production
Construction Aggregates	$640.6	46.0%
Precious Metals	271.9	19.5%
Base and Ferrous Metals	215.0	15.4%
Coal and Natural Gas	110.6	7.9%
Industrial Minerals	152.4	11.0%
Gemstones	2.2	0.2%
TOTAL	$1,392.7	100.0%

Table 1. Value of 1993 Mineral Production in The Pacific Northwest**

* Values are measured at the point of production and do not include shipping. Metal ore production is based upon the market prices of the recoverable contained metals. Phosphate production is based on the price of phosphate rock. The table does not include downstream products such as cement and lime.

** From an unpublished analysis by the Oregon Department of Geology and Mineral Industries.

IDAHO

The state of Idaho has a favorable geologic setting for several different types of minerals. The panhandle, to the north, is characterized by highly deformed sedimentary rock and large granitic intrusions. Some of the world's richest lead and silver ores are found here. A large area of the center of the state contains mostly granites of late Cretaceous to early Tertiary age. These are important sources of gold, molybdenum, and other metallic elements. Young volcanic rock covers a wide band from the southwestern corner of Idaho up to Yellowstone Park. Although this rock is a potential source of geothermal power, few significant metallic ore deposits are located in it. The southeast corner of the state, however, is covered by thick layers of Paleozoic and Mesozoic sedimentary rocks which hold much of the world's phosphate resources.

In 1993, $269.7 million worth of minerals were produced in Idaho. The five highest ranking minerals in terms of production values were phosphate, construction aggregate, gold, silver, and lead.

Idaho also has cobalt, yttrium, and tantalum resources, although none of these are being mined today. These rare metals are strategically important to the United States, and Idaho is one of the few states known to have significant quantities of them. While foreign sources of these metals are more competitive, Idaho's resources are considered strategic because if foreign sources are cut off in the event of political upheaval, they can be exploited.

Idaho does not produce any coal, oil, or natural gas, primarily because it is covered by thick layers of volcanic rock. While commercial quantities of energy materials may be present, exploring and drilling through the volcanic rock would be both difficult and prohibitively expensive.

Mineral Commodity	Value of 1993 Production ($Mn.)
Construction Aggregates	$69.0
Precious Metals	67.2
Base and Ferrous Metals	31.9
Coal and Natural Gas	0.0
Industrial Minerals	101.4
Gemstones	0.2
TOTAL	$269.7

Table 2. Value of 1993 Mineral Production in Idaho

Historically, Idaho was one of the world's leading sources of lead and silver. Today, the production values of these metals are exceeded in Idaho by phosphate, construction aggregate, and gold.

Much of the silver from Idaho was mined in the Coeur d'Alene district. This area in northern Idaho is 30 miles long and ten miles wide, consisting of highly faulted sedimentary rock that was invaded by granitic intrusions. Thermal solutions from these intrusions carried metallic compounds with them. When these precipitated out along the faults in the sedimentary rocks, rich veins of lead, zinc, antimony, and silver ores were created.

Production began in the Coeur d'Alene district in the 1880s. Unlike the short-lived placer mines in other districts around the West, the rich and extensive underground vein deposits of Coeur d'Alene ensured many years of production. About five times more silver was produced from Coeur d'Alene than from the famous Comstock Lode in Virginia City, Nevada.

By 1891 there were 40 mines in the Coeur d'Alene district. From 1892 to 1899, the region was the site of labor disruptions, violent confrontations, and bombings. In 1892 a strike was called by the miners' union over wages and other issues. Later a group of miners resorted to blowing up an ore mill. The governor of Idaho responded by declaring martial law and bringing in federal troops to restore order.[*]

Labor unrest continued off and on throughout the 1890s as mines hired non-union workers. In 1899, a large gang of men took possession of a passenger train, stole dynamite, and went on to Kellogg while picking up volunteers enroute. They shot their way into the office of the area's largest mine. The office was destroyed, and they proceeded to dynamite the mine's concentrator. Once again martial law was declared, and federal troops returned. Eventually, the mines became unionized.

Today, mining continues in the Coeur d'Alene district, although output is less than in the past. Silver and lead rank fourth and fifth, respectively, in value of production from Idaho's mines. While there is still very rich ore, it is often quite far below the surface and expensive to produce. Over 95 percent of the world's silver is now being produced as a minor byproduct of copper, lead, and gold refineries or is coming from recyclers. Silver is far less expensive to produce from these sources than from primary silver mines. As a result, production from primary silver mines such as those in Idaho has been declining.

Gold is the third most important mineral product in Idaho. Much of it comes from mines in central and western areas of the state where there are extensive granitic rock bodies. This region hosts several large lode deposits of gold. Unlike placer deposits, where gold has been eroded from bedrock and carried by water into deposits along streams, lode deposits consist of minerals inside hard rock. The region also contains deposits of molybdenum and cobalt.

---

[*] *Mining Highlights of the Coeur d'Alene District,* presentation by Henry Day to the 60th anniversary of the Idaho Mining association, July 18, 1963.

The first significant gold discovery in Idaho was made in the Boise Basin where a placer deposit was uncovered in 1862. Over the next two decades numerous placers were discovered in the state. Only a few, such as those found in Idaho City, remained active for many years.

The future of gold mining depended on the development of lode mining. Lode mining was delayed, however, until the 1880s when rail service was introduced. Production received a large boost after several important technological improvements such as the invention of dynamite and advances in pumping equipment. All this played an important role in the economic development of Idaho.*

There is an extensive phosphate field in southeastern Idaho, which in the early 20th century was the largest known source of phosphate in the world.** Since then large phosphate reserves were developed in Florida, North Carolina, and Africa. Still, Idaho is a significant producer. In 1993, $78.4 million worth of phosphate rock was mined in Idaho and the state produced $568.2 million dollars worth of phosphoric acid and elemental phosphorous in plants and smelters.

Over 90 percent of the world's phosphate is used as fertilizer. Much of Idaho's phosphate rock, however, is smelted in energy-intensive electric furnaces and made into elemental phosphorous. The elemental phosphorous is then made into chemical products used in laundry detergent and food processing. Producing elemental phosphorous is economical in Idaho because electric power costs are much lower than in other phosphate mining regions of the world.

The Idaho Phosphoria Formation spreads over 130,000 square miles. Around Pocatello and Soda Springs, Idaho, thick beds of this formation are near the surface. Millions of years of weathering by groundwater and exposure to the atmosphere have acted to concentrate the phosphate minerals. These high-grade deposits are targeted by mining companies.***

Idaho is also a major producer of vanadium pentoxide, a metal used in steel production and catalysts. It is extracted from waste materials that are taken from an elemental phosphorous smelter in Soda Springs.****

OREGON

All but a few parts of Oregon are covered with thick layers of relatively young volcanic rock and sediments. However, many mining companies believe Oregon has an anti-mining business climate. As a consequence, Oregon's output of metallic

---

* Merle Wells, *Gold Camps and Silver Cities,* Idaho Bureau of Mines and Geology, 1963.
** Virgil Kirkham, *Phosphate Deposits of Idaho and Their Relation to the World Supply,* Idaho Bureau of Mines and Geology, March, 1925.
*** Mike O'Driscoll, *U.S. Pacific Northwest—An Outpost of Industrial Mineral Wealth,* Industrial Minerals, April 1989.
**** Earl H. Bennett and Virginia S. Gillerman, *Mining, Minerals, and the Environment in Idaho, 1994,* Idaho Geological Survey, October 1995.

minerals lags far behind the other states of the Pacific Northwest. In 1993, Oregon produced $240 million worth of minerals with less than one percent coming from metals. Ranked in order, the five most important minerals produced in the state were construction aggregate, natural gas, diatomite, limestone, and pumice.

Construction aggregate accounts for 89 percent of all the mining in Oregon.* Aggregate is produced in all 36 counties and is usually consumed within 25 miles of where it is mined. For this reason, most of the mining in Oregon occurs around urbanized areas. In 1993, 52 percent of the aggregate tonnage produced in the state came from just seven urban counties: Columbia, Jackson, Lane, Linn, Marion, Multnomah, and Washington. (Figure 1, p. 369.)

Areas of igneous intrusions and metamorphic rock are found in parts of northeastern and southwestern Oregon. Over 95 percent of the gold and silver mined in Oregon came from those parts of the state.

Gold was discovered in the Klamath Mountains of southwestern Oregon in 1850. In 1862, discoveries were made in the Blue Mountains of Baker and Grant Counties. Waves of prospectors from California moved into Oregon. Following them were farmers, ranchers, and loggers. Gold mining was thus the principal force behind the early development of the logging, construction, and cattle industries in Oregon.

Within a few years, almost every placer gold district in Oregon was being mined. In parts of eastern Oregon where water was scarce, long ditches that brought water to placer deposits were built. The Eldorado Ditch, extending over 100 miles and

Mineral Commodity	Value of 1993 Production ($Mn.)
Construction Aggregates	$214.0
Precious Metals	0.8
Base and Ferrous Metals	0.6
Coal and Natural Gas	7.1
Industrial Minerals	16.2
Gemstones	1.3
TOTAL	$240.0

Table 3. Value of 1993 Mineral Production in Oregon

---

* R. Whelan, *Oregon's Mineral Industries, An Assessment of the Size and Economic Importance of Mineral Extraction in 1993*. Oregon Department of Geology and Mineral Industries, December 1994.

taking ten years to build was the longest. Long abandoned, some of these ditches are now being used by municipal water districts and for irrigation.

Sparking the gold rush was the legend of the Blue Bucket Mine. The mine was said to be a stream filled with gold pebbles discovered by accident by pioneers in eastern Oregon. The gold was so plentiful that it was fashioned into fishing weights. The mine got its name because, according to the legend, children filled a blue bucket with gold pebbles but later tossed them aside as their wagon train proceeded west. Stories of the discovery, which was never proven, were widely believed by prospectors who feverishly combed eastern Oregon and parts of Idaho.*

Many of the best gold deposits were quickly played out, and by 1882 gold production in Oregon dropped precipitously.** This changed in the 1890s when new technologies allowed miners to greatly expand output from lode deposits. Gold and byproduct silver production surged. By 1910, output fell as many of the high-grade lode deposits were depleted. Placer mining, however, was experiencing a resurgence as steam and electric powered systems started replacing hand-operated placer mining equipment. When, in 1913, a large scale dredge was introduced in Sumpter Valley, placer gold output more than doubled in Oregon from what it was in 1912.

Gold production slowly fell once again, and by 1928 it was 90 percent below what it was thirty years before. The Great Depression stimulated gold mining by reducing the cost of labor and equipment at the same time as the price of gold was fixed by the U.S. Government to $35 an ounce in 1934. Output jumped from 10,931 ounces in 1928 to 113,402 ounces in 1940. Production collapsed when the federal government passed War Production Board Order L-208 in 1942. In contrast to the country's immediate needs during World War II, gold mining was seen as a non-essential activity. Output in 1943 was only 1,097 ounces. Gold mining in Oregon never recovered, and today only about 2,000 ounces of gold are mined in the state each year.

Presently no coal is mined in Oregon. However, coal mining did play a role in the early development of Coos Bay. The area developed into a major port in the late 19th century, in part because the coal mines there were major suppliers to San Francisco and the steamships which ran up the coast. When the Southern Pacific Railway was extended to Coos Bay, the railroad developed the Beaver Hill Mine which supplied coal to the railroad's locomotives for many years.

Seams of coal made from ancient sediments are contained in the shale, sandstone, and other sedimentary rocks which underlie much of the Coos Bay area. Although the total quantities of coal are large, the seams tend to be thin and surrounded by weak shale layers which make mining difficult. The coal also contains many impurities.

---

*   Howard Corning, *Oregon End of the Trail,* Oregon State Board of Control, Salem, OR: 1940.
**   Howard Brooks and Len Ramp, *Gold and Silver in Oregon,* The State of Oregon Department of Geology and Mineral Industries, Portland, OR: 1968.

Coal mining peaked in Coos County in 1905 when 110,000 tons were mined. When the discovery of oil in California and the introduction of diesel eliminated the best markets for the coal, eventually all the mines closed.

Chromite and nickel minerals occur in ultramafic rocks which crop out in parts of southwestern and northeastern Oregon. Ultramafic rocks, intrusives that are usually formed at great depths, are magnesium and iron-rich igneous rocks that have little or none of the quartz or feldspar that are characteristic of granite.

Because ultramafic rocks are relatively uncommon, deposits of their associated chromium and nickel minerals are rare. Unfortunately for Oregon, the ores found in the state are marginal by world standards. Mining of chromium and nickel has occurred off and on over the past century in Oregon. The chromium deposits were important during the First and Second World Wars. In Douglas County, the Riddle nickel mine was a major producer for many years. The ore that remains is low grade by world standards and the nickel smelter relies mostly on ore imported to Oregon from New Caledonia in the South Pacific.

Natural gas is produced from the Mist Gas Field in Columbia County, the only operating field in the Pacific Northwest. The gas comes from marine sediments contained in thick and highly porous sandstone trapped between layers of volcanic rock which act to contain the natural gas. Heat from volcanic activity may have accelerated the creation of the natural gas.

The Mist Gas Field was discovered in 1979. Reichhold Chemical Company, which had a nearby fertilizer plant, needed a low-cost supply of natural gas. At the same time, the local utility, Northwest Natural Gas, was seeking a gas storage site. Together they financed the exploration which led to the discovery of the Mist Gas Field.

The government of Columbia County earns substantial royalties from natural gas production. During the Depression, when many private timberland holders went bankrupt, the county took over the title for vast acreage of private forests. In later years, the county sold the land for timber but retained the mineral rights.

While the gas field is still a significant producer of natural gas, its future value is in its use for storage. During the summer months when natural gas prices are low, Canadian pipeline gas is pumped underground into depleted areas of the Mist Gas Field. The gas is then extracted during cold spells in the winter when local demand for gas exceeds the ability of pipelines to deliver.

Oregon used to be a major producer of mercury, with production peaking in the early 1940s when demand for mercury declined because of environmental concerns, mines in Lane, Douglas, Jackson, and Jefferson Counties eventually closed down. Mercury ore is frequently found near the surface and is deposited by hot volcanic solutions.

Central and southeastern Oregon are covered by young volcanic rock. A number of valuable non-metallic minerals come from these areas. Much of the country's pumice is mined in central Oregon. Pumice is a volcanic glass that contains many gas bubbles and is formed from rapidly cooling, frothy masses of lava. It is used in lightweight concrete and as landscaping stone. Perlite, also of volcanic origin and used in lightweight concrete, is mined in Lake and Baker Counties. Mined perlite is heated in a process that makes the mineral puff up into very light pellets. Perlite is a common ingredient in household plant soil mixtures.

Diatomite, used as a filtering medium and as an ingredient in cat litter,* is produced in southern and eastern Oregon. This mineral is formed from the silica-rich skeletons of single-celled algae which settled at the bottom in ancient bodies of water. Dried lake beds are the setting for diatomite deposits in Oregon.

Oregon's production of gemstones is increasing. Some volcanic rock sediments in central Oregon contain precious opal. Oregon sunstones, a gem variety of the mineral feldspar, are found in Lake and Harney Counties. High quality sunstones are brilliant red and contain minute particles of copper.

WASHINGTON

The western and southern parts of Washington are covered by much of the same young sedimentary and volcanic rocks that are found over most of Oregon. Other than construction aggregate and coal, relatively few minerals are produced from these sections of Washington. Northeast of Seattle, however, there are extensive granitic intrusions which stretch all the way to the Idaho border. These are host rocks for most of the valuable metallic mineral deposits in Washington.

Mineral production equaled $536.8 million in Washington during 1993. Sixty percent of the total came from construction aggregate which was produced for the state's fast-growing population. Ranking second in production was coal, followed by gold, silica, and silver.

Mineral Commodity	**Value of 1993 Production ($Mn.)**
Construction Aggregates	$324.8
Precious Metals	87.8
Base and Ferrous Metals	1.7
Coal and Natural Gas	103.5
Industrial Minerals	19.0
Gemstones	0.1
TOTAL	$536.8

Table 4. Value of 1993 Mineral Production in Washington

---

*   *Mineral Facts and Problems,* U.S. Bureau of Mine, 1985.

Coal-bearing rock occurs along the western margins of the Cascades and in a few places along the eastern flank. The coal beds were formed from organic sediments that were covered by volcanic rock.

The first coal mine in Washington was started in 1854 in Whatcom County. About that time, a large coal discovery near Centralia provided the incentive for the building of the first railroad linking Seattle with the Columbia River.*

By 1877, coal mines around Seattle were major suppliers to San Francisco. Production in the state grew steadily until 1918, peaking at four million tons that year. Coal mining declined because of the growing availability of oil, the introduction of hydroelectric power, and competition from coal mines in Utah and Wyoming. By 1968, Washington's coal output totaled only 55,000 tons.

In 1970, Pacific Power and Light started a mine-mouth coal power plant in Centralia. This boosted coal production to nearly 4.5 million tons a year. Today, the Centralia Mine and the much smaller John Henry Mine in King County account for all the coal production in the state.

The first significant gold placer was discovered in 1859 along the Similkameen River near the Canadian border. Hundreds of miners stampeded into the area, causing the population of Okanogan City to swell to nearly 3,000. The rush was short-lived, however, as news of much richer discoveries in Canada drew miners northward.**

Many small placer gold discoveries were made throughout Washington. In 1868, a major gold strike was made in Kittitas County on Swauk Creek. Placer gold production from this area continues on a small scale to this day.

The first lode deposits of gold were uncovered in 1871 in Okanogan County. This was soon followed by discoveries around Swauk Creek in 1874 and the Wenatchee gold belt in 1885. Large settlements and capital investments were made. Lode mining had a far greater importance to the industrial development of the state than placer mining.

Prospectors looking for gold found rich base metal and silver ores. The first of these was the discovery in 1871 of the Ruby-Conconully District in Okanogan County. In Pend Oreille County, massive lead-zinc sulfide deposits were found in the Metaline District. Over 19 million tons of ore have been produced there since 1890. Large deposits of silver, lead, and zinc were found in Stevens County in 1894. Two years later the gold and silver Republic deposits of neighboring Ferry County were discovered. Mines in this region continue to produce large amounts of gold and silver.

Farther south in Chelan County, lode deposits of copper and gold were uncovered. Production began on the nine-mile-long Wanatchee gold belt in 1894. In recent years, ore from this deposit was extracted through the Cannon Mine.

---

*　G. Beikman, K. Gower, and D. Dana, *Coal Reserves in Washington,* Washington Division of Mines and Geology, Bulletin 47, 1961.

**　Wayne Moen, *The Mineral Industry of Washington—Highlights of Its Development, 1853–1980,* Washington Department of Natural Resources, Information Circular 74, 1982.

Most of the gold and silver produced in recent years in Washington came from underground mines. Unlike Nevada's gold industry, which relies mostly on heap leaching low-grade ores with cyanide solutions, Washington's mines extract ore from high-grade vein deposits. High-grade ore from the mines is milled into concentrate, which is then shipped to smelters as far away as Japan.

With so much mining of complex ores in northern and central Washington, downstream investments were needed. In 1887, the first smelter in Washington was built in Colville. It produced lead and silver. Eventually six smelters were constructed. Washington was a major producer of nonferrous metals and silver throughout the first half of the 20th century.

Non-ferrous metal mining peaked in 1955. Because of declining ore supplies and rising costs, all of the smelters in Washington closed. By 1993, the last base metal mine suspended operations. Washington still has substantial base metal resources, however, many of them are deep underground and too costly to mine given current economic conditions.

Stevens and Okanogan Counties have large sedimentary rock formations that contain unusually pure deposits of limestone and dolomite. The limestone is used by the paper and construction industries.* Dolomite is used in agriculture and as a source of magnesium metal. Washington has the only thermal magnesium smelter in the United States. The process is commercially viable because of the availability of low-cost electric power and dolomite. Magnesium metal is used as an alloying ingredient by aluminum smelters in the Pacific Northwest.

There are several large occurrences of ultramafic rock which have been thrust up into the Western Cascades. One occurrence, a massive olivine deposit, is on Twin Sisters Mountain, between Skagit and Whatcom Counties. Olivine is a magnesium-rich mineral that is made into refractories for solid waste incinerators and steel foundries. Washington and North Carolina are the only states that produce olivine.

WESTERN MONTANA

Western Montana consists mostly of sedimentary rock and igneous intrusions. One igneous formation, the Boulder batholith, is of particular importance to Montana's mining history. A batholith is a huge domed intrusion that has no known bottom and covers an area in excess of 40 square miles. In the southern section of western Montana, some of the sedimentary deposits have been so extensively deformed and eroded that metamorphic rocks appear near or at the surface of the earth.

In 1993, the 17 counties that comprise western Montana produced minerals worth $345.8 million. The five leading minerals produced were copper, gold, construction aggregate, molybdenum, and silver.

---

*   *Washington Geology*, Washington State Department of Natural Resources, Olympia, WA: March 1995

Mineral Commodity	Value of 1993 Production ($Mn.)
Construction Aggregates	$32.8
Precious Metals	116.1
Base and Ferrous Metals	180.8
Coal and Natural Gas	.0
Industrial Minerals	15.8
Gemstones	0.2
TOTAL	$345.8

Table 5. Value of 1993 Mineral Production in Western Montana

The mines around Butte, Montana once dominated the world copper industry. The ore deposits were part of the Boulder batholith which was formed deep underground about 80 million years ago. Cooling and external forces caused the granite batholith to fracture. Over time mineral solutions migrated up through these fractures and deposited various metallic minerals. What resulted were rich veins of ore several feet wide and over a mile deep.

Anaconda expanded into a number of new businesses such as timber. As insurance against union work stoppages, it opened copper mines in South America. Unions were once so powerful in Butte that non-unionized businesses were prevented from locating in the city.

By the 1940s, technological advances at Anaconda Copper were no longer offsetting the combination of failing ore grades and greater mining depths. Costs started rising, and by 1955 the company switched to low-cost open pit mining. In 1983, all operations in Butte ceased.

Currently, two large mines operate around Butte. The Continental Pit is a large-scale producer of low-grade copper, silver, and molybdenum ores. The open pit mine is run by Montana Resources. The operation is a financial success because it employs modern mining methods on a large scale at the site of an old mine. Reflecting the radical change in the labor force of this region in the past 30 years, the 325 miners at the Continental Pit are non-union. West of Butte, the Beal Mountain gold and silver mine is a similar operation, a large-scale, open pit mine with only 110 miners who are exploiting a mine site discovered in the 19th century.[*]

Some of the old Anaconda properties are environmental clean-up sites. Butte copper ore contained large amounts of arsenic, selenium, and cadmium and smelter

---

[*] Robin McCulloch, *Montana Mining Directory—1993,* Open File Report 329, Montana Bureau of Mines and Geology, 1994.

technology used in the past did little to keep these toxic elements from being released into the environment. Cleaning the extensive pollution from the old mine and smelter wastes will provide Butte with a future source of employment.

Helena, Montana, began as a mining town. In 1864, gold was discovered in Helena. A rush started, and Helena became a supply and transportation hub for many of the outlying mining camps in the area. When lode deposits of silver and base metals were discovered near Helena, the city's access to capital and equipment allowed it to exploit the resource. In 1890, Helena had over 50 millionaires among its 10,000 residents. The silver panic of 1893 destroyed much of the silver mining and smelting in Helena. The loss was cushioned by the city's success at being named the state capital four years earlier.* Today, there is one large underground mine in the Helena region which produces gold, silver, lead, and zinc concentrates.

As is typical of most base metal mining districts, the first discoveries around Butte involved gold. Numerous placer gold claims were filed in 1864. The mines, however, were not very productive, and some prospectors turned their attention to veins of silver ore. Butte, like other western gold camps, lacked the capital, technology, and transportation to process complex ores. The mines had to ship their silver ore as far away as New Jersey because there were no local smelters.

It wasn't until 1876 that the first successful smelter for silver ore was opened in Butte. It sparked a sharp increase in silver mining, much of which was financed by British and French investors. Three years later a copper smelter was completed.

In 1881 the railroad was extended into Butte, and with it came a means for bringing in mining equipment and taking out metal. That same year an Irish immigrant named Marcus Daley sold his interest in a silver mine and reinvested his money on an undeveloped copper deposit in nearby Anaconda. But without a large ore processing facility, his deposit was worthless. Daley turned to investors in San Francisco, New York, and Boston for money to build a mine, mill and smelter. His company, Anaconda Copper, became the largest copper producer in the world. It also produced byproducts such as silver, gold, cadmium, manganese, and zinc.

Anaconda flourished. Demand for copper was skyrocketing with the country's expanding use of electricity and telegraphs. With access to low-cost hydroelectric power, Anaconda was able to produce low-cost copper at its Great Falls refinery. In 1893, the repeal of the Sherman Silver Production Act devastated Montana's silver mines. Anaconda continued getting stronger both financially and politically. The company's power was enhanced in 1899 when Daley teamed up with William Rockefeller and associates of the Standard Oil Trust to reorganize Anaconda into the Amalgamated Copper Company.**

---

*   *1987 Guidebook of the Helena Area,* Special Publication 95, Montana Bureau of Mines and Geology, 1987.
** Brian Shovers et al., *Butte and Anaconda Revisited,* Butte Historical Society, 1991.

Many of the miners came to Butte from Italy, Finland, Ireland, and England. Irish miners migrated from the Michigan copper range where ore deposits were being depleted. A collapse of tin prices in England brought in experienced underground tin miners from Cornwall, England. Largely because of copper mining in Butte, western Montana became a melting pot. It was a highly industrialized region with a unionized workforce. This contrasts strongly with the more conservative, less ethnically diverse eastern half of the state.

Anaconda used its wealth for gaining political influence in Montana. It was viewed by many Montanans as a heavy handed political force in the state. This spawned an anti-mining legacy which is reflected in the high severance taxes in Montana.

The United States is a leading producer of talc, and almost one-half of it comes from Montana. Talc mines located in southwestern Montana are important to the economies of Beaverhead and Madison Counties. Talc is found in metamorphic rocks along with marble and graphite* The construction industry uses talc in ceramic tiles, paints, sanitary ware, roofing materials, and joint compounds. Large amounts are also consumed by the paper, plastics, and cosmetics industries. Talc from Montana is favored by automotive companies around the world for its value in ceramics used in catalytic converters.

Western Montana has numerous active placer mines that operate seasonally. Collectively, they produce between 5,000 and 15,000 ounces of gold a year. Some placers also recover sapphires. There are several sapphire placer mines on the Missouri River near Helena.

There has been considerable interest in a number of precious and base metal exploration sites in Montana in recent years. Some of these are in environmentally sensitive areas. Although some of these sites could be mined safely, grassroots opposition to mining has been fierce. Especially vocal are newer residents of western Montana who moved to the state for its rural atmosphere and recreational opportunities. Development of several large projects has been blocked.

---

*   Eugene Perry, *Talc, Graphite, Vermiculite, and Asbestos in Montana*, Montana School of Mines, Butte, Montana: 1948.

Economic Geography

THE ECONOMY Chapter 11

Steven R. Kale
Oregon Department of Transportation

Historically, dependency on natural resources has characterized the economy of the Pacific Northwest. In recent years, however, the region's economy has become increasingly diversified, especially in larger urban areas. Despite this increasing diversification, more remote parts of the region remain dependent on resource-based industries.

This chapter highlights selected aspects of the economy of the Pacific Northwest, beginning with a brief sketch of the historical development of the regional economy. A portrait is then presented of the Northwest's economy in the early 1990s, along with a discussion of recent trends for the region as a whole and in terms of metropolitan-nonmetropolitan differences. This is followed by summaries of regional economic restructuring, of issues affecting the region's economy in the 1990s, and of state and regional economic development efforts. An appendix at the end of the chapter identifies sources, definitions, and other considerations for analyzing regional economic change in the Pacific Northwest.

HISTORICAL OVERVIEW

Pre-European Settlement

Before European settlement, the economy of the region's native populations could best be described as subsistence (Table 1). Most native groups relied on fishing, hunting, and gathering. Coastal Native Americans were especially dependent on products from the ocean, while native groups in the Columbia Plateau and between the Cascade Mountains and the Coastal Ranges subsisted on fish, wild game, roots, and nuts. Fishing was far less important in the Great Basin, where native groups hunted small animals and gathered roots, seeds, nuts, and whatever else they could find to eat.

Pre-European Settlement: (Pre-1800s)
Native American Subsistence Economy with Limited Local and Regional Trading

Early European Explorers and Fur Trading: (1780s–1840s)
Maritime and Overland Fur Trading

The Pioneer Economy: (1840s–1880s)
Agriculture and Mining

The Rail Era and the Timber Products Industry: (1880s–1930s)
Lumber and Wood Products, Agriculture, Fishing

The New Deal and World War II: (1930s–1945)
Government Construction, Military Projects, Aircraft, Shipbuilding

Post-World War II and Emerging Suburbia: (1945–1960s)
Lumber and Wood Products, Aircraft, Diversification

Prosperity, Recession, Recovery, and Rapid Growth: (1970–The Present)
Lumber and Wood Products, Diversification, High Technology, Service Industries

Table 1. Periods of Economic Change in the Pacific Northwest.
Source: Based in part on Schwantes (1996)

Many native groups were involved in trading. Before introduction of the horse, the exchange of goods was limited by how far traders could walk or travel by canoe or dugout. Trading along the coast and elsewhere was further limited by rugged terrain. After the Spanish introduction of the horse, trading areas expanded. The region's first horses are believed to have been obtained by Native Americans in the Great Basin in the late 1600s or early 1700s. Ownership of large herds was among the first signs of material wealth among native populations.

Early European Explorers and Fur Trading

Most of the first European explorers to the Pacific Northwest came not to settle or obtain riches but to search for a Northwest Passage between the Atlantic and Pacific Oceans. Failure to find the fabled passage, the often-present fog, and the absence of good harbors contributed to a general lack of interest in the region's economic resources. This changed when early explorers made contact with native populations along the coast and discovered an abundance of furs that could be readily traded.

The peak of the maritime fur trade lasted for about 25 years from the 1790s into the early 1800s, and was followed by a period in which British and American traders competed to claim territory where they would have exclusive rights to vast supplies of "soft gold." Competing claims for exclusive rights occurred mostly in the early 1800s when it was as yet unclear which nations had territorial jurisdiction over the Pacific Northwest.

The territorial dispute was settled in 1846 when Great Britain and the United States agreed the 49th parallel would be the latter's northern border. The mid- to late 1840s also corresponded with the end of the region's fur trading era and the beginning of widespread Euro-American settlement.

The Pioneer Economy

Overtrapping, changing tastes in fashion, and a desire to establish a more permanent territorial presence contributed to the transition from fur trading to an agricultural economy. Many of the region's first farmers, especially in the Willamette Valley, were former fur traders. Further spurring the early agricultural economy were the thousands of settlers who arrived in the 1840s and 1850s after taking the Oregon Trail across the Great Plains, Rocky Mountains, and the Columbia Plateau to the final leg down the Columbia River or over the Cascades.

Discovery of gold in California stimulated exporting from the Pacific Northwest. Much of the Willamette Valley's early agricultural production was sent first to California and later to mining centers within the region. Among the earliest mining centers in the Northwest were 1) Jacksonville, Oregon and Colville, Washington in the 1850s; and 2) the Clearwater, Florence, Boise Basin, and Owyhee districts in Idaho; the Powder River and Canyon City districts in Oregon; and the Virginia City, Bannock, and Helena districts in Montana in the 1860s and later. Gold was the primary mineral mined, although silver was produced at several mining centers. Additionally, as early as 1853, coal was mined near Coos Bay, Oregon, mainly for export to California. Larger coal mines opened in western Washington during the 1870s and 1880s.

The Rail Era and the Timber Products Industry

The first east-west railroad across the region was completed in 1883, followed in 1887 by the first north-south railroad from the Northwest to California. Completion of transcontinental rail lines and construction of numerous spur lines opened up much of the Northwest to settlement and resource development and exploitation. Access to rail lines resulted in expanded markets for agricultural producers who formerly had relied on waterborne or more rudimentary means of transportation. Farmers east of

Photo 1. Idaho City, Idaho.

the Cascades especially profited from the ability to ship their products more cheaply to markets in California and the eastern United States.

Though not directly stimulated by the completion of transcontinental railroads, the fisheries industry grew dramatically in the late 19th and early 20th centuries. The supply of salmon seemed to be endless, and technological advances in fishing and canning enabled vast quantities to be harvested and processed. Major canning centers were located along the Columbia River, the Puget Sound, and coastal rivers.

Perhaps the greatest growth in the late 1800s and early 1900s occurred in the Northwest's timber industry. Prior to the rail lines, most timber production occurred close to waterways where logs could be transported in "rafts" or aboard ships. Construction of rail lines opened vast amounts of formerly inaccessible forests and was a major factor in the emergence of the Northwest's timber industry.

During the first few decades of the 20th century, lumber and wood products became the region's dominant industry. Timber processing was labor intensive, and in some parts of the Northwest, population increased dramatically. Along with population growth came demands for market-based goods and services. Rapid expansion in the early decades, however, came to an abrupt halt by the Depression of the 1930s.

The New Deal and World War II

The federally sponsored New Deal helped the Pacific Northwest adjust to the economic distress of the 1930s. Several New Deal programs made major contributions to the region's economic development. Construction of bridges, dams, and other projects resulted in thousands of jobs; moreover, the dams stimulated further development through the generation of electricity and provision of water for irrigation. Construction of dams and subsequent development provided numerous jobs for new and existing residents, but also contributed to a dramatic reduction in the region's salmon migration.

World War II brought much of the Northwest out of the Great Depression. Thousands of residents and in-migrants found jobs in defense-related industries. Most notable were the aircraft industry in Seattle and nearby communities and shipbuilding in Portland and the Puget Sound area. The war accelerated growth in the region's emerging aluminum industry and led to the creation of the Hanford Nuclear Reservation in central Washington as well as numerous military bases.

Post World War II and Emerging Suburbia

After 1945, employment declined in shipbuilding, aircraft production, and other industries that had geared up for wartime efforts. Lumber and wood products once again became the region's leading industry as families moved to emerging suburban communities. The aircraft industry continued strong with the increasing demand for commercial planes in the 1950s and 1960s and military planes during the Vietnam conflict.

By the end of the 1960s, resource-based industries were far less important in the region's major urban centers than they had been earlier in the century. In many remote communities, resource-based industries remained dominant, although in some, services-producing sectors tempered this dominance. Tourism, government, retail trade, medical and business services, and other non-resource based sectors gained importance as economic restructuring occurred both nationally and regionally.

Prosperity, Recession, Recovery, and Rapid Growth

By the end of the 1970s, production in the lumber and wood products industry slowed. The recession of the early 1980s hit the Pacific Northwest especially hard, and thousands of workers lost their jobs. At about the same time, high technology industries were experiencing rapid growth, leading some observers to conclude that the region was trading timber jobs for high technology jobs. Other analysts, however, speculated that a more likely scenario was that low-paying jobs in trade and services were taking the place of relatively high-paying timber products jobs.

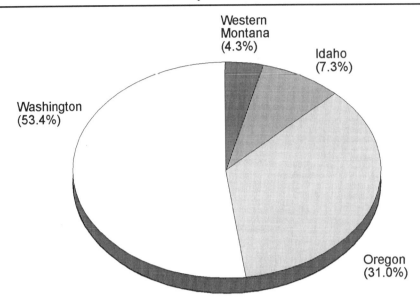

Figure 1. Employment, 1990 (in percent).
Source: United States Bureau of Economic Analysis, Regional Economic Information System.

By the late 1980s, the Pacific Northwest had recovered from the recession earlier in the decade. The region's economic growth in the early 1990s was among the strongest in the nation. Moreover, unlike the early 1980s, the region continued to grow in the midst of a national recession. By the mid-1990s, regional economic growth had slowed but was still above the national rate. The following sections provide further detail about the region's economy since the 1970s.

REGIONAL EMPLOYMENT AND INCOME

Employment

In 1994, about 5.7 million persons (four percent of the national total) were employed in the Pacific Northwest. Washington accounted for 53 percent of the regional total, while western Montana accounted for only four percent (Figure 1).

During most of the 1970s, employment grew faster regionally than nationally (Figure 2). The slower regional rate of growth in the early 1970s was due to declining employment in Washington where the Boeing Company and its suppliers laid off large numbers of workers. During the first half of the 1980s, the nation outperformed the region in employment growth, but by the latter half of the 1980s, the region once again grew faster than the nation. This trend continued into the 1990s, though by 1994 the difference between regional and national growth rates had narrowed somewhat from earlier in the decade.

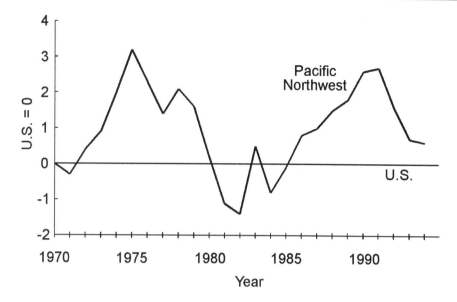

The graph shows the number of percentage points that year-to-year regional employment growth or decline was above or below the rate of year-to-year growth or decline for the nation. For example, if employment growth for a year was 1.0 percent nationally and 3.0 percent regionally, then regional growth would be 2 percentage points higher than national growth.

Figure 2. Regional Employment, 1970–1990.
Source: United States Bureau of Economic Analysis, Regional Economic Information System.

Income

Regional income-per-person was $21,194 in 1994 while nationally, income-per-person was $21,696. Income-per-person ranged from $17,535 in western Montana to $22,526 in Washington (Figure 3).

During much of the 1970s, income-per-person in Oregon, Washington, and western Montana increased faster than it did for the nation as a whole, and was higher than the national average for part of the decade in Oregon and all of the decade in Washington (Figure 4). In the early 1980s, regional income dropped substantially in comparison to the nation. By the late 1980s and into the early 1990s, income-per-person in all three states and western Montana was growing slightly faster than in the United States as a whole, but remained lower relative to the national average than during the 1970s.

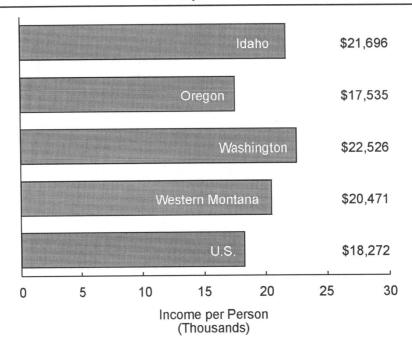

Figure 3. Income per Person, 1990.

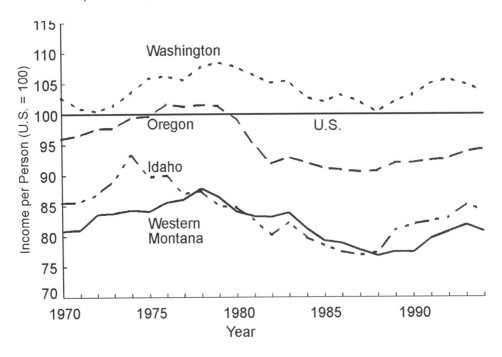

Figure 4. Income per Person by State, 1970–1990.
Source: United States Bureau of Economic Analysis, Regional Economic Information System.

METRO/NONMETRO DIFFERENCES

Employment

Of the 136 counties in the Pacific Northwest, 22 are metropolitan (metro), 22 are nonmetropolitan (nonmetro) adjacent to metro areas, and 92 are more remote or "nonadjacent" nonmetro (Figure 5). Eleven of the region's metro counties are in Washington, nine are in Oregon, and two are in Idaho. None of the 17 counties in western Montana is classified as metro.

The region's 22 metro counties comprise 13 metro areas (Table 2). In terms of employment the Seattle metro area made up of King and Snohomish Counties is the region's largest, accounting for just over one-fourth of the regional total. Second

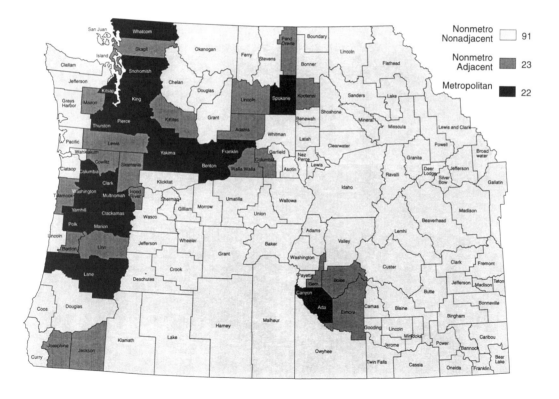

Figure 5. Metro and Nonmetro Counties.
(See this chapter's appendix for the definitions of metro areas, nonmetro adjacent counties, and nonmetro nonadjacent counties.)

largest is the Portland-Vancouver metro area (Clackamas, Columbia, Multnomah, Washington, and Yamhill Counties in Oregon, and Clark County in Washington), which accounts for 18 percent of the region's employment. Among metro areas, Bellingham and Medford-Ashland employ the fewest. About 59 percent of the region's total workforce is employed in the Seattle, Portland, and other metro areas along the Puget Sound and in the Willamette Valley.

In general, Boise and several Puget Sound metro areas grew the fastest after 1970; Tacoma and Yakima were the slowest growing metro areas. Considerable variation in growth occurred for several metro areas. In the 1970s, the Richland-Kennewick-Pasco area ranked first in percentage growth, primarily due to the construction of nuclear power plants at the Hanford Nuclear Reservation. During the 1980s, however, Richland-Kennewick-Pasco was the region's slowest growing metro area as power plant construction ended and several facilities closed at Hanford. In the early 1990s, the pattern reversed itself again as employment grew in response to accelerated waste-cleanup activities.

Photo 2. Seattle Skyline.

	Employment, 1994	Average Annual Change (percent)			
		1970–80	1980–90	1990–94	1970–94
Seattle	1,465,712	4.9	4.6	1.1	5.2
Portland-Vancouver	1,015,757	4.6	2.6	2.8	4.4
Tacoma	302,747	1.2	3.3	2.2	2.6
Boise City	226,199	6.3	3.1	6.0	6.9
Spokane	224,287	3.6	2.1	3.3	3.6
Eugene-Springfield	166,283	5.3	1.8	1.9	3.9
Salem	161,141	5.2	2.4	2.7	4.6
Yakima	110,762	3.1	2.5	1.9	3.1
Bremerton	105,841	5.0	4.5	2.1	5.6
Richland-Kennewick-Pasco	99,957	9.2	0.9	4.4	6.1
Olympia	96,222	5.8	5.3	3.6	7.3
Medford-Ashland	85,059	6.3	2.9	3.2	5.7
Bellingham	83,335	4.5	4.7	3.4	5.9

Table 2. Employment Change in Metro Areas, 1970–1994.
Source: U.S. Bureau of Economic Analysis, Regional Economic Information System

Bellingham, which ranked tenth among the 13 metro areas in percentage employment growth during the 1970s, ranked second in the 1980s and first from 1985 to 1990. Bellingham's growth has been attributed to several factors, including increased demand for consumer goods and services associated with shoppers from British Columbia. Bellingham's growth moderated in the early 1990s as changes in the currency exchange rate contributed to fewer bargains for Canadians shopping south of the border.

From 1970 to 1994, nonmetro employment increased the fastest in San Juan County, Washington, in Blaine County, Idaho, in Deschutes County, Oregon, and in Kootenai County, Idaho (Table 3). San Juan County has become a haven for recreational sailors and others who prefer the amenities of playing, living, and working in an island setting. Two of the region's most popular ski resorts (Sun Valley and Mt. Bachelor) are in Blaine and Deschutes counties, which also attract numerous summer tourists. Coeur d'Alene Lake has become an important recreational attraction for Kootenai County, which is adjacent to metro amenities in Spokane.

Photo 3. Sun Valley, Idaho.

Other rapidly growing nonmetro counties generally are located in amenity-rich areas along lakes, coastal areas, and ski resorts. For example, in Bonner and Valley counties in Idaho, attractions include Lake Pend Oreille, Payette Lake, and several ski areas. Flathead County, Montana, contains several lakes, a ski area, and the western entrance to Glacier National Park. Gallatin County, Montana contains several ski areas as well as Montana State University. Port Townsend in Jefferson County, Washington is a popular coastal tourist attraction.

The greatest nonmetro employment declines or slowest growth since 1970 generally have been experienced in resource-dependent counties. The greatest percentage losses were in two counties historically dependent on the mining industry: Deer Lodge County, Montana and Shoshone County, Idaho. The Montana county was the location of a large copper smelter that closed during the 1970s; in the Idaho county the silver mining industry experienced hard times during the 1980s and into the 1990s. Other counties with declines or slow growth include several agriculture- and timber-dependent counties such as Clearwater, Lincoln, and Oneida Counties in Idaho; Gilliam, Morrow, and Wheeler Counties in north central Oregon; and Columbia County in southeastern Washington. Most of these counties are sparsely populated and remote from large population centers and scenic amenities.

Photo 4. Coeur D'Alene, Idaho.

Several nonmetro counties have undergone booms and busts since 1970. For example, Skamania County, Washington and Morrow County, Oregon were among the fastest growing during the 1970s, but were among those experiencing the greatest percentage declines during the 1980s. For Skamania County, this reversal was largely due to in-migration in the late 1970s and out-migration in the 1980s of workers building a second electricity generating powerhouse at the Bonneville Dam. For Morrow County, the reversal was due to in- and out-migration associated with construction of a coal-fired power plant.

In contrast, Gilliam County, Oregon was among the region's slowest growing counties for the 1970–1994 period, but from 1990 to 1994 was the 10th fastest growing. This turnaround was due largely to employment generated by the opening of several landfills for solid waste and low-level hazardous materials.

About 28 percent of the Pacific Northwest's employment is in nonmetro adjacent and nonadjacent areas, but the proportion varies considerably regionally (Table 4). The proportions are 100 percent in western Montana and 65 percent in Idaho, but are much lower in Oregon (27 percent) and Washington (16 percent).

1970-1980		1980-1990		1990-1994		1970-1994	
Counties with the Most Growth	Percent	**Counties with the Most Growth**	Percent	**Counties with the Most Growth**	Percent	**Counties with the Most Growth**	Percent
San Juan, WA	13.2	Blaine, ID	8.0	Boise, ID	11.0	San Juan, WA	13.8
Morrow, OR	12.8	San Juan, WA	6.4	Kootenai, ID	8.9	Blaine, ID	12.4
Deschutes, OR	12.0	Custer, ID	5.7	Teton, ID	7.7	Deschutes, OR	12.1
Blaine, ID	9.9	Jefferson, MT	5.6	Ravalli, MT	7.5	Kootenai, WA	11.9
Skamania, WA	9.3	Kootenai, ID	5.3	Gallatin, MT	6.5	Bonner, ID	8.4
Kootenai, ID	8.6	Deschutes, OR	5.0	Flathead, MT	5.8	Gallatin, MT	8.1
Bonner, ID	8.0	Island, WA	5.0	Bonner, ID	5.6	Jefferson, WA	7.9
Josephine, OR	7.6	Jefferson, WA	5.0	Valley, ID	5.3	Ravalli, MT	7.1
Ferry, WA	7.3	Gallatin, MT	4.4	Pend Oreille, WA	5.2	Valley, ID	6.8
Stevens, WA	7.3	Skagit, WA	4.4	Gilliam, OR	5.1	Flathead, MT	6.5
Counties with the Greatest Declines or Least Growth	Percent	**Counties with the Greatest Declines or Least Growth**	Percent	**Counties with the Greatest Declines or Least Growth**	Percent	**Counties with the Greatest Declines or Least Growth**	Percent
Deer Lodge, MT	-3.5	Skamania, WA	-3.7	Shoshone, ID	-2.5	Deer Lodge, MT	-1.7
Wheeler, OR	-2.4	Garfield, WA	-2.9	Butte, ID	-1.5	Shoshone, ID	-1.4
Clearwater, ID	-1.4	Shoshone, ID	-2.9	Wheeler, OR	-1.5	Garfield, WA	-1.3
Columbia, WA	-0.8	Sherman, OR	-1.9	Powell, MT	-1.0	Wheeler, OR	-1.0
Gilliam, OR	-0.8	Caribou, ID	-1.8	Mineral, MT	-0.5	Sherman, OR	-0.9
Sherman, OR	-0.7	Deer Lodge, MT	-1.5	Columbia, OR	-0.5	Clearwater, ID	-0.6
Lincoln, WA	-0.4	Sanders, MT	-1.5	Douglas, OR	-0.2	Columbia, WA	-0.3
Oneida, ID	-0.4	Morrow, OR	-1.4	Harney, OR	0.1	Gilliam, OR	0.3
Garfield, WA	-0.3	Camas, ID	-1.1	Garfield, WA	0.2	Lincoln, ID	0.3
Lincoln, MT	-0.2	Lincoln, WA	-1.0	Lincoln, ID	0.2	Oneida, ID	0.1

Table 3. Nonmetro Counties, Most Growth and Greatest Declines in Employment, 1970-1994 (average annual percentage change). *Source:* United States Bureau of Economic Analysis, Regional Economic Information System.

Photo 5. Abandoned Mining Facility, Shoshone County, Idaho.

Over twice as many of the Pacific Northwest's nonmetro employees work in nonmetro nonadjacent counties (20 percent) as work in nonmetro adjacent counties (8 percent). By far the greatest proportions of nonmetro nonadjacent employees are in western Montana (100 percent) and Idaho (55 percent).

Although nonmetro workers comprise a much larger portion of total employment in Idaho and western Montana than in Oregon and Washington, they are nonetheless more numerous in the latter two states than in Idaho or western Montana. Nonmetro employment in 1994 was 421,000 in Idaho, 479,000 in Oregon, 486,000 in Washington, and 245,000 in western Montana. These totals suggest that rural economic prosperity or distress affects about the same number of workers in Oregon and Washington, but has a greater overall impact in Idaho and western Montana because metropolitan areas dominate the economies of Oregon and Washington.

As occurred elsewhere in the U.S. during much of the 1970s, employment in the region's nonmetro adjacent and nonadjacent areas increased more rapidly than in the region's metro areas (Figure 6). By the late 1970s, metro areas began to grow faster than nonmetro areas. In the early 1980s, the Pacific Northwest's metro and nonmetro areas experienced a major recession. Metro areas recovered from the recession much

Photo 6. Landfill, Gilliam County, Oregon

more quickly than did nonmetro areas while remote nonmetro areas recovered more slowly than did nonmetro adjacent areas.

Income

Income-per-person is higher in the region's metro areas than in its nonmetro areas (Table 4). For the region as a whole and in Idaho and Oregon, incomes were higher in nonmetro adjacent areas than in nonmetro nonadjacent areas. In Washington, however, incomes were greater in more remote nonmetro areas than in those near metro areas. Higher incomes in nonmetro adjacent than in remote areas likely reflect a spillover of higher wage jobs from metro to nearby nonmetro areas. This spillover apparently occurs to a greater extent in Idaho and Oregon, where considerably fewer counties are in the nonmetro adjacent category than in the nonmetro nonadjacent category.

Income-per-person is higher in Seattle and Portland-Vancouver than in the region's other metro areas (Table 5). Except for Yakima, Washington's small metro areas have higher incomes than Oregon's small metro areas. Seattle and Portland-Vancouver were the only metro areas where income-per-person was greater than the national average for metro areas in 1970, 1980, 1990, and 1994. Most metro areas

	Employment (percent)	Income-per-person (dollars)
METRO		
Idaho	35	21,511
Oregon	73	21,649
Washington	84	23,416
Western Montana	0	n.a.
REGION	72	22,770
NONMETRO		
Idaho		
Adjacent	10	18,053
Nonadjacent	55	16,601
Oregon		
Adjacent	8	17,917
Nonadjacent	19	17,626
Washington		
Adjacent	8	18,145
Nonadjacent	8	18,318
Western Montana		
Adjacent	0	n.a.
Nonadjacent	100	14,447
REGION		
Adjacent	8	18,063
Nonadjacent	20	17,417

n.a.: not applicable

Table 4. Metro and Nonmetro Employment and Income Per Person, 1994.
Source: U.S. Bureau of Economic Analysis, Regional Economic Information System

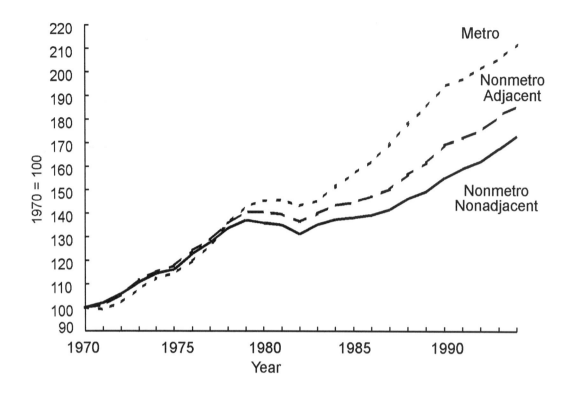

Figure 6. Metro and Nonmetro Employment Change, 1970–90.
Source: United States Bureau of Economic Analysis, Regional Economic Information System.

were nearer the national average in 1970 and 1980 than in 1990 or 1994. The Boise metro area improved the most in ranking, going from seventh in 1970 to fourth in 1990. Boise's improvement appears to have been fueled by growth in durable manufacturing such as high technology.

Income-per-person is higher in a number of nonmetro counties than in metro areas (Table 6). Most nonmetro counties with high incomes are sparsely populated and dependent on agricultural production, primarily wheat or other labor-intensive crops. Several high-income nonmetro counties (i.e., Blaine in Idaho and San Juan in Washington) have attracted wealthy in-migrants. The ten highest-income nonmetro counties are about evenly split between Idaho, Oregon, and Washington. None is in western Montana. Lewis and Clark County, home of the Montana state capitol in Helena, has the highest income-per-person in western Montana.

Nearly all of the ten lowest-income nonmetro counties are in Idaho and western Montana; none is in Oregon. In most counties, low incomes are due to the lack of year-round jobs in agriculture or timber industries. In Madison County, Idaho,

1970		1980		1990		1994	
Metro Area	Income	Metro Area	Income	Metro Area	Income	Metro Area	Income
Seattle	4,616	Seattle	12,369	Seattle	22,962	Seattle	27,097
Portland-Vancouver	4,307	Portland-Vancouver	10,921	Portland-Vancouver	19,188	Portland-Vancouver	22,890
Olympia	4,266	Richland-Kennewick-Pasco	10,746	Olympia	18,073	Boise City	21,511
Bremerton	4,231	Bremerton	10,026	Bremerton	17,312	Olympia	21,202
Tacoma	4,005	Olympia	9,924	Boise City	17,200	Richland-Kennewick-Pasco	20,798
Richland-Kennewick-Pasco	3,946	Tacoma	9,707	Tacoma	16,999	Richland-Kennewick-Pasco	19,870
Boise City	3,899	Spokane	9,445	Richland-Kennewick-Pasco	16,783	Spokane	19.565
Bellingham	3,746	Boise City	9,442	Bellingham	16,720	Bremerton	19,264
Spokane	3,742	Salem	9,156	Spokane	16,317	Bellingham	19,190
Salem	3,533	Eugene-Springfield	9,127	Medford-Ashland	15,918	Eugene-Springfield	19,167
Eugene-Springfield	3,500	Bellingham	9,038	Eugene-Springfield	15,908	Medford-Ashland	18,892
Yakima	3,394	Yakima	8,837	Yakima	15,496	Salem	18,234
Medford-Ashland	3,314	Medford-Ashland	8,713	Salem	15,454	Yakima	17,760
U.S. Metro Average	4,047	U.S. Metro Average	9,940	U.S. Metro Average	19,885	U.S. Metro Average	21,696

Table 5. Income Per Person in Metro Areas, 1970, 1980, and 1990 (in dollars).
Source: United States Bureau of Economic Analysis, Regional Economic Information System.

Photo 7. Montana State Capitol, Helena.

low-paid or unpaid students at Ricks College comprise a substantial portion of the county's total population.

ECONOMIC RESTRUCTURING

Like the nation as a whole, the Pacific Northwest is undergoing economic restructuring within industry groupings and companies, regionally, and among sectors of the economy. Restructuring within industry groupings and companies occurs as employers substitute capital for labor and close production facilities or lay off employees. Regional restructuring occurs when employers move production. Sectoral restructuring consists of shifts in the economy from goods-producing activities (e.g., manufacturing and construction) to services-producing activities (e.g., government, trade, and services).

Changes occurring in the timber products industry illustrate industrial restructuring. Regional employment in the timber products industries declined substantially from a peak employment of 190,000 in 1979 to 140,000 in 1982, a loss of 50,000 jobs in less than five years (Figure 7). Timber products employment recovered somewhat during the latter part of the 1980s, but not to pre-recession levels. Timber harvest

1970		1980		1990		1994	
Highest Income Counties	Income	**Highest Income Counties**	Income	**Highest Income Counties**	Income	**Highest Income Counties**	Income
Clark, ID	6,283	Sherman, OR	18,750	Clark, ID	31,370	San Juan, WA	28,992
Sherman, OR	5,930	Garfield, WA	13,226	San Juan, WA	24,108	Blaine, ID	27,712
Lincoln, WA	5,552	Gilliam, OR	12,882	Blaine, ID	22,782	Clark, ID	27,154
Garfield, WA	5,529	Lincoln, WA	12,832	Sherman, OR	20,193	Sherman, OR	22,886
Camas, ID	5,404	Morrow, OR	12,789	Lincoln, WA	19,601	Chelan, WA	21,176
Gilliam, OR	5,255	Clark, ID	11,622	Garfield, WA	18,895	Deschutes, OR	20,341
Columbia, WA	1825, Columbia, WA	Power, ID	11,504		18,795	Benton, OR	20,327
San Juan, WA	5,026	Blaine, ID	11,384	Adams, WA	18,217	Skagit, WA	20,177
Adams, WA	4,644	San Juan, WA	11,281	Skagit, WA	17,768	Valley, ID	20,103
Lewis, ID	4,483	Adams, WA	10,959	Deschutes, OR	17,608	Jefferson, WA	19,885
Lowest Income Counties	Income	**Lowest Income Counties**	Income	**Lowest Income Counties**	Income	**Lowest Income Counties**	Income
Madison, ID	2,351	Franklin, ID	5,906	Madison, ID	9,685	Madison, ID	11,085
Lake, MT	2,513	Madison, ID	6,370	Franklin, ID	10,627	Mineral, MT	11,707
Bear Lake, ID	2,695	Ferry, WA	6,565	Bear Lake, ID	11,039	Franklin, ID	12,361
Sanders, MT	2,746	Owyhee, ID	6,573	Mineral, MT	11,118		12,841
Granite, MT	2,817	Pend Oreille, WA	6,573	Sanders, MT	11,482	Bear Lake, ID	13,032
Franklin, ID	2,845	Jefferson, ID	6,576	Oneida, ID	11,514	Sanders, MT	13,095
Madison, MT	2,851	Boundary, ID	6,666	Boundary, ID	11,562	Oneida, ID	13,369
Bonner, ID	2,859	Lake, MT	6,718	Teton, ID	11,647	Owyhee, ID	13,401
Gem, ID	2,874	Payette, ID	6,749	Lincoln, MT	12,027	Jefferson, ID	13,651
Powell, MT	2,888	Mineral, MT	6,784	Deer Lodge, MT	12,128	Minnedoka, ID	13,688
Income		Income		Income		Income	
U.S. Average	4,047	U.S. Average	9,940	U.S. Average	18,666	U.S. Average	21,696

Table 6. Nonmetro Counties, Highest and Lowest Incomes per Person, 1970, 1980, 1990, and 1994 (in dollars).
Source: United States Bureau of Economic Analysis, Regional Economic Information System.

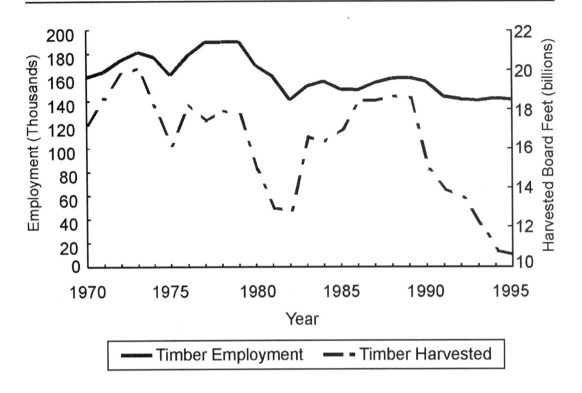

Figure 7. Socioeconomic Specialization of Nonmetro Counties.

* See Table 8 and this chapter's appendix for more information on socioeconomic specialization.

during the late 1980s, however, was greater than just before the recession, showing that the industry was becoming less labor intensive.

In the 1990s, employment once again declined to levels of the recession years in the early 1980s. Similarly, timber harvest rebounded during the latter 1980s but by 1995 had declined to levels below those of the recession years.

In some parts of the United States, regional economic restructuring has been represented by the movement of production facilities from metro to nonmetro areas, and more recently, to foreign locations. In the Pacific Northwest, the opening of high technology and other types of branch plants is an example of regional restructuring that is based partly on lower costs and partly on other factors such as the availability of skilled labor, quality of life amenities, and tax incentives or other government-sponsored inducements.

With rapid growth in the services sector along with slow growth in manufacturing and to a lesser extent in government, sectoral restructuring is occuring in the Pacific

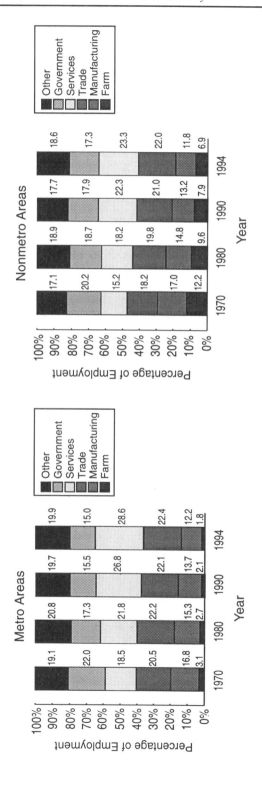

Figure 8. Timber Industry Harvest and Employment, 1970–1992*.
Source: United States Forest Service (1980, 1982, 1992).

* Includes data for all of Montana. Most of the state's timber production occurs in the 17 counties included in western Montana.

Northwest's metro and nonmetro areas. In 1970, for example, the services sector's share of employment was 18.5 percent in metro areas and 15.2 percent in nonmetro areas (Figure 8). By 1994, services' share had risen to 28.6 percent in metro areas and 23.3 percent in nonmetro areas.

By contrast, between 1970 and 1994, manufacturing's share of the Pacific Northwest's economy fell from 16.8 percent to 12.2 percent in metro areas, and from 17 percent to 11.8 percent in nonmetro areas. For metro areas, the greatest decline in share occurred for the government sector, which dropped from 22 percent of employment in 1970 to 15 percent in 1994. For nonmetro areas, the greatest decline in share occurred for the farm sector, which dropped from 12.2 percent in 1970 to 6.9 percent in 1994. Overall, the farm and government sectors account for a greater share of total employment in nonmetro areas than in metro areas, services are relatively more important in metro areas than in nonmetro areas, and manufacturing and trade are similar in relative importance in metro and nonmetro areas.

Photo 8. Abandoned Farm, Klickitat County, Washington.

THE REGION'S ECONOMY IN THE 1990S

Issues Affecting the Region

Federal macroeconomic, fiscal, and trade policies will continue to have important implications for the region. High interest rates, for example, have in the past adversely affected the region's ability to export its products. Free trade agreements, such as the North American Free Trade Agreement (NAFTA) between the U.S., Canada, and Mexico, also are having beneficial impacts on some sectors of the economy and adverse impacts on others. The net effect of these agreements is not yet clear.

Other efforts to restrict or promote trade with other Pacific Rim countries could adversely or positively affect the Northwest. One such effort is the Pacific NorthWest Economic Region (PNWER), established in 1989 by governmental leaders in Alaska, Alberta, British Columbia, Idaho, Montana, Oregon, and Washington (Bluechel 1993). The purpose of the PNWER is to encourage public-private cooperation in the promotion of the Pacific Northwest's economy in the global marketplace. International trade is encouraged through the PNWER's working group, Export NorthWest, whose major accomplishment to date is the establishment of an information system for matching Northwest products with prospective buyers.

Federal and state laws for threatened or endangered species will continue to be of great concern to many northwesterners. The June 1992 federal listing of the northern spotted owl as a threatened species has extended the shadow cast on the region's timber industry. Restricting timber harvest in spotted owl habitat, much of which is in old-growth forests, has contributed to a reduction in timber-based jobs. Timber-harvest proponents and opponents generally were unable or unwilling to compromise in the 1980s and thus far in the 1990s. Optimism about resolving conflicts between the timber industry and environmental groups grew with the Forest Conference convened in Portland by President Clinton in March 1993. Congressional and legal actions since then show that conflicts are far from over.

The listing of other species as threatened or endangered could raise even more concerns about the region's economic future, especially in nonmetro areas. One example is the September 1992 listing of the marbled murrelet as a threatened species, which has resulted in restrictions on the harvesting of old growth timber within a 50-mile band inland from the Oregon and Washington coasts.

A potentially more significant issue is the listing of various species of salmon as threatened or endangered. Estimates indicate that 100 years ago, 10 million to 16 million salmon and steelhead trout migrated annually into the Columbia River Basin (Northwest Power Planning Council 1987). The number in the early 1990s was down to about two million and falling, largely due to construction of dams, overfishing, and destruction of habitat for spawning.

Declining numbers of salmon in the Columbia-Snake River basin led the federal government in the early 1990s to list as endangered the Snake River sockeye salmon as well as Snake River spring/summer chinook and fall chinook. Further listing could

Photo 9. Shutdown Sawmill, Paisley, Oregon.

lead to dramatic increases in electricity prices and shipping rates for agricultural and other commodities due to the need to direct more of the Columbia Basin's water to the survival of salmon.

Listing of coastal salmon and other migratory fish raise similar concerns. Of special interest are coastal runs of coho salmon and steelhead trout. The July 1996 listing of another migratory species—the Umpqua River cutthroat trout—as endangered is viewed by some as a precursor to the listing of coastal salmon and steelhead. Listing of additional species likely would lead to further restrictions on logging and other uses in the Coast Range, Cascades, and elsewhere in the western part of Pacific Northwest.

Efforts to Diversify

To reduce cyclical effects and dependency on one or a few sectors, local and regional groups throughout the Northwest will continue efforts to diversify their economies. Diversification occurs primarily in two ways: 1) as declines occur in dominant (basic) economic sectors without corresponding declines in support sectors (Type One), and 2) as communities or regions become more attractive for economic activities that previously were less significant there (Type Two). Type One diversification is most notable in the region's resource-based communities, especially those dependent on timber and wood products. Type Two diversification is the type on which most economic development efforts focus.

Declining employment in resource-based sectors does not necessarily lead to corresponding declines in other economic sectors, especially if owners of retail and service businesses believe that the resource-based sectors will recover. Type One diversification occurs more or less by default as employment declines in the resource-based sector. Such diversification, however, is short-lived if recovery occurs or if local retail and service businesses eventually close in the absence of recovery.

Efforts to achieve Type Two diversification include attracting branch plants or relocating businesses, encouraging expansion of existing businesses, or starting-up new businesses. Establishing a favorable economic climate for business expansion and start-ups is receiving increased attention due to the realization that few businesses begin new branch plants or relocate, and that many communities in the U.S. and elsewhere are competing for only a few branch plants and relocations.

Economic diversity is greater in the region's metro areas than in its nonmetro areas (Beyers 1991), and future efforts to diversify likely will be more achievable in larger communities than in smaller ones. In the short-term, growth rates may be higher in communities specializing in one or a few sectors. Over the long-term, economies of areas able to diversify through new start-ups and business expansions generally will perform better than economies tied to resource-based industries subject to cyclical and structural changes.

Summer and fall tourism and recreational activities contribute to diversification in selected mountainous, coastal, and otherwise scenic parts of the region. Efforts will increase to reduce seasonal variations in tourism by promoting wintertime activities such as skiing and storm and whale watching along the Pacific Coast. Nature-based tourism, sometimes termed "ecotourism," will receive greater emphasis in a few nonmetro areas.

Establishing gaming centers for bingo and other forms of legalized gambling has become an increasingly popular development strategy, especially for Native American reservations. Passage of the Indian Gaming Regulatory Act in 1988 created new opportunities for Native Americans seeking to diversify economies of reservations and other tribal lands. The act allows tribes to establish facilities offering games not expressly forbidden by state laws.

Four years after the act's passage, the Cow Creek Band of the Umpqua Tribe opened the Pacific Northwest's first gaming center in Canyonville, Oregon. This facility initially allowed bingo only, but subsequently was expanded to offer a wide variety of games as well as lodging facilities. By Fall 1996, there were two tribal casinos in Idaho, six in Oregon, and 15 in Washington (Figure 9). Although several tribal casinos operated in eastern Montana, none was located in the state's 17 western counties.

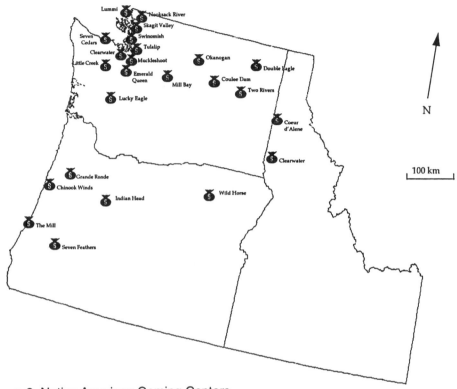

Figure 9. Native American Gaming Centers
Source: Based on Bourie (1996)

Establishment of casinos has not been without controversy. For example, several tribes in Washington are operating casinos without having a state-tribal compact as required by the Gaming Regulatory Act. In Oregon, opening of two casinos in a heavily traveled highway corridor has contributed to congestion and highway safety concerns. In Idaho, controversy was so great that voters in 1992 passed a constitutional amendment preventing Native Americans from opening casinos. This was not enough, however, to stop casino development because Idaho later implemented a state-run lottery, thereby clearing the way for reservation gambling.

Controversy also has arisen due to community discontent with several tribes building or seeking to build casinos on off-reservation tribal property. As sovereign nations, tribes are not required to negotiate with local officials, who in some cases have vigorously tried to keep casinos away from their communities. Despite these and other issues, more casinos are likely to open because of the lucrative revenues generated for economic development and other tribal programs.

To some residents, gaming centers are but one type of "locally unwanted land use" (LULU) or activity they do not want "in their backyard" (NIMBY). Other examples include landfills, chemical waste-disposal sites, and prisons. While some communities view these activities as highly undesirable, others welcome them. Several landfills have opened in remote parts of eastern Oregon and Washington, primarily to handle solid waste generated in Seattle, Portland, and other metropolitan areas, including from as far away as the Napa Valley in California (Burke 1995). A number of state prisons also have been built in eastern Oregon and Washington, but have run into considerable opposition in western parts of these states.

Other efforts to diversify the region's nonmetro areas include attracting or starting high-technology or other "non-traditional" companies, especially in the larger communities with universities. Adding value to, rather than exporting, raw materials will continue to be part of an overall economic development strategy in some areas.

Attracting retirees will also likely receive greater attention. Increasing numbers of retirees will be tempered by concerns about greater demands for certain types of services, health care for example, and about reduced support for others, for example, public education.

A potentially emerging trend is the movement of "lone eagles" to locations where information technologies allow them to conduct business from their homes or small offices (Beyers and Lindahl 1996). A recent study in Washington state revealed that lone eagles accounted for about three percent of in-migrants statewide in 1994 (Salant, Carley, and Dillman 1996). A growing number of lone eagles may help diversify economies of remote areas, especially those with scenic amenities and the necessary technological infrastructure.

REGIONAL ECONOMIC DEVELOPMENT EFFORTS

Private Sector

Hundreds of private sector groups are working to diversify or otherwise strengthen the Pacific Northwest's economy. Most private sector groups are community-based organizations, such as local chambers of commerce or private economic development organizations. Other private sector groups, including utilities, railroads, and financial institutions, provide assistance to users of their services.

Private foundations and environmental groups have recently become more involved in community economic development. The Northwest Area Foundation, for example, works to promote economic revitalization and improve the standard of living for residents of Idaho, Iowa, Minnesota, Montana, North Dakota, Oregon, South Dakota, and Washington. Headquartered in St. Paul, Minnesota, the foundation provides grants to alleviate rural and urban poverty and promote sustainable development throughout its eight-state service area.

Photo 10. Prison-Siting Protest, Linn County, Oregon.

Several environmental groups have recognized the need to resolve problems associated with employment declines in resource-based industries. Examples of such groups include Ecotrust, which has provided assistance and funding in the Willapa Bay area of Washington; the Pacific Rivers Council, which has helped several Oregon communities to identify economic development potential associated with watershed restoration; and The Wilderness Society, which has assisted in efforts to diversify timber-dependent communities (The Wilderness Society 1992; Johnson 1993; Bryant 1997).

A number of groups have actively argued against locking up natural resources on federal lands. The Sagebrush Rebellion of the 1980s has been followed by the wise-use movement of the 1990s. Leaders of these movements have argued for the protection of private property rights and against excessive restrictions on the extraction of resources from public lands. Such groups have become increasingly vocal in the Pacific Northwest with its high percentage of publicly owned lands.

Public Sector

Numerous public sector organizations are working to diversify economies at the local or regional level. City or county economic/community development departments are probably the most widespread. Other groups include port districts and authorities, which facilitate business development in their service areas, and multi-county economic development districts and councils of government, which provide assistance in some parts of the Northwest.

Through numerous agencies and programs, the federal government directly and indirectly influences regional economic development. Some programs of federal agencies adversely affect the economies of communities in the Northwest; other programs are intended to encourage economic development. Examples of the latter include programs of the Department of Agriculture, Economic Development Administration, Department of Housing and Urban Development, and the Small Business Administration. The Dislocated Worker Program, which receives funding through the U.S. Department of Labor, has become one of the most well-known federal assistance programs in some of the Pacific Northwest's timber-dependent communities.

Another example of a federal assistance program is the Northwest Economic Adjustment Initiative, established as an outcome of the Clinton Administration's 1993 Forest Conference. The Initiative works with regional leaders in California, Oregon, and Washington to combine federal, state, and other resources that will help revitalize economies of timber-dependent communities, including those adversely affected by listing of the northern spotted owl as an endangered species.

State governments influence economic development in a variety of ways. The most obvious are through levying or forgiving taxes and fees, and through expenditures of revenue. Government investments in human and infrastructural development probably have the greatest impact on long-term economic development. The former includes investments in K-12 schools, and in public universities and colleges, which train future workers and provide assistance through extension programs, business development centers, and other outreach services. Infrastructural investments include those in public works such as transportation, sewers and treatment facilities, and water lines and related facilities.

State economic development agencies provide direct assistance. Publishing economic and demographic data and providing information about available sites and buildings are among the most basic services. State development agencies also recruit businesses, run lending programs, provide grants, and offer technical assistance and training. Historically, development efforts have focused on manufacturing, but more recently, they have diversified to target many different types of businesses, including those in the travel and tourism industry and film making. State development agencies operate international offices to encourage reverse investment, facilitate exports from the Northwest, and otherwise promote international trade and investment.

During the remainder of the 1990s, regional economic development strategies will continue to evolve and will include efforts to enhance international trade, tourism, targeted recruiting, film and video production, development of infrastructure, worker training, and community assistance, especially for those affected by changes in natural resource industries. Programs to create an economic environment that supports "growing your own business" will no doubt continue.

Private and public groups in the Pacific Northwest may become more involved in efforts to evaluate economic and environmental tradeoffs. Emphasis on ensuring economic vitality through environmental protection, conservation, and natural landscape preservation could lead to innovative economic development programs and strategies which have received little attention to date (Power et al. 1995; Power 1996; Rasker 1995).

APPENDIX

ANALYZING THE REGIONAL ECONOMY: A BRIEF PRIMER

Overview

Books, magazines, newsletters, and similar published sources provide information about historical and recent trends in regional economies. Many such documents depend wholly or in part on data from state governmental agencies. Unfortunately, because information is not compiled uniformly among states, data for multi-state regional analyses often must be sought from other sources.

Among these other sources are federal agencies such as the U.S. Department of Commerce and the U.S. Department of Labor. The Bureau of the Census, a sub-agency of the Department of Commerce, provides the most comprehensive economic data for states, counties, cities, and other geographical units. Every 10 years, the Bureau conducts a census of the population; 1990 is the most recent year for which data are available from the Census of Population, Census of Housing, and Census of Population and Housing.

The Bureau of the Census also produces *County Business Patterns*, an annual publication released to the public about three years after the most recent year for which data are available. This publication provides relatively detailed data for most sectors except agriculture and state and local governments. Because of concerns about disclosing information about individual companies, data are sometimes withheld, especially for small counties where a single business may account for all the employment in an industry grouping.

The U.S. Bureau of Economic Analysis (BEA), another sub-agency of the Department of Commerce, is a frequently used source of annual data for two important economic indicators: income and employment. These data are provided in published and unpublished formats that allow comparison among states, counties, and other geographical units. For example, the BEA publishes the *Survey of Current Business*, a monthly magazine that provides numerous tables of changes in national, state, and regional economic indicators. The BEA also produces the Regional Economic Information System (REIS). Data from the REIS are released to the public about two years after the most recent year of data in the system.

Analysis of regional economic change can focus on a multi-state region, on the region's component states, on counties, or on other geographical sub-units. Analysis can further focus on parts of a region defined by characteristics such as urban or rural. Decennial censuses from the Bureau of the Census provide the most detailed economic data for urban and rural areas.

To examine economic changes for years between those for which the Bureau of the Census collects data, researchers rely on other sources of data and geographical units of analysis. Data for metropolitan (metro) and nonmetropolitan (nonmetro) areas are sometimes used when data for urban and rural areas are not readily available.

Metro areas, except those in New England, contain one or more central counties or similar geographical units with 1) a central city of 50,000 population or more, and 2) an urbanized population of at least 50,000 population in one or more metro counties with at least 100,000 population (U.S. Office of Management and Budget 1990). Outlying counties may be included as part of a metro area if they are contiguous and meet specified requirements of commuting to the central county(ies) and of metropolitan character, such as population density and percent of the population that is urban. All other counties are nonmetro.

Based on results of the 1990 Census, there are 22 metro and 114 nonmetro counties in the Pacific Northwest (U.S. Office of Management and Budget 1992). The number of metro and nonmetro counties represents a change from the 1980 Census when 20 counties were metro and 116 were nonmetro. The changes are the addition of Columbia County, Oregon to the Portland-Vancouver metro area, and the addition of Canyon County, Idaho to the Boise metro area. Other changes include 1) the addition of Ashland to the title of the Medford metro area (i.e., Medford-Ashland), and 2) the deletion of Vancouver as a separate metro area and its inclusion with the Portland metro area (i.e., Portland-Vancouver).

In late 1996, the U.S. Office of Management and Budget designated an additional metro area in the Pacific Northwest: Pocatello, Idaho (Miller 1996). The analysis of economic trends in this chapter was conducted before the designation of Pocatello (Bannock County) as a metro area.

To differentiate economic changes in metro areas from those in nonmetro counties near metro areas and those in more remote nonmetro counties, researchers have developed the concepts of nonmetro adjacency and nonadjacency. Nonmetro counties near metro areas often are economically integrated with the metro counties and are termed "nonmetro adjacent." These are counties that are physically adjacent to one or more metro counties; at least two percent of the employed labor force in the nonmetro county commutes to a central county (Hines, Brown, and Zimmer 1975; Butler 1990). All other nonmetro counties are nonadjacent.

The time period for examining economic data depends on the analyst's objectives. The analyst may want to portray the economy at one or more points in the past or during a recent year, or may want to look at change over time. Unfortunately, data may not be available for the geographical units and time periods the analyst wants to examine; thus choices must be made about geographical units, data sources, and time periods over which to portray a region's economy.

The choice made for this chapter was to use county-level employment and income data from the BEA's Regional Economic Information System. These data were used to show regional changes for metro and nonmetro areas from 1970 to 1994, the most recent year for which data are available. Data were obtained on-line from the University of Virginia Social Sciences Data Center at http//www.lib.virginia.edu/socsci/reis/reis1.html.

Information about Recent Economic Changes

Most sources of data for viewing broad-scale economic changes in regions such as the Pacific Northwest are several years old before they become available to the public. Analysis based on these sources thus is "dated" and may not capture significant recent changes. Two examples illustrate. In the early 1990s, unlike most of the 1970s and 1980s, economic growth was stronger in Idaho than in Oregon and Washington. Additionally, substantial layoffs at the Boeing Company contributed to a less robust economy for Seattle and the Puget Sound area in the early 1990s than was suggested by trends of the 1980s.

To understand more recent changes, analysts consult a variety of sources. Major newspapers, such as the *Idaho Statesman*, *The Oregonian*, *The Seattle Times*, the *Seattle Post Intelligencer*, and the *Spokane Spokesman Review* are very useful daily sources of information about the regional economy. *Marple's Business Newsletter*, a Seattle-based publication, provides information about the regional economy every two weeks. *Northwest Economic Indicators*, published bimonthly in Portland, covers economic trends in Alaska, Idaho, Montana, Oregon, and Washington. *U.S. Bank's Economic Barometer*, published semi-annually in Portland, covers economic trends in Idaho, Nevada, Oregon, Utah, Washington, and 34 counties of northern California.

U.S. Bancorp, in cooperation with the Northwest Policy Center at the University of Washington, produces *Northwest Portrait*, an annual publication that covers economic trends in Alaska, Alberta, British Columbia, Idaho, Montana, Nevada, Oregon, Utah, Washington, and northern California. The Northwest Policy Center works with economic policy in Alaska, Idaho, Montana, Oregon, and Washington, and produces numerous documents, including *The Changing Northwest*, a quarterly newsletter addressing strategies for a vital economy. *Communities in the Lead: The Northwest Rural Development Sourcebook* is an example of one the Center's publications designed to help the region's communities plan and manage economic development efforts.

State employment security agencies provide published and unpublished data about state economies. In the Pacific Northwest, these agencies are the Idaho Department of Employment in Boise, the Oregon Employment Department in Salem, the Washington Employment Security Department in Olympia, and the Montana Department of Labor and Industry in Helena. Various other state government agencies, such as agriculture, economic development, and transportation, produce newsletters or other information about regional economic changes.

REFERENCES

Beyers, William. "Trends in Service Employment in Pacific Northwest Counties: 1974–1986" *Growth and Change* 22(4)(1991):27–50.

Beyers, William and David Lindahl. "Lone Eagles and High Fliers in Rural Producer Services" *Rural Development Perspectives* 11(3)(1996):2–10.

Bluechel, Alan. "The Pacific NorthWest Economic Region: A Multi-State, Multi-Province Regional Approach to Economic Development" *Economic Development Review* 11(3)(1993):27–29.

Bourie, Steve. *1997 American Casino Guide*. Dania, FL: Casino Vacations, 1996.

Bryant, Rebecca. "Earth Tones" *Planning* 63(1)(1997):19–21.

Burke, Jack. "First California Trash Exported by Rail" *Traffic World* (July 3, 1995):22–23.

Butler, Margaret. Rural-Urban Continuum Codes for Metro and Nonmetro Counties. Staff Report No. AGES 9028. Washington, D.C.: U.S. Department of Agriculture, Economic Research Service, April 1990.

Hines, Fred, David Brown, and John Zimmer. *Social and Economic Characteristics of the Population in Metro and Nonmetro Counties, 1970*. AER-272. Washington, D.C.: U.S. Department of Agriculture, Economic Research Service, March 1975.

Johnson, Kirk. "Emerging Lessons for Reconciling Community and Environment" *The Changing Northwest* 5(1)(1993):4–5. Seattle: University of Washington, Northwest Policy Center.

Miller, Berna. "Metro Status" American Demographics 18(1)(1996):26–27.

Northwest Power Planning Council. Compilation of Information on Salmon and Steelhead Losses in the Columbia River Basin. Portland, 1987.

Power, Thomas Michael et al. *Economic Well-Being and Environmental Protection in the Pacific Northwest*. A Consensus Report by Pacific Northwest Economists. Missoula: University of Montana, 1995.

Power, Thomas Michael. *Lost Landscapes and Failed Economies*. Washington, D.C.: Island Press, 1996.

Rasker, Ray. *A New Home on the Range: Economic Realities in the Columbia River Basin*. Washington, D.C.: The Wilderness Society, 1995.

Salant, Priscilla, Lisa Carley, and Don Dillman. "Estimating the Contribution of Lone Eagles to Metro and Nonmetro In-Migration." Technical Report no. 96–19. Pullman: Washington State University, Social and Economic Sciences Research Center, 1996.

Schwantes, Carlos. *The Pacific Northwest: An Interpretive History*. 2nd Edition. Lincoln: University of Nebraska Press, 1996.

The Wilderness Society. *From Dreams to Realities: Diversifying Rural Economies in the Pacific Northwest.* Washington, D.C., 1992.

U.S. Forest Service. *Production, Prices, Employment, and Trade in Northwest Forest Industries.* Portland, Oregon: Pacific Northwest Research Station, 1980, 1981, 1992, and 1996.

U.S. Office of Management and Budget. "Revised Standards for Defining Metropolitan Areas in the 1990's" *Federal Register* 55 (March 30, 1990):12154–12160.

U.S. Office of Management and Budget. *Revised Statistical Definitions for Metropolitan Areas.* OMB Bulletin no. 93–05. Washington, D.C., 1992.

MANUFACTURING IN THE NORTHWEST   Chapter 12

William A. Rabiega with John Hall

INTRODUCTION

Lumber, pulp, and paper from the forest. Canned and frozen fruits and vegetables from the farm. Aluminum ingots refined using cheap, hydroelectric power. These are probably the first images which come to mind when manufacturing in the Northwest is mentioned. A close link between the factory and the land is presumed.

Next, someone might remember his or her Pendleton jacket or Jantzen swimwear. The link to the land becomes more remote, and the role of the region's entrepreneurs and innovators begins to emerge. If you are from Seattle or Portland, the high technology of Boeing and Tektronix and the people who founded those firms will come to mind. You yourself might be one of the highly skilled people who work for those companies. You will almost certainly know someone who is in their employ.

All of these images are valid as far as they go, but they do not give a complete picture of manufacturing in the Northwest. The intent of this chapter is to elaborate and sharpen these pictures and to present new ones. Because this presentation uses some of the graphic and statistical techniques of analytical geography, the reckoning from it will be substantially more explicit than the image based on the popular impressions depicted in the opening paragraphs. Once this new reckoning is in place, it is hoped that not only will your knowledge of the Northwest be expanded, but you will also have a better idea of how an economic geographer might view the naively given landscape.

MANUFACTURING LOCATION THEORY

Knowing nothing else about a region, an economic geographer might begin her or his analysis of the region by reviewing some basic location theory tenets. What are the economics of location choice? What are the factors considered in location decision-making? How are those factors ordered and weighted by individuals with prominent roles in the decision process and companies with varying organizations?

Ultimately, the importance and distribution of manufactures in the region and the value of the resources which support that distribution must be assessed.

Both micro-economic and regional economic theory pertain to this analysis. Micro-economic theory provides that entrepreneurs locate firms so as to maximize utility. Classic location theory equates maximizing utility with minimizing costs. Hence the basic task of the decision maker is to find the place with the lowest aggregate costs for transportation, labor, taxes, power, loan interest, etc., and build a plant there.

Even if one follows these classical notions, it is unlikely that a perfect location will be selected. To do so the decision maker must have perfect knowledge of his or her product, market, competitors, and all the factors of location, and must not be constrained by limited funds or other considerations. The decision maker must also subordinate any personal goals, such as living in a pleasant climate or having quality, inexpensive housing, to the company's institutional goals. Such personal goals are external to the company's interests.

Current thought, while accepting that manufacturing locations must have a low enough cost to return a reasonable target profit, allows that other attributes than positive increments on the company balance sheet may add to the utility of a place. These include the amenity factors of climate, schools, housing, cultural institutions, and recreational opportunities. Most of these factors do not directly affect costs or revenues, but enhance the personal utility of the firm's employees. They can indirectly figure into revenues in that the firm is in a better position to attract highly qualified and mobile staff, or that in the long run, they might discount wages and salaries because of the high external amenities.

Current thought also holds that corporate goals, including those for targeted profit levels, may be affected by the organization of the company, or even by individuals in key positions within the organization. A small family-owned and managed firm may thus behave quite differently in its location decisions than a public multinational conglomerate. In the family firm a limited market may be given and personal assets, such as a localized businesspersons' network, may strongly color decisions. With large public firms, these elements are likely to have little effect.

As will be detailed subsequently, the presence of the lumber and wood products industry in the Northwest can be explained largely in classic location theory terms. Overall costs are minimized by processing near the raw material site. On the other hand, the presence of the aircraft and aerospace and electronic instruments industries must involve utilities other than pure cost minimization. There are, in fact, few if any industries wishing to serve the United States/North American market that would find a minimal cost location in the Northwest. Even lumber milled from Northwest forests is primarily distributed in the West Coast and Mountain states. The strongest assets the region holds for contemporary manufacturing locations are the quality of its workforce and its external amenities.

THE REGIONAL ECONOMIC BASE AND MANUFACTURES

The principal area of regional economic theory of importance to this discussion concerns the local economic base. The economic base is comprised of all the industries—agricultural, mining, manufacturing, or service—which bring income in from outside the region in which they are located. These firms can either sell outside the region or attract consumers into the region. In either case they serve to draw new wealth, often termed "clean dollars," to the region and are termed "basic" industries.

These "clean dollars" are then recirculated through the region when they are used to buy raw materials and support services locally and when the employees of the firm spend their earnings. Hence a single dollar in basic income may represent two or even three dollars in business transactions within the regional economy. Industries that primarily recirculate dollars derived from basic firms are termed "non-basic."

Manufactures are routinely considered to be basic industries, and it is often true that the goods they make must be sold beyond the localities in which they are made for the firms producing them to be economically viable. But some manufacturing activity is clearly non-basic. For example, a producer of computers with a worldwide market may consume virtually the total output of a local supplier of silicon chips. That local supplier, though a manufacturer, is thus part of the non-basic sector of the region; it does not bring in clean dollars, but is a recirculater of those brought in by the computer manufacturer. Its production, and all the transactions associated with that production, are part of the multiplier effect of the computer firm. The computer firm is part of the basic sector, and the chip firm, though closely linked to the computer firm, is included with non-basic activities.

In general, the economic health and character of a region is tied to the health and character of its basic industries. They are the motor that powers the economy. When production and profits are high in the basic industries, this boom ripples through the non-basic sectors and the personal incomes of those in the workforce. When business is bad for basic industries, the detriments of that weakness spread through the non-basic firms and, inevitably, show up in the personal circumstances of individual households.

Such negative effects can be mitigated by having basic industries which are not linked into local purchasing chains, or by having a large array of basic activities which sell in varied markets or have cyclical highs and lows at different times. Regions with such diversified sectoral structures differ substantially in their economic character from those wherein a few basic firms are linked to almost everything that happens. Thus, viewing and analyzing the revenues and/or the employment among sectors of a local economy can reveal much about its character. This is true of manufacturing compared to other major sectors, e.g., agriculture and services, and of manufactures compared among themselves.

TOOLS FOR SPATIAL AND SECTORAL ANALYSIS

Two fairly straightforward statistics are used here to aid in sectoral and spatial analysis, the shift/share index and the location quotient. The intent of both of these techniques is to give context to employment data for any industry. They are used most frequently when reviewing manufacturing in a locality and give context by comparing the current employee numbers in manufacturing categories to some base standard.

With shift/share indexes, the proportion of change in a subregion, here the northwestern United States, from a given base year to the current period is compared to the change in a broader region, here the total United States, over the same time. This is stated formally as:

$$SS_i = \frac{e_{ik}/e_{ij}}{E_k/E_j}$$

where: SS_i = the shift-share index for a given industry in region i over the time period j through k.
e_{ik} = current employment in the industry in region i.
e_{ij} = base year employment in the industry in region i.
E_k = current national employment in the industry.
E_j = base year national employment in the industry.

The numerator shows the regional shift in the industry, and the denominator shows the national shift. After the division of the denominator into the numerator, the index shows whether the industry is growing or shrinking in the region compared to the national development pattern. If the SS is less than 1.00, the industry is shrinking; if 1.00, it is following the national trend; and if greater than 1.00, the regional growth is greater than that nationally or the shrinkage is less than the national rate.

The location quotient is useful to show spatial concentrations of a manufacture and depicts a rough definition of a localized economic base. It is calculated by comparing the proportion an industry has of subregional employment compared to that for national employment. Formally:

$$LQ_{ij} = \frac{e_{ij}/te_i}{E_j/TE}$$

where: LQ_i = the location quotient for the ith place
e_{ij} = employment in the jth industry in the ith place
te_i = total employment in the ith place
E_j = national employment in the jth industry
TE = total national employment

When a LQ_i is 1.00, the locality has exactly the same proportion of its workforce in an industry, the jth industry, as the nation; when it is 0.00 to .999, the locality shows a dearth compared to the national standard; and when the score is greater than 1.00 an abundance compared to that standard is indicated. This index is also roughly tied to the economic base of a region. An abundance of an industry implies that some of its production must be exported, thus bringing in income from outside the region. Thus high LQ_i's are associated with basic industries.

However, a single firm which sells goods nationally or even internationally may not in itself employ enough workers in its county of location to cause an L.Q. of greater than 1.00. This is particularly true in metropolitan areas with large workforces and an overall dominance of service industries in the economy. Similarly, metropolitan counties may show less concentration in manufacturing in general compared to more rural places because in metropolitan areas business is primarily in service industries. The L.Q. is more about showing the importance of an industry locally than the absolute amount of it that is present.

In the analyses of manufacturing geography which follow, the shift/share analysis is used to show changes in the importance of manufacturing in the Northwest. The locational quotients will show where selected manufactures are concentrated and give some idea of how basic these activities are to the various local economies.

THE CHANGING STRUCTURE OF NORTHWEST MANUFACTURING

Since 1960 the proportion of United States workers in manufacturing has continually decreased. Table 1 shows the decade by decade changes in proportion, and the percentage shift associated with those changes. Note the precipitous drop between 1980 and 1990 and the overall drop between 1960 and 1990. Employment in manufacturing as a proportion of jobs has clearly diminished in the United States over the last three decades, with the greatest declines during the corporate restructuring of the

Manufacturing Share of Employment

1960	1970	1980	1990
0.2709	0.2445	0.2244	0.1769

Shift

1960–1970	1970–1980	1980–1990	1960–1990
−9.75 %	−8.22 %	−21.17 %	−34.7 %

Table 1. Shifts in Manufacturing Employment in the United States, 1960 to 1990

	1960	1970	1980	1990
Manufacturing Employment	451,379	486,282	649,735	726,193
Total Employment	2,104,861	2,539,634	3,644,747	4,408,347
Share	0.2144	0.1915	0.1783	0.1647
Shift	1960–1970	1970–1980	1980–1990	1960–1990
	−10.68 %	−6.9 %	−7.63 %	−23.18 %

Table 2. Manufacturing shift/share for the Northwest 1960–90

1980s. This is not to say the absolute number of manufacturing workers has decreased. In fact, that number has gone up, but it has not gone up in proportion to the overall growth of employment.

In the Northwest absolute employment in manufacturing has increased over that same time period; however, following the national trend, it has not grown as much as the rest of the economy. Hence, the share of jobs in manufacturing has decreased, but not as sharply as in the United States as a whole. Table 2 shows these changes. Hence, while manufacturing has become a less significant proportion of employment in both the United States as a whole and the Northwest, it is relatively stronger in the Northwest (see Table 3). Thus, the Northwest is now, by comparison to the rest of the country, more of a manufacturing region than it was in 1960. This change has occurred primarily in the last decade.

Percentage of U.S. Manufacturing	1960	1970	1980	1990
Employment in the Northwest	2.58	2.58	2.96	3.55
Shift/Share	1960–1970	1970–1980	1980–1990	1960–1990
	0.9897	1.0144	1.1718	1.1764

Table 3. Northwest Shift/Share of Manufacturing

MANUFACTURING LOCATION IN THE NORTHWEST

Location Quotients for manufacturing employment in Pacific Northwest counties are shown on Map 1. Three categories are represented, L.Q.'s of 1.2 or greater, L.Q.'s from 0.80 to 1.2, and those below 0.8. The data to support this, and all subsequent maps and numbers, comes from the Employment by Industry Tables of the 1990 United States Census of Population.

The highest category includes counties notably above the U.S. average in manufacturing employment, the middle group includes those within a random range of it, and the lowest group includes those counties which are not manufacturing-oriented in their employment structure. While no county in the region actually reports zero persons employed in manufacturing, the manufacturing workers in the lowest group of counties can most likely be classified in the non-basic sector.

Looking at Map 1, the home counties of major metropolitan areas, e.g. Multnomah County for Portland, Oregon, King County for Seattle, Washington, and those counties around them, show up in the top two categories, as should be expected. More notable are the number of rural counties that exhibit higher L.Q.'s, e.g. clusters in Northern Idaho and Montana. Manufacturing appears as much a rural employment as an urban one in the Northwest.

Food and Kindred Products

Persons employed in the manufacturing of food and kindred products are depicted in Map 2. This category of manufactures includes canning, freezing, and packaging foodstuffs, meat packing houses, bottling plants, dairies, etc.—familiar components of the industrial landscape. Kindred products include such things as packaging grass seed or nursery seedlings for shipment. Five groups of L.Q.'s are depicted where counties with scores of greater than 5.00 signify manufacturing specialization and an export-based (basic) sector of the local economy.

In general, the high L.Q.'s appear in rural counties, close to the agriculture which grows their raw materials. The orchard areas of southcentral Washington and northwest Oregon show these high numbers, as do the field crop and irrigated areas of Idaho. Clearly food processing is carried out near the sources of raw product, and periods of high employment likely follow seasonal harvests.

Textile Manufacture

There is not a single county in the Northwest where the L.Q. for employment in textile manufacture equals or exceeds 1.00. Thus, compared to the national standard, this industry is a relatively minor part of the Northwest's economic structure. Paradoxically, two homegrown textile companies, Jantzen, Inc. and Pendleton Woolen Mills, produce products that are recognized worldwide. However, both are known in niche rather than mass markets. Thus the volume of manufacture, and the employees

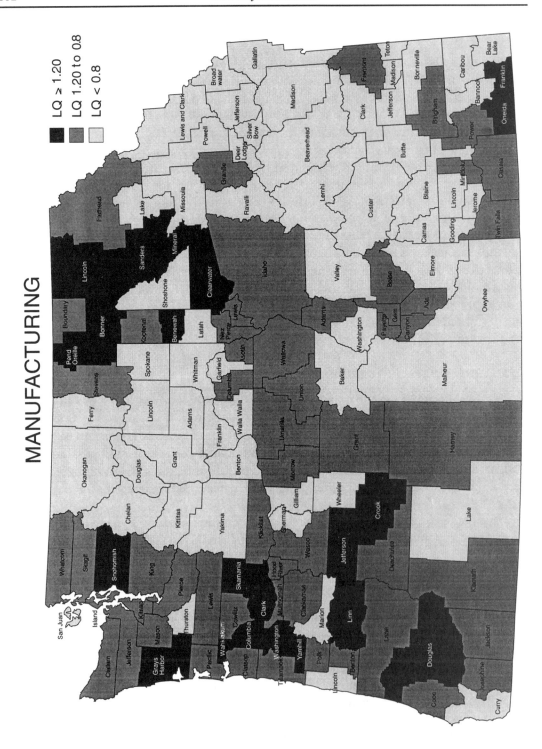

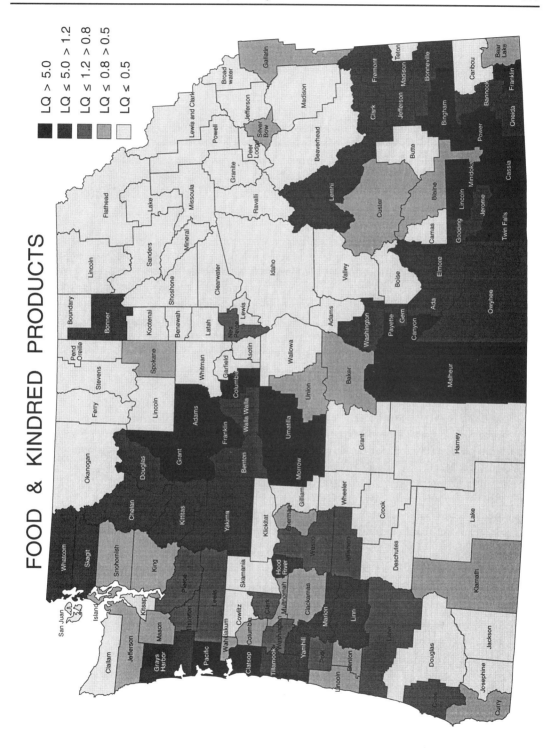

needed to support it, is relatively small. Also, particularly in the case of Jantzen, much of the actual fabrication is not done in the Northwest.

Jantzen, Inc.

Jantzen, Inc. was founded in Portland, Oregon as Portland Knitting Mill in 1910. One of the three founding partners, John A. Zehnbauer, managed the company until 1956. In its earliest days, the firm was both a manufacturing and retailing operation. The manufactures included heavy knit sweaters and hosiery and retailing was done over a wider range of apparel products. The company designed and manufactured its first swimsuits in 1913 for the Portland Rowing Club.

The company began operating under the Jantzen name in 1918. Through the 1920s swimwear was its major product, and the Red Diving Girl logo gained international recognition. By 1930 operations had also become international, with company-owned plants in Australia and England and a licensee in Canada. In the late 1930s Jantzen diversified its lines to include knit sweaters, foundations (bras, etc.), and sportswear (under the name of Sun Clothes). These basic product lines were continued into the 1960s, and new plants were opened in South Carolina to support them. The Canadian plant was purchased by Jantzen, further multi-nationalizing its structure.

Since then, company activities have centered on modernization of technology, rationalizing operations and internal management, and acquisitions and mergers. Jantzen acquired Breton Mills in North Carolina in 1966. In 1970 Jantzen shares were publicly listed on the New York Stock Exchange. Blue Bell, Inc. headquartered in North Carolina, purchased Jantzen in 1980. By that time, the intimate apparel division and the Australian subsidiary had been divested. Blue Bell, in turn, was purchased by VF Corporation in 1986. Jantzen now operates as an independent subsidiary, with about one-quarter of its sales volume in swimsuits and new lines in backpacking and mountaineering gear under the JanSport label.

Printing and Publishing

Like textile manufacture, the printing and publishing industry is not found in great concentration in the Northwest. A scattered handful of counties show L.Q.'s edging over 1.00, but there is no place with a notable focus of employment in this industry. Again, there may be a press or two with national and international distribution, but it remains that printing is a relatively limited and non-basic employment in the Northwest.

Chemical Manufacture

With the notable exception of several Idaho counties, Bear Lake, Bingham, Bonneville, Butte, Custer, Franklin, and Power, L.Q.'s for employment in chemical manufacture are well under 1.00. Within the Idaho counties, the actual numbers employed in even the concentrated places are in the low hundreds. This implies a single specialized plant. Thus the chemical manufacture in these counties is likely linked to localized resources, including the advantage of confining hazard to remote areas. As a region, however, the Northwest is not characterized by chemical manufacture.

Furniture, Lumber and Wood Products

As Map 3 indicates, the importance of manufacturing furniture, lumber, and wood products in the Northwest cannot be overstated. The number of counties with L.Q.'s of over 5.00 is remarkable. The great majority of counties in Oregon have 5.00's or more. Six counties, two each in Oregon, Idaho, and Montana, show numbers above or near 20.00. Grays Harbor county in Washington has an L.Q. of 53.62; better than half of all employed persons in that county work in this industry.

This is also a rural industry. In many remote counties, the employment structure better fits the stereotype for a gritty manufacturing town than a sylvan wilderness. In these places the health of the local economy is more tied to the industrial cycle than the harvest season. Discontinuities between common perception and economic realities are clearly implied.

Further, the regional importance of this industry should not be confused with its health. As detailed in other sections of this text, forest-linked industries are struggling to find a new equilibrium in the Northwest, and that eventual equilibrium will almost certainly be at a lower level of production than that historically seen. Pope and Talbot, a firm that grew up in the Northwest, starts as a rough and tumble frontier partnership and ends, for now, as a public multi-national with only a small proportion of its interests in the region of its youth.

Pope and Talbot in the Northwest

Pope and Talbot is a lumber, pulp, and paper products company headquartered in Portland, Oregon with a 140-year history in the Northwest. It began operations as a retailer of lumber in San Francisco in 1850. The founding partners also ran a "lighterage" service (using longboats to move cargo to and from ships moored in the harbor). By 1854 they were operating the Puget Mill Company at Port Gamble, Washington and acquiring a fleet of ships to move their lumber to San Francisco and other ports of opportunity.

In addition to lumber manufacture and shipping, the Mill Company was involved in import and export trade, retail marketing, land development, and

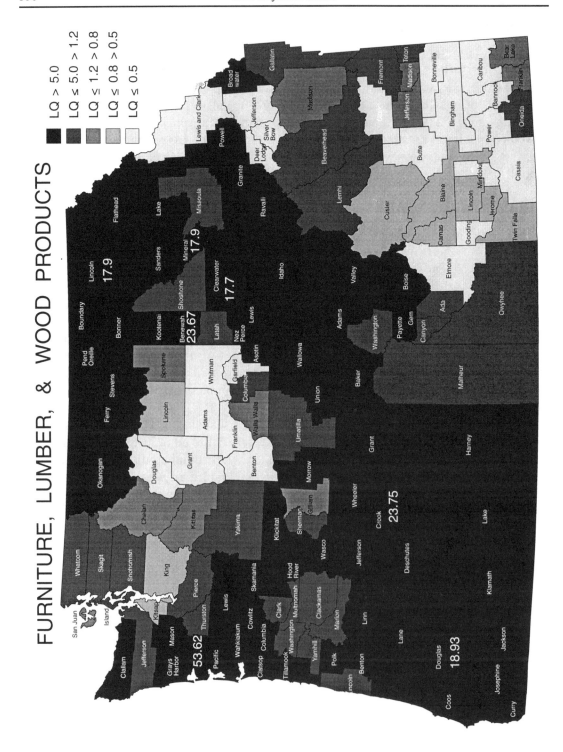

tugboat operations well into this century. The last of their maritime fleet, the Pacific Argentine Brazil Line, was divested in 1956. They withdrew from intercoastal trade, including shipping to San Francisco in 1963. Thus the vertical integration wherein the company held its own forest properties, cut, milled, and marketed its lumber, and was a merchant to its workers was also ended. Until this time the main market for its lumber was the northwestern U.S.A. and California, with episodic trade in virtually all points of the Pacific Rim.

By 1990, the company operated only a single sawmill in the Northwest at its historic beginning place in Port Gamble. This mill primarily supports the California market and operates or closes as building there prospers or declines. Its Northwest and Japanese markets are served by mills in British Columbia, Canada. It reaches new midwestern markets from mills in South Dakota and Wyoming. The primary reason cited by Pope and Talbot for divesting itself of sawmills and timberlands in the Northwest and acquiring the mills in British Columbia and other states is increasing environmental regulation and limitation of timber harvests in Oregon and Washington.

Pope and Talbot's single pulp plant at Halsey, Oregon supplies material to local newsprint and consumer product companies. However, the great bulk of its operations are now in diaper plants and tissue mills in California and the eastern U.S.A. Thus, while timber supplies were abundant and cheap in the Northwest, Pope and Talbot was primarily a regional lumber miller and wholesaler. As supplies decline and become more expensive, the company is shifting its operations away from lumber to nondurable goods and away from its historic home.

It is notable that, except for the decade of the 1920s, Pope and Talbot has always been managed by direct descendants of founding officers of the company. The movement away from the Northwest cannot thus be attributed to an outside takeover and management restructuring. In its 1990 Annual Report, Pope and Talbot reduced the asset value of their last Northwest mill, citing inadequate timber supplies to produce at previous, and assumedly more profitable, levels in the future. It seems unlikely that this milling company which grew up in Washington and Oregon will be sawing any timber in either of these states within a few years.

Primary Metal Manufactures

Primary metal manufacturing includes benification, smelting, refining, and some foundry functions. While the Northwest has been noted for its production of aluminum and various rare and valuable metals, employment concentration in these activities is found most frequently in Washington, where eleven counties show in the greater than L.Q. of 1.2 categories. Oregon shows eight such counties, and Idaho and Montana seven.

These are nearly all rural counties with a single plant employing possibly a few hundred workers. The historical linkage to hydroelectric power is important in these locations, however it is much less important than it formerly was, as the cost for this electricity is now relatively high by the international standards of this industry. What is again clear is the importance of yet another manufacture in the rural economies of the Northwest.

Fabricated Metals

Almost any firm which manipulates metal as its single raw material group to manufacture products can be characterized as a metal fabricator. Much metal fabrication is done as part of the non-basic component of a local economy. This is true in the Northwest where only a handful of counties show L.Q.'s for this manufacture in the 1.00 and 2.00 range. This is not a characteristic industry in the region.

Machinery Including Computers

This is a broadly inclusive category of manufactures including various traditional machinery, e.g. motors, and high technologies such as computers. As an industry, machinery is not widespread in the Northwest; only a few counties show scores above 1.00 and even these do not exceed 2.00. Again, some niche companies may ship widely, but this is in the main a non-basic component of the economy. Software development and distribution, an industry of note in the region, is a service industry, and not part of the fabrication of computing machines as a manufacture.

Other Electrical Equipment, Instruments, and Components

Insofar as the Northwest is identified with high technology electronics, it is probably in the manufacture of electrical equipment, instruments, and components. While employment in these manufactures is concentrated in only a handful of places, and in all but one instance, is between an L.Q. of 1.00 and 2.00, that single case has drawn enough notoriety to be named, "The Silicon Forest." The L.Q. for that case, Washington County, Oregon is 4.0126.

The local firm around which the "Forest" was seeded is Tektronix. Its history provides a good picture of this manufacture in this region.

Tektronix in Portland

Tektronix was formed in 1946 with a retail electronics as well as manufacturing operations. The founding partners were Portland residents, one of whom, Howard Vollum, had conceived a design for an improved oscilloscope while working with radar during World War II. The scope Tektronix manufactured was very high speed and accurate for its time and built to be used as a measurement and research instrument. Because the partners expected the worldwide market for such specialized instruments to be small—only around 100 per year—they expected to be a small firm with little specialization of jobs and commensurately little separation between employees and management. The retailing store split away in 1947, although much of the early support for the firm as a whole came from it.

The first permanent location for Tektronix was the second story of Hawthorne Electronics, the retailing operation. In 1947 there were six employees; by 1991 there were 11,636 in five U.S. states and four overseas locations. Several other firms' founders were originally "Tek" people, so the company's local impact on employment goes beyond its own plants.

From its inception, Tek remained heavily committed to research and development, hence it continued and continues to produce new products at the leading edge of technology. It virtually re-invented the cathode ray tube (CRT) for its scopes and is constantly involved in innovation.

Currently Tektronics has three divisions: test and measurement, television systems, and computer graphics. The company entered the television field in the middle 1950s, producing scopes to monitor equipment during broadcasts, as it does now. In the 1980s it began to produce computer graphic work stations and printers.

Transportation Equipment Manufacture

No manufacturing firm is more closely identified with the Northwest than Boeing Aircraft and Aerospace. The only notable concentration of employment in this category is King County (Seattle), and the counties around it. Quotients of 3.00 to 5.00 appear in these places. While some other companies, e.g., Freightliner Trucks in Portland, have major reputations and national distribution, the story of Boeing is the essential story of manufacture of this type in the Northwest.

Boeing Aircraft

The Boeing Company was founded in Seattle and began its operations as Pacific Aero Products in 1916. Its founder, William E. Boeing, a timberman turned flight enthusiast, remained its president and/or chairman of the board until 1934. Single engine seaplanes were the first product of the company with the U.S. Navy as the targeted customer.

In its early years the firm's financial performance was uneven. Orders won during World War I were cut back with the Armistice in 1918. Much of Boeing's early manufacturing work was as a subcontractor to build planes of other designers. Some stability was added by operating airlines for the limited passenger traffic at that time, and more importantly, as the contracted carrier for U.S. Airmail.

Economic security through both vertical and horizontal industrial integration was aggressively pursued by William Boeing. By the end of the 1920s, he was operating United Aircraft and Transportation Corporation, which held not only Boeing Airplane Company, but also Pratt & Whitney (aircraft engines), Hamilton (propellers), Chance Vought, Stearman, Sikorsky (other aircraft builders), Pacific Air Transport, and Boeing Air Transport. In 1934, in response to Federal anti-trust regulations, this diversified company was divided into the present day United Airlines, United Technologies, and the Boeing Company.

As a producer of military aircraft, Boeing is best known for its large bombers, the B-17 and B-29 of World War II, and the B-52 which is still in use. However, most people today primarily associate this company with its 700 series of civil passenger craft, the 707, 727, 737, 747, 757, 767, and the new 777. These are extremely successful products and account for a notable proportion of the foreign export manufacture income of the United States.

Yet, in the early 1970s, as the 747 was poised for introduction, the Boeing Company nearly failed. In that period, following a general depression in the airline industry, Boeing had few new orders (none domestically for 18 months), and was deeply involved in the ill-fated SST (SuperSonic Transport) program. The company reduced its workforce from 80,000 in 1970 to 37,000 by the end of 1971.

At that time Boeing diversified its activities to include spacecraft for NASA programs, air launched cruise missiles (from B-52 frames), and computer technology development. It was also involved in experiments with hydrofoils and mass transit vehicles. Some commercial jetfoils, light rail vehicles in Boston and San Francisco, and the demonstration Personal Rapid Transit system in Morgantown, West Virginia remain as remnants of these corporate and public transit experiments.

Today Boeing is a strong force in the aerospace industry. As was the case in the 1970s, the Seattle area's economic fortunes and those of the Northwest in general, are closely linked to the vitality of this firm whose prosperity is somewhat cyclical in nature.

SUMMARY

The Northwest is the home of diverse manufactures. It is correctly identified with the production of aircraft and some categories of high technology electronics. These industries are in the Northwest because the founding innovators of what would prove to be highly successful companies in new areas of technology made this region their home. It is properly noted for the wood, metal, and agricultural manufactures it produces. Here the link to resources from the forest, land, and rivers is evident. Yet another group of companies, like Jantzen, have identity with the region even though the employment they support is modest.

Manufacturing is becoming more important in the Northwest than it has been, at least in comparison to the national economy. Firms are being aggressively recruited to this region, particularly in the high technology electronics groups. Internationally based firms locating plants in the Northwest are among those supporting this manufacturing employment base, and their products are tied to international markets. In some sense, the Northwest is like a third-world nation emerging from being a producer and exporter of raw materials into a manufacturer. At the same time the service sector is growing even more quickly.

That manufacturing is strong and growing in this region in these times of the "post-modern" service-oriented economy is notable, even remarkable. But what is even more striking is the virtual spatial ubiquity of high concentrations of manufacturing employment across both urban and rural counties in the Northwest. This is very unlike an emergent third world nation, and perhaps unique to the region. This may be the abode of lumberjacks and farmers. It may be blessed by a physical geography that makes it a mecca for tourists. But it is at the same time very much a large and diverse manufacturing town.

REFERENCES

A Brief History of Jantzen, Inc. Portland, Oregon: Jantzen, Inc., 1990.

Annual Report, Pope and Talbot. Portland, OR: Pope and Talbot, 1990.

Bureau of Census. *1990 Census of Population: Social and Economic Characteristics.* Washington, D.C.: United States Government Printing Office, 1993.

Coman, Edwin T. and Gibbs, Helen M. *Time, Tide, and Timber; Over a Century of Pope and Talbot.* Portland, Oregon: Pope and Talbot, Inc., 1978.

Company Profile. Seattle, Washington: The Boeing Company, 1991.

Corporate Backgrounder. Portland, Oregon: Tektronix, Inc., 1991.

Flight Path: A History of the Boeing Company. Seattle, Washington: The Boeing Company, 1990.

Isard, Walter. *Methods of Regional Analysis.* Cambridge, MA: The M.I.T. Press, 1960.

Lee, Marshall M. *Winning with People: The First 40 Years of Tektronix.* Portland, Oregon: Tektronix, Inc., 1986.

Smith, David M. *Industrial Location.* 2nd. ed. New York: John Wiley and Sons, 1981.

AGRICULTURE   Chapter 13

James W. Scott
Western Washington University

INTRODUCTION

With some 97,000 farms and approximately 56 million acres of farmland—including 12 million acres of harvested cropland and 7.3 million acres of irrigated cropland and pasture—the states of Idaho, Oregon, and Washington, together with the sixteen western counties of Montana, comprise a major region of agricultural activity in the continental United States.

Farmed on an extensive scale only in the past one hundred years, the Pacific Northwest since the beginning of this century has been making notable contributions to the nation's larder with such crops as wheat, barley, hay, sugar beets, potatoes, and dried peas, and beans; a miscellany of fruits, particularly apples, pears, and cherries; and a wide variety of specialty crops such as asparagus, mint, hops, and cranberries. Pastoral farming is only slightly less important: from the region come large quantities of beef and beef products, milk and milk products, poultry and eggs. And although at the present time agriculture employs only a tiny fraction of the adult labor force, as it does in most other parts of the country, it nonetheless makes a notable contribution to the gross domestic product of all the northwestern states.

Long separated from the settled East and Midwest by vast distances and sparsely populated regions with few developed resources, the Pacific Northwest's agriculture was a largely local—or at best a regional—activity until the completion of the various transcontinental railroads and the ensuing drop in freight rates, which made the transportation of its produce economically possible. Even then, transportation costs remained a vitally important and often restraining fact of life that dictated that agriculture be commercially super-efficient, highly specialized, and dependent on a steady supply rather than an occasional surplus. Further, improved technologies in packaging and in food preservation helped insure more effective participation in the agricultural commerce of the nation and in the development of international markets.

This chapter examines the physical bases of agriculture, the structure of the farm and the organization of today's farming in the region, the changing methods of working the land, and the productivity of Pacific Northwest agriculture. The final section identifies and briefly examines the major agricultural regions of the Pacific Northwest.

THE PHYSICAL BASES OF AGRICULTURE

Climate, soils and drainage, water supply, topography, ease of access, and the presence or absence of a wide variety of soil organisms, pests, and diseases constitute the major physical factors responsible for the success or failure of any farming operation. Human ingenuity and technological innovations have, of course, long helped to moderate and even to offset the limitations posed by these physical factors, while economic considerations have both contributed to the adoption of technologies or to their abandonment.

Initially in most, if not all, agricultural regions of the world it has been those areas best endowed with good soils, abundant water supply, and an attractive climate that were the first areas settled and those most rapidly developed. The Pacific Northwest proved no exception. Portions of western Oregon, the Willamette Valley in particular, western Washington, the Puget Sound Lowland especially, and the Lower Mainland of British Columbia, with their mild maritime climate and generally heavy, but fertile soils, became the earliest agricultural regions of the Pacific Northwest. Accessibility also was a significant factor. By contrast, the arid and semi-arid lands located east of the Cascades and the Rockies were only slowly, if at all, brought into agricultural production, and in large part only as economic incentives and political opportunities afforded or as relevant technological changes were initiated.

A century and a half of farming experience has led to the emergence of clearly articulated agricultural patterns and to the creation of a variety of agricultural regions that have become increasingly responsive to the range of physical controls affecting the Pacific Northwest. Where the latter have been less than favorable, suitable modification in farming practices and techniques—largely made possible by a wide range of government and privately sponsored research and development—have generally sufficed to insure at least some modest and often phenomenal success.

It is not necessary here to do more than briefly review the physical factors identified above as they affect or influence the region's agriculture. Other chapters provide more complete coverage of climate, soils, topography, water quality and water availability, as well as examining a wide range of other environmental factors.

Sufficient quantities of water was the most important prerequisite for the pioneer farmer, as it remains for his modern counterpart. West of the Cascades this has seldom been a problem, with precipitation averaging upwards of 30 inches a year—predominantly in the form of rain except in the higher elevations of the Cascades, the Olympics, and the Oregon Coast Range—water availability is, in most seasons, an assured factor in farming. Coupled with exceedingly rare periods of drought, moderate year round temperatures, (the annual range of temperature is seldom more than 25°F), a long frost-free growing season with only an occasional late spring frost and hardly ever an early fall frost, as well as various other *positive* climatic factors, the so-called "wet west" of the Pacific Northwest was the earliest region of agricultural development, boasting an astonishing range of crops and livestock. The emphasis throughout the first six or seven decades of farming was on mixed crop/livestock activities—what has come to be referred to as *general farming*.

Precipitation east of the Cascades, by contrast, drops to less than 10 inches a year in most areas close to the mountains and only a few inches more in areas just west of the Rockies. Accompanying the low precipitation are the more extreme temperatures of both winter and summer and thus a far higher annual range of temperature, higher evapotranspiration rates due to long periods of summer drought and extreme heat, a shorter frost-free growing season—a period occasionally of as little as sixty days in parts of northeast Washington, northern Idaho, and western Montana—and greater susceptibility to such extreme climatic conditions as hailstorms and hot, desiccating winds. Consequently, the "dry east" of the Pacific Northwest has been forced over the past century and more to develop strategies that can offset these many *adverse* climatic phenomena.

Dry farming techniques and the use of semi-arid grassland for extensive, seasonal livestock farming were among the earliest strategies adopted, and they are still extremely important in most areas of the "dry east" today. Requiring considerable capital investment and effective organization are the irrigation systems and techniques that have, almost literally, caused "the desert to bloom." Irrigation has also enabled intensive use to be made of large parcels of land in all the northwestern states, as well as in British Columbia. The water for irrigation is derived in large part from the many rivers that rise in the Rocky Mountains and flow westward to the Pacific, the Columbia and Snake rivers in particular. Other strategies developed to respond to low rainfall total and high evapotranspiration rates include the cultivation of specially developed drought-resistant seeds and various drought-resistant plants.

No less important than water for successful agriculture is the quality of the soil for both crops and pasture. Especially important are the soil's natural fertility, its workability, and its qualities, particularly ease of drainage.

The Pacific Northwest has a wide range of soils that includes inceptisols, ultisols, mollisols, aridosols, entisols, alfisols, vertisols, spodosols, and histosols. Not all of these are used, or indeed are useful, for farming. Histosols, for example, which are found in boggy areas, though exceedingly high in organic content, are poorly drained and difficult to work. By contrast, others such as *aridisols* and *mollisols* are lower in organic content and have very limited water-retention capacity, yet they can be easily worked provided sufficient water is available.

The soil's natural fertility and the availability of an adequate water supply were the initial requirements for pioneer farmers in the Pacific Northwest. Later on—generally after about 1900—the adoption of irrigation on a wide scale and the increased use of inorganic fertilizers and lime, together with the application of pesticides and herbicides, enabled farmers to modify and even improve the quality of their soils. Hence, over the past century and a half, extensive acreages in most of the Pacific Northwest have been utilized for crop farming and the raising of livestock. Notable exceptions are the higher reaches of the Olympics, Cascades, and Rockies and those areas where the original forest cover has either remained undisturbed or where cut-over forest has been replanted.

Among the soils that have been farmed most successfully and continuously since the mid-19th century are the *ultisols* that cover much of western Oregon from the California border through the length of the Willamette Valley to the Columbia River. These soils continue through parts of western Washington as far north as southern Puget Sound. Ultisols have a generally high organic content, are dark in color, and are easily drained.

Farmed for almost as long are the *inceptisols* of the Puget Sound Lowland north to the Canadian border, and along the lowerlying areas of the Pacific Coast Lowland from central Oregon to the Strait of Juan de Fuca. These soils, however, are less well drained and have a high clay content, although they have a moderate to high organic content and may contain varying amounts of only partially weathered parent material.

Covering most of the Columbia Basin, virtually all of eastern Oregon, and a large part of southern Idaho are *mollisols* and *aridisols*. Mollisols occur widely in areas of steppe or shrub-steppe. They are generally high in bases and organic matter and are well-drained. Aridisols, which usually occur in areas of very low rainfall of ten inches or less annually, are lower in organic content than mollisols, although they may have high concentrations of calcium carbonate. With the ready adoption of dry farming methods, as in the Palouse, and with the widespread use of irrigation, as in the Columbia, Snake, and Yakima valleys, these soils are some of the most productive soils in the Pacific Northwest and indeed the world.

Limiting the success of farming in virtually every section of the Pacific Northwest region have been numerous pests and diseases. Among the former are a variety of insects both above ground and in the soil. Fences and barbed wire have helped control the depredation of crops caused by such herbivores as deer and elk, as well as livestock losses caused by coyotes, foxes, and other small carnivorous mammals. Traps and various poisons have also been employed by farmers to control losses although in recent years, they have been used ever more cautiously as environmental groups and animal rights groups have protested.

Insects and various soil organisms have long posed, as they continue to pose, problems for the farmer. Their number are legion. Some, such as grasshoppers and locusts above ground and cutworms and wireworms in the soil, may be a threat to virtually all crops; others, by contrast, are *crop specific*. These include the Hessian fly which attacks wheat, the beet leaf hopper that long threatened beet farmers in the Snake River Valley, the caddis moth that is a particular threat to apple and other tree fruit orchards, the alfalfa weevil and various smuts and rusts that affect cereal crops, and a host of others. Research carried out at state agricultural stations and the land grant universities in all four states, as well as by a number of commercial chemical companies, has led to the development of a wide range of insecticides, pesticides, and herbicides, and a range of hormones and chemicals to combat these environmental threats to agricultural productivity. Despite the progress made, the "battle" is never won, as new diseases and more resistant strains continue to make their presence felt. At the same time the dire warnings sounded first by Rachel Carson in her book *Silent Spring* are having some effect and *organic farming*, which shuns the use of insecticides, pesticides, and herbicides, is conducting counterattacks on the various pests and diseases in quite different ways, and with a modest degree of success.

Plant and animal diseases—bacterial and viral—and various fungi pose additional problems for the farmer. Fungicides and a wide range of antibiotic and other medicines have provided protection against many of these, but at no small cost to the farmer. Nevertheless, outbreaks of such diseases as hoof-and-mouth, bovine tuberculosis, and equine encephalitis still occur periodically, usually requiring mandatory slaughter of large numbers of animals.

FARMS AND FARMERS

The changes that have occurred in American agriculture in the past half-century have been little less than revolutionary. They affect virtually all aspects of farming, and perhaps none more than farm management and the farm labor force. Fifty years ago close to one-fourth of the American population lived on a family farm. Today, barely more than five percent live there while a scant three percent of the country's labor force works the land devoted to agriculture.

Increasingly farms are corporately-owned, and such farms make up a much higher percentage of the total farm land. With such corporate ownership and management, it is hardly surprising that a high proportion of the nation's crops and livestock products come from a small percentage of farms. Meanwhile the number of family farms continues to drop, and these contribute ever smaller percentages of the total farm output. And, as Fraser Hart reminds us in his recent book *The Land That Feeds Us*, in order to survive the family farm has been forced to become the *family business*.

Analysis of recent Censuses of Agriculture will show that the Pacific Northwest has been as greatly affected by such changes and trends as any other agricultural region of the country.

Land in Farms

Among the three Pacific Northwest states, Oregon has the largest total acreage of farmland. However, as a percentage of the state's total land area, Oregon is well behind Washington, which, while barely 70 percent the size of Oregon, has almost as large an area of farmland. Further, much of Oregon's farmland is in pasture or range and less than one-sixth is harvested cropland. Western Montana has an even smaller percentage of harvested cropland. Both Idaho and Washington have much higher percentages of harvested cropland than either Oregon or western Montana, Idaho's being almost twice that of Oregon and more than two and a half times that of western Montana.

	Idaho	Oregon	Washington	W. Montana
Land in farms	13,931,875 ac.	17,809,165	16,165,568	8,323,607
Farmland as a % of total land	26.0%	28.6%	36.7%	—
Harvested cropland	4,349,122	2,832,663	4,597,476	1,031,138
Cropland as a % of total farmland	31.1%	15.8%	28.4%	12.4%

Table 1. Land in Farms, 1987.
Source: Census of Agriculture, 1987.

There are, of course, wide variations in each state between counties in the amount of farmland devoted to harvested crops (Figure 1). In 1987 Idaho had seven counties in which more than 50 percent of the total land in farms was harvested cropland: Canyon, Jerome, Latah, Lewis, Madison, Minidoka, and Teton. In Oregon there were only five such counties: Hood River, Linn, Marion, Polk, and Washington; And in Washington but three counties: Skagit, Whatcom, and Whitman. In western Montana no county had much more than 10 percent of its farmland used for harvested crops, and eight of its sixteen counties had 10 percent or less: Deer Lodge, Granite, Jefferson, Lewis & Clark, Missoula, Powell, Saunders, and Silver Bow. Low percentages of harvested cropland were recorded in only one Idaho county: Adams, where only seven percent was harvested cropland; while in Oregon eight counties—Baker, Crook, Curry, Douglas, Grant, Harney, Wasco, and Wheeler—and in Washington two counties—Ferry and Okanogan—also fell below ten percent. In two Oregon counties—Curry and Wheeler—the harvested cropland constituted little more than two percent of the total land in farms. It should be noted that of the nineteen counties listed above with low percentages of harvested cropland only two of them—Curry County and Douglas County, both in Oregon, are located west of the Cascades (Figure 1).

Number and Size of Farms

The long-term trends toward fewer and larger farms were quickly established in the years following the Second World War. In the three Pacific Northwest states there were close to 160,000 farms still operating in 1954, but by 1987 the number had dropped to less than 90,000, the decline being most rapid in Washington, which had a 48 percent reduction, compared to Oregon's 41 percent and Idaho's 37 percent. And although there was also a modest reduction in the total amount of farmland, particu-

Agriculture

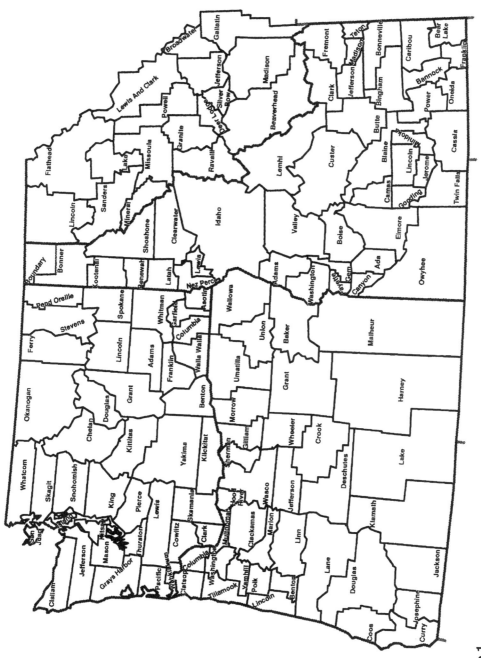

Figure 1.

	Idaho		Oregon		Washington	
	1954	1987	1954	1987	1954	1987
Number of farms	38,740	24,142	54,441	32,014	65,175	33,559
# of farms under 50 acres	11,439	8,289	27,113	16,924	37,660	17,402
Av. farm size	371	577	389	556	271	480

Table 2. Number and Size of Farms, 1954 and 1987.
Source: Census of Agriculture, 1987.

larly in western Oregon and western Washington, there was commensurate increase in most areas in the average size of farms, as Table 2 shows.

A county-by-county analysis of farm size in the three states and the sixteen counties of western Montana shows wide variations. The largest average size of farm is found in Wheeler County, Oregon, where farms average 5,896 acres, that is approximately nine square miles. The smallest average farm size is found in Kitsap County, Washington, where farms average only 24 acres. (Figure 1)

Eleven of Idaho's forty-four counties have average farm size in excess of 1,000 acres; Clark County being the largest with an average size of 3,520 acres. Thirteen of Oregon's 36 counties have farms that average more than 1,000 acres, as do nine of Washington's 39 counties, where only one, Ferry County, exceeds 2,000 acres in average farm size. In western Montana nine of the sixteen counties there have an average farm size of more than 1,000 acres; Beaverhead County at 4,090 acres is the largest. (Figure 1)

In marked contrast are those counties where the average farm size is under 100 acres. They include four counties in Oregon—Clackamas, Hood River, Josephine, and Multnomah—and no fewer than fourteen in Washington—Clark, Clallam, Chelan, Cowlitz, Island, Jefferson, King, Kitsap, Mason, Pierce, Skamania, Snohomish, Thurston, and Whatcom; all but Chelan County are located in western Washington. Neither Idaho nor western Montana has any county with an average farm size of less than 100 acres.

The counties with farms of small average size are almost all, in fact, in western Oregon and western Washington, and for the most part are farms engaged in intensive crop farming, truck farming, berry or fruit farming, dairy farming, or poultry farming. By contrast, those counties with the largest average farm sizes—upwards of 2,000 acres—are counties largely engaged in range farming of beef cattle or occasionally sheep; while those of medium size—from 1,000 to 2,000 acres—are often involved

	1954	1974	1987
Idaho	$33,466	$204,091	$336,615
Oregon	$30,178	$170,145	$299,755
Washington	$31,018	$198,154	$355,576

Table 3. Average Value of Farms, 1954, 1974, and 1987.
Source: Census of Agriculture, 1987.

in wheat and other small grain production, although many others of this size are also engaged in range farming.

Dramatic increases in the value of farmland and farm buildings in the past few decades reflect not only the general inflation of prices, but also the increased competition for land for other uses in many parts of the Pacific Northwest region, in particular the heavily urbanized areas of western Washington and western Oregon. Between 1954 and 1987 there was approximately a tenfold increase in the value of the average farm in Idaho, a close to tenfold increase in Oregon, and an almost twelvefold increase in Washington, as Table 3 shows.

Likewise there was rapid increase in the value per acre of agricultural land between 1954 and 1987. In Idaho land values rose from $90 to $572; in Oregon from $78 to $542; and in Washington from $115 to $739.

The average value of farms and their value per acre, of course, vary greatly from one county to another depending on size, location, type of farming, availability of irrigation, and other factors. In Clark County, Idaho, Ferry County, Washington, and Beaverhead County, Montana, the average value of a farm is more than $1 million, while in 27 others—six in Idaho, five in western Montana, nine in Oregon, and six in Washington—the average value exceeds a half-million dollars. Only two counties, Shoshone, Idaho, ($113,913) and Kitsap, Washington ($148,751), have average farm values of less than $150,000. Another fifteen counties—six in Oregon, six in Washington, and three in Idaho—have average values of less than $200,000. However, in all but three of these, the Idaho counties of Bear Lake, Franklin, and Gem, the average price of land per acre is above the state average, and in a number of counties this is markedly so. (Figure 1)

The highest average value per acre is found, not surprisingly, in the Seattle area: in Kitsap County, where it averaged $7,526, and King County, where it averaged $6,131 in 1987. Pierce County, adjacent to Tacoma, averaged $4,553, and in Oregon two counties adjacent to Portland, Multnomah with $3,840 and Clackamas with $3,754, had averages about seven times that of the state average while Hood River, a little to the east of Portland had the state's highest average of $4,553. In Idaho only 20 of its 44 counties exceeded the state's average of $572 per acre, and only five of these had averages of more than $1,000, the highest being Canyon County at $1,438. Likewise in western Montana, the average value of farmland is much below that of Oregon and Washington, and although all 16 counties of western Montana exceeded the state average of $205, only Flathead County with $1,024 and Ravalli County with $1,054 had average values of more than $1,000 per acre. (Figure 1)

Farms by Type of Organization and Ownership

Individual or *family farms* continue to predominate in absolute numbers of farms, although the actual number of such farms continues to decline year by year. Between the 1982 Census of Agriculture and that of 1987 Washington witnessed a decrease of close to ten percent and Oregon a decrease of seven percent in the number of farms owned by individuals and families. Idaho's decrease, however, was much less, being only a little less than three percent. During the same period partnerships increased only very modestly in Washington, maintained their numbers in Idaho, and actually decreased in numbers in Oregon. In all three states, by way of contrast, there was an increase in the numbers of corporate farms, most markedly in Washington, where the increase was close to ten percent, and in Oregon, where it was more than 13 percent.

Many factors have been responsible for these trends in farm organization and ownership; Among them, the economic factors are the most obvious and, in most instances, the most significant. Success in farming rests in no small measure on the efficiency of the operation and the proper investment in land and equipment. The operator's "know how," experience, diligence, ability to take responsibility, and make the right decisions are all involved, as are the size of the unit—which obviously will be widely different from, say, a truck farm to a cattle ranch—the unit's physical

Farm by type		Idaho	Oregon	Washington
Individual/family	1982	20,950	29,802	31,107
	1987	20,368	27,766	28,289
Partnership	1982	2,312	2,807	2,748
	1987	2,314	2,603	2,850
Corporate	1982	1,175	1,311	2,043
	1987	1,207	1,490	2,248
Other: cooperative	1982	277	167	182
institutional, etc	1987	253	155	172

Table 4. Number of Farms by Type of Organization, 1982 & 1987.
Source: Census of Agriculture, 1987.

characteristics of soil and terrain, the efficient use of machinery, automation and computerization, and the availability and dependability of suitable markets. Not all of these are constants, and on the ability of the operator to respond to seasonal changes or to interpret correctly short-term and long-term trends depends the success or the failure of the farm operation as a whole, whether it be a family farm, partnership, cooperative, or corporation.

For thousands of family farmers trends toward increased mechanization and computerization, as well as the need to expand operations to compete effectively in regional or national markets, have been thwarted by lack of capital and the inability to borrow sufficient money to upgrade operations. For others, particularly those in marginal agricultural districts, a reluctance to radically change the social fabric and abandon a long-established way of life has resulted in economic failure, and eventually the enforced sale of the farm. The latter has sometimes been offset or delayed by the owner taking a job off the farm in some nearby town and becoming a part-time rather than a full-time farmer.

In the Pacific Northwest the family farm has been facing most rapid decline in Oregon and Washington, where thousands of small farms have disappeared in recent decades, some as a result of swelling suburbia and exurban developments, others to become integral parts of large corporate farms. In Idaho and western Montana the decline, though less evident, has also occurred. Even so, in all four states the individual or family farm remains the typical unit, accounting for approximately four-fifths—from 79 percent to 84 percent—of the total number of farms. However, the average size of such farms and the average value of their farm production are markedly below those of the corporate farm.

Corporate farms are increasing rapidly in number and even faster in their share of total farmland. They are especially important in many areas east of the Cascades, where corporate farms control vast acreages of cropland, pasture, and rangeland. In Idaho's Ada, Blaine, Camas, Elmore, and Gem counties, Oregon's Lane and Deschutes counties, and Washington's Benton, Franklin, and Walla counties, more than one-third of the total farmland is now controlled by large corporations. And in many of the adjacent counties in all three states, large amounts of land, ranging from one-quarter to one-third of the total farmland, is now in corporate farms. (Figure 1) In Montana large corporate farms are concentrated in central and eastern parts of the state; nonetheless, in all sixteen counties of western Montana corporate farms control sizeable proportions of the total land in farms.

The evolution of a large corporate land holding and the multitude of specialized activities associated with this are particularly well described in Alexander McGregor's *Counting Sheep*, which follows the fortunes of the McGregor family through many decades "from open range to agribusiness on the Columbia Plateau."

THE CHANGING FARM OPERATION

Many of the changes that characterize today's farming operations can be traced back to the introduction of such machines as a steel plow and the Cyrus McCormick combine harvester in the early to mid-nineteenth century, as well as the many successful experiments in plant and animal genetics and the development of artificial fertilizers at that time or shortly thereafter. However, the changes that truly revolutionized farm operations, changes that brought about exponential increases in productivity and radically modified the farmer's way of life, are indubitably the products not of the nineteenth but the twentieth century. The introduction of machines to replace animal power and human labor, the advent of rural electrification, the exponential increase in the use of fertilizers and agricultural chemicals, the regional spread of irrigation, and the changing nature of the farm labor force are specific aspects of the farm operation that caused sweeping and permanent change.

Machinery and Equipment

Although hand tools, such as the spade, the pitch fork, and a wide array of traditional hand tools and equipment, are likely to be found on every size of farm, machines, mechanized tools, and electrical or gas-powered equipment have largely replaced them in most farm operations. Most numerous and in use on virtually every farm are trucks or pickups and wheel tractors. (Table 5)

The multipurpose tractor with suitable attached equipment is used for a wide assortment of tasks from plowing and seeding to weeding and harvesting of crops, as well as a host of other chores around the farm formerly done by animals or people. Over the past several decades ever more powerful and efficient machines have become available, although on many farms tractors purchased three or four decades ago may still be in use.

	Number of farms	Farms with trucks/ pickups	Farms with 4 or more	Farms with wheel tractors	Farms with 4 or more
Idaho	24,142	22,096	5,058	20,575	4,831
Oregon	32,014	28,179	3,963	26,045	3,636
Washington	33,559	28,503	5,139	28,666	4,256

Table 5. Trucks and Pickups and Wheel Tractors, 1987.
Source: Census of Agriculture, 1987.

	Combines		Mower Conditioners		Balers	
	# of farms	total machines	# of farms	total machines	# of farms	total machines
Idaho	7,162	9,495	8,376	9,093	10,391	12,101
Oregon	4,406	6,442	8,106	9,215	8,359	9,565
Washington	5,193	7,996	8,601	9,930	8,390	9,783

Table 6. Other Farm Machinery and Equipment, 1987.
Source: Census of Agriculture, 1987.

Other machines and equipment designed for more specific uses and for particular crops have also been developed. These range from huge combine harvesters that tower a dozen or more feet above the ground complete with air-conditioned cabins and other accoutrements to smaller machines for harvesting raspberries, cherries, and other easily bruised crops. The Census of Agriculture provides detailed information on only a select few of these, including for the Pacific Northwest region grain and bean combines, mower conditioners, and pickup balers (Table 6).

The assessed value of these machines and other farm equipment has risen sharply in recent decades while the cost of purchasing new machines and equipment or replacement of worn-out ones has added greatly to the burdens of the family farmer, who, often unable to raise or borrow the necessary working capital, may be squeezed out and forced to give up the land to a bank, developer, or corporate farm.

In Idaho the average value per farm of machinery and equipment in 1987 was a little less than $49,000, in Washington close to $46,000, and in Oregon approximately $38,000. These averages, however, hide the wide range of assessed values that exist within each state from one county to another. In Idaho there are six counties—Adams, Boise, Bonner, Clearwater, Gem, and Valley—with average machinery/equipment values of less than $30,000, but none under $20,000. In seven other Idaho counties—Cassia, Clark, Fremont, Lewis, Madison, Power, and Teton—the values are all above $80,000, with Power County having the highest county average in the Pacific Northwest of $137,287. In Oregon, however, the range is greater than in Idaho: Three counties—Lincoln, the lowest at $12,631, Deschutes, and Douglas—have average values of less than $20,000, while two counties—Gilliam and Sherman—have values well above $100,000. But it is in Washington where the range of assessed values is greatest. Kitsap county has an average of only $9,656, while ten other counties have averages between $10,000 and $20,000—Clallam, Clark, Cowlitz, Island, Jefferson, Mason, Pend Oreille, San Juan, Skamania, and Wahkiakum. At the other extreme, four Washington counties—Columbia, Franklin, Lincoln, and Whitman—have average equipment values of more than $100,000, the highest is Whitman at $118,479. The range of values in western Montana is much less, although most of the 16 counties have values below the state average of more than $60,000, from Sanders County at $21,012 to Broadwater County at $81,483. (Figure 1)

Rural Electrification

As far as the domestic side of farming is concerned, no development in this century has brought about so many changes as the spread of rural electrification. In dairying and many other farm operations, the availability of electricity has made possible a vast array of important changes. Between May 1936, when President Roosevelt signed the Rural Electrification Act aimed at providing electricity to every area of the nation, and 1955 the number of farms across the country with electricity increased almost sixfold, from under 800,000 to close to 4,500,000. Public Utility Districts (PUDs) came into existence in all parts of the country and the number of electricity generating stations—thermal and hydro—jumped accordingly. In the Pacific Northwest the establishment by the Congress of the Bonneville Power Administration and the completion of dozens of dams and power plants on the Columbia and Snake rivers and many of their tributaries helped provide the power that was transmitted to virtually every farm and rural settlement in the three northwest states and to western Montana by the 1960s. Details of these developments are provided elsewhere in this volume.

Commercial Fertilizers and Agricultural Chemicals

A huge, highly diversified and profitable industry has developed over the past few decades to provide aid to farmers in expanding the farm's productivity and in combating the many bacterial and viral diseases and the numerous pests and weeds that threaten the survival of agriculture in some regions of the country and the overall success of farming in others.

The use of fertilizers and a great range of other chemicals—herbicides, pesticides, and insecticides—is widespread in the Pacific Northwest. And although today there is mixed evidence as to the employment or the avoidance of agricultural chemicals and fertilizers, the overall trend is still upward, rather than the reverse (Table 7 provides information on their use for the three states of Idaho, Oregon, and Washington).

The total cost of these fertilizers and chemicals is considerable: more than $270 million for chemicals, half of which was spent in Washington, and more than $400 million for fertilizers, with Idaho and Washington each spending about $150 million. The application of lime to "sweeten" highly acidic soils is found almost entirely in western Oregon and western Washington.

	Commercial Fertilizers		Agricultural Chemicals	Lime	
	farms	acres	farms	farms	acres
Idaho	13,109	3,314,974	14,634	18	364
Oregon	17,254	2,385,614	18,044	1,749	74,626
Washington	18,031	3,893,955	19,811	1,546	49,707

Table 7. Commercial Fertilizers and Agricultural Chemicals, 1987.
Source: Census of Agriculture, 1987.

Hybrid Seeds and Genetically-Enhanced Plants

Great advances have been made in the past half century in the development of a wide range of hybrid seeds. Hybrid corn, which was first introduced in the Midwest in the late 1920s, is now grown in virtually every region of the country, and particular species have been developed to suit a variety of climates and soils. Wheat, barley, and many other crops are now largely dependent on hybrid seeds. As a result yields per acre have risen dramatically—sometimes six or eight times. Offsetting these advances and economic inducements are the additional costs incurred—the necessity for annual purchases of hybrid seed and the need for heavier application of fertilizers—while the loss of native varieties has led to less diversity and, according to some experts, an even greater loss of taste and other once distinctive characteristics.

In the past decade great advances have been made in genetically enhanced plants: the promotion of specific desired characteristics such as size, shape, and color, and the elimination of less desirable ones, including easy spoilage and susceptibility to frost damage. But relatively few of these are as yet in wide use, and in the Pacific Northwest none is much beyond the stage of laboratory experiment. However, once proved successful in the region, there is little doubt that quick diffusion of such genetically enhanced species will take place, just as it did a half century or more ago with hybrid seeds.

Irrigation

Although irrigation had been attempted on a small scale in various places in the West prior to 1850, it was not until the 1860s, following the successful efforts of Mormon farmers in Utah working cooperatively and under the aegis of their bishops, that irrigation became a rapidly more important factor in regional agriculture. By the 1880s cooperative irrigation had diffused northward and northwestward, first into the

Photo 1. An aerial view looking southeast over Grand Coulee Dam, with Banks Lake in the upper right of the picture. (U.S. Department of the Interior, Bureau of Reclamation, Pacific Northwest Region).

valleys of southern Idaho, occupied then, as now, by Mormon farmers, and eventually into areas of eastern Oregon and the Yakima and Spokane valleys of Washington.

The costs of irrigation farming were high, and the technical problems, even with gravity flow irrigation, many. Moreover, sufficient capital was not always available, and real progress on a more than local scale could not be made until larger sums of capital were secured. In the 1880s such sums began to be provided by the Northern Pacific Railroad and other land grant railroads, which were eager to dispose of their land to prospective Euro-American settlers. One such investment by the Northern Pacific Railroad was the Yakima Canal Company organized in 1889. But it was not until the federal government got involved that irrigation was extended to many areas of the Pacific Northwest.

The federal government's first attempt to expand irrigation, in consort with state governments, was, in Donald Worster's words, "a dismal and discouraging failure." (Worcester, 1985, p. 157) The Carey Act of 1894 empowered the "desert" states of the West to preempt up to one million acres for sale to would-be settlers and to provide them with water for up to 160 acres at reasonable cost. Only Idaho, however, took any great advantage of the act, but even then much less than one million acres were settled and watered. It was not until passage of the National Reclamation (Newlands) Act of 1902 that the impact of the federal government began to be felt across the region, as projects were undertaken in valley after valley of Idaho, Oregon, Washington, and western Montana. Progress was especially significant following the initiation of the Columbia Basin Project in the Roosevelt Era of the 1930s and the bringing on line of Banks Lake reservoir, just south of Grand Coulee Dam, and its distributing canals around 1950.

The huge economic outlays for such federal projects and the environmental consequences of these have been discussed endlessly throughout the past half century. Recent critics have included not only environmentalists, but economists, political leaders, academics, and others. Particularly informative but critical works are two recent books: *Rivers of Empire: Water, Aridity and the Growth of the American West* by the historian Donald Worster, and *Cadillac Desert: The American West and Its Disappearing Water* by the journalist Marc Reisner.

In 1987 every county, without exception, had some acres of irrigated farmland, a large proportion of which was harvested cropland. The number of farms provided with irrigation and the total acreage irrigated are shown in Table 8.

Two regions in particular stand out as the areas where irrigation has made possible most, if not all, of the agricultural success: the valley of the Snake River and the lower reaches of many of its tributaries in Idaho, and Washington's Big Bend Country, south and east of the Columbia River and west of the Palouse Hills.

Twelve of Idaho's counties, nearly all of them in the Snake River basin, had farms that in 1987 were provided with an average of more than 200 acres of irrigated land,

	Farms Irrigated	Irrigated Acreage	Average per farm
Idaho	16,620	3,219,192	194
Oregon	14,411	1,648,205	114
Washington	15,437	1,518,648	98
Western Montana	9,520	963,154	101

Table 8. Farms and Irrigated Acreages, 1987.
Source: Census of Agriculture, 1987.

Photo 2. Center pivot irrigation, Columbia Basin Project between Othello and Pasco. (U.S. Department of the Interior, Bureau of Reclamation, Pacific Northwest Region).

six others had more than 300 irrigated acres and one, Power County, averaged more than 400 acres. Bingham County, one of those with an average of more than 200 acres, had more than 300,000 irrigated acres. Twin Falls and Cassia counties each had well over 200,000 irrigated acres. In Washington only three counties exceeded averages of more than 200 acres and one county, Lincoln, 300 acres. However, Grant County, with 369,179 acres irrigated, exceeded all other counties in the Pacific Northwest, while Yakima County had 247,313 irrigated acres. In Oregon two counties, Harney and Lake, had averages of more than 500 acres per farm and two others averages of more than 200 acres. Klamath County, however, had the largest irrigated acreage, 213,183 acres, with Malheur County close behind with 193,271 acres. Beaverhead County in western Montana with only 288 irrigated farms averaged more than 800 acres per farm, and Powell County more than 400 acres. Only Beaverhead and Madison counties, however, exceeded 100,000 acres of irrigated land. (Figure 1)

Whatever the long-term prospects for such continued levels of irrigation, and some experts are decidedly pessimistic, there is no gainsaying the fact that much of the century's agricultural successes in the Pacific Northwest, like those in the Central Valley of California, can be credited to irrigation, for which the methods for providing have undergone great changes in recent decades. Gravity-fed ditch irrigation has largely given way to carefully monitored drip irrigation, while extensive areas have seen the emplacement of center pivot irrigation.

Farm Labor

Although it is an exaggeration to claim that American agriculture in the past half-century has been transformed from a labor intensive activity to a largely machine-dependent industry, there are certainly sectors, like poultry farming, in which mechanization, computerization, and even a degree of automation have gone far toward eliminating a large number of jobs. In fact, the number of jobs on the farm today is but a small fraction of what it was 50 years ago. In addition, fewer of them are filled by permanent hired labor, and in many specialized operations, such as fruit farming or truck farming, seasonal labor, some of it schoolchildren but most of it Mexican-American, has become the standard practice. And even in many specialized farm operations machines have been invented and introduced to cut down on, if not totally eliminate the need for, large armies of seasonal workers.

The only permanent labor present on a high proportion of full-time family farms may be that of the owner-operator and spouse, aided seasonally, if need be, by other members of the immediate family. And on most part-time farms the need for even seasonal labor is likely to be small or even non-existent. Table 9 provides information on farm operators in 1987 in the three Pacific Northwest states and western Montana.

	Farms with full-time owner/operators	**Farms with part-time owner/operators**	**Farms with operators 200 days or more off farm**
Idaho	14,550 (60%)	9,592 (40%)	7,555 (31.3%)
Oregon	15,359 (48%)	16,655 (52%)	12,646 (39%)
Washington	17,654 (53.6%)	15,905 (46.4%)	12,330 (36.7%)
Western Montana	3,973 (55%)	3,238 (45%)	2,451 (34%)

Table 9. Farm Operators, 1987.
Source: Census of Agriculture, 1987.

Photo 3. Unpainted Barn, Stevens County, Washington. (J.W. Scott photo)

A county-by-county analysis of these statistics indicates great variations, some of which might be accounted for by the close proximity of urban areas where there are numerous job opportunities. However, even in some of the more remote counties of eastern Oregon, eastern Washington, central and northern Idaho, and western Montana there are large numbers of part-time farmer-operators. As Table 9 indicates, Oregon leads in the percentage of part-time operators, who are more numerous than full-time operators, and in the numbers of part-time operators working more than 200 days off the farm. By contrast, Idaho has the largest percentage of full-time farm operators, who outnumber the part-time operators three to two. Although the trend toward more and more part-time operators is likely to continue for some time, a slowing down of the trend has been apparent for the past decade or so.

Replacement of Animal Power and the Disappearing Barn

One of the most obvious changes in the farm operation in the past century has been the replacement of horses and other animals (mules and oxen) by tractors and other mechanized machinery. As a result the demand for draft horses such as Clydes-

Photo 4. Two barns in one, Berthusen Park, nr. Lynden, Washington. Hans Berthusen's first barn in the foreground was later incorporated in a much larger barn built at the turn of the century. The Berthusen homestead is now part of the Lynden parks system. (J.W. Scott photo)

dales and Percherons has dwindled to near zero, although a few continue to be raised for display purposes and plowing contests at agricultural fairs. In the Pacific Northwest mules, which were never very numerous, have virtually disappeared from the farm scene.

With the declining numbers of draft animals to be fed and housed has come a much decreased need for the stable and the barn; the latter in particular was once the most noticeable building on any farmstead. Hay lofts being no longer needed, barns are becoming little more than "white elephants" on many farms. Large numbers of barns are being left to decay, their roofs unrepaired and their sides unpainted. And when eventually the barn if no longer of any use for the storage of machines and equipment, or needed for the occasional housing of animals, it is likely to be abandoned in favor or more efficient sheds, garages, and workshops, most of metal construction. The demise of the barn, so regretted by the romantically inclined and the folklorist (who see it as yet another indicator of the decay of rural life), is a hard fact of life for today's farmer. His livelihood depends as never before on the efficiency of the farm operation rather than on the visual appearance of the farm and its various outbuildings.

FARMING TYPES AND PRODUCTION

A mix of physical and human factors enter into the choice and the eventual success or failure of the types of farming practiced in any region. Delayed settlement and therefore economic development of the Pacific Northwest, combined with its relative isolation from the rest of the nation, helped propel the region into early specialization and heavy dependence on external markets. While these external markets continue to play a major role in northwest agriculture, specialization has not led to an agricultural picture that is in any significant way monocultural. Diversity remains a notable characteristic of most, if not all, agricultural regions of the Pacific Northwest.

Although many farms continue to rely on a variety of crop and livestock products, the percentage of such farms has dropped dramatically since the 1950s when Otis Freeman reported on the widespread distribution of *general farming* in the region (Freeman, 1954, 287–90). In the past 40 years, large numbers of small family farms that were formerly engaged in general farming have disappeared altogether, other farms have been amalgamated into larger units, and specialized corporate farms have become significantly more important in terms of the total farm output. However, while general farming has declined in importance, regional diversity in crop and livestock production is as great as it ever was.

Today an astonishing range of crops and a wide variety of livestock products come from farms across all three northwestern states and the western portion of Montana. In 1989 Idaho ranked first in the nation in the production of potatoes, raising more than a quarter of the nation's crop, and it was third in the production of barley, sugar beets, hops, and three other crops. By contrast, Idaho ranked among the ten leading states in only one livestock product: American cheese. Oregon likewise ranked high the same year in the production of a variety of crops, but among the first ten states only in the production of mink, sheep and lambs, and wool. As in many earlier years, Oregon led in the production of rye grass, fescue, and Kentucky bluegrass, as well as in the production of five berry crops, blackberries, boysenberries, loganberries, and black and red raspberries. In addition, it ranked among the top five states in 17 other crops, including hops, potatoes, onions, beans, and pears. Washington, like its neighbors, also ranked higher in crop production than in livestock products. In 1989 it ranked first in the nation in the production of ten crops: hops, lentils, dry edible peas, spearmint oil, apples, Concord grapes, sweet cherries, pears, carrots, and red raspberries. And it was among the top five states in eleven others, including asparagus, wheat,

	Crops	Livestock Products
Idaho	$1,648,412,000 (60%)	$1,074,276,000 (40%)
Oregon	$1,716,920,000 (70%)	$ 727,724,000 (30%)
Washington	$2,404,486,000 (69%)	$1,251,379,000 (31%)

Table 10. Agriculture by Value of Production in 1989.
Source: Agricultural Statistics, various states.

barley, and cranberries. However, only in the production of mink and milk products did Washington rank in the top ten livestock-producing states.

In value of production, agriculture in the three northwestern states totalled close to $9 billion. Approximately 65 percent of this, or $5.7 billion, came from crops and the remainder, $3.05 billion, from livestock products (Table 10 provides details).

It is important to note that despite what has been said above about the lower rankings for livestock products than for crops, a livestock product ranked first in each state's list of leading agricultural products in 1989 (Table 11).

However, as prices fluctuate or as productivity changes due to climatic and other factors, the rankings of the top products frequently change places among themselves.

Cash-Grain Farming

Although various grains were grown in the Pacific Northwest as early as the 1820s, it was not until the advent of commercial wheat farming in the Palouse Hills in the 1870s and 1880s that the Pacific Northwest became a major producer of small grains. Today, almost one-fourth of the barley grown in the United States and close to one-eighth of the wheat comes from the three northwestern states, Washington being the leading producer of wheat and Idaho of barley.

Large acreages are planted to wheat in all three states, although from year to year the totals may vary considerably. There was, for example, a marked decline in the total acreage between 1982 and 1987, from more than 5.3 million acres to about 4.2 million acres. Production, which totalled 275 million bushels in 1982, had dropped to 248 million bushels by 1987.

Wheat production is heavily concentrated in the Palouse region of Washington and adjacent regions of Idaho and Oregon, although there is some production in virtually all counties in all three states as well as western Montana, where more than seven million bushels was produced in 1987. Whitman County, Washington, where in 1987 almost 29 million bushels was raised, leads the region. Four other Washington counties, Lincoln, Walla Walla, Grant, and Adams, each produced more than ten million bushels, and these together with Whitman County accounted for well over 60 percent of the state's total wheat production. In Oregon, Umatilla County led with

	Product	Value
Idaho	Cattle & calves	$661,000,000
	Potatoes	$560,000,000
	Wheat	$334,000,000
	Milk	$322,000,000
	Sugar beets	$176,000,000
	Barley	$130,000,000
	Hay	$114,000,000
Oregon	Cattle & Calves	$330,000,000
	Greenhouse & Nursery	$279,000,000
	Hay	$245,000,000
	Wheat	$215,000,000
	Milk	$205,000,000
	Potatoes	$136,000,000
	Ryegrass seed	$102,000,000
Washington	Milk	$559,000,000
	Apples	$466,000,000
	Wheat	$446,000,000
	Cattle & Calves	$402,000,000
	Potatoes	$402,000,000
	Hay	$251,000,000
	Greenhouse & Nursery products	$140,000,000
	Pears	$140,000,000

Table 11. The Leading Agricultural Products, 1989.
Source: Agricultural Statistics, the various states.

	1988			1989		
	Winter	Spring	Irrigated	Winter	Spring	Irrigated
Idaho	770,000	380,000	551,000	810,000	560,000	680,000
Oregon	660,000	95,000	133,000	815,000	105,000	172,000
Washington	1,300,000	970,000	305,000	2,100,000	280,000	317,000

Table 12. Wheat Acreages: Winter Wheat, Spring Wheat & Irrigated.
Source: Agricultural Statistics, the various states.

more than 16 million bushels, followed by Morrow County with almost eight million. Idaho's Bingham County was first with just less than 11 million bushels, and with Latah and Cassia counties each producing more than six million bushels. In Montana, where the total production exceeded that of Washington by more than 30 million bushels, the sixteen western counties raised only a small fraction of the state's total and with more than four million of the seven million bushels raised coming from two counties: Gallatin and Broadwater. (Figure 1)

Not only do total wheat acreages planted vary from year to year, the proportions of the total planted to *winter wheat* or *spring wheat* may vary even more. In some parts of the region severe weather conditions may destroy large amounts of the winter wheat which then has to be replaced by spring varieties. This happened in early 1991 in much of eastern Washington and in parts of eastern Oregon and Idaho. However, devastating weather occurs on a wide scale only rarely, and the advantages gained in planting winter wheat—fields prepared and seed sown in fall, a ground cover that protects the soil against winter winds, and a crop that can be harvested weeks earlier than spring wheat—more than offsets the occasional loss of much of the planted seed. But there are marked variations in the proportions planted to winder and spring wheat in each state and from one year to the next, as there is also in the acreages that are irrigated (see Table 12).

Yields per acre may also vary greatly. In Washington, for instance, in 1989 the average yield was 48.7 bushels per acre while in the same year in Oregon it averaged 58.5 bushels and 66.7 bushels in Idaho. On irrigated land the yields are considerably higher. In 1989 these averaged 83.1 bushels per acre in Idaho, 90.0 bushels in Washington, and 92.8 bushels in Oregon.

Like wheat, barley is grown in most counties of all three states, but it is the Palouse region of Washington, Whitman County in particular, and nearby parts of Idaho and Oregon, together with the valley of the Upper Snake River, where barley plays an especially crucial role. Both feed and malting varieties of barley are grown in all three states, but whereas in Idaho they are grown in more or less equal amounts, in Washington and Oregon feed varieties far exceed malting varieties in importance.

Fluctuations in the acreages planted to barley, however, are more pronounced than they are for wheat. In Washington, for example, a record 1,200,000 acres was planted in 1985, but five years later only 350,000 acres was planted—a decline of approximately 70 percent. In Idaho the area planted to barley in the 1980s has varied much less dramatically, ranging from a high of 1,370,000 acres in 1984 to a low of 840,000 acres in 1987.

Much smaller acreages are planted to oats, mostly in places where wheat and barley are not the major crops. In 1989 Idaho had a total of 90,000 acres planted to oats, most of this in northern and eastern counties of the state, with Kootenai and Bonneville being the leaders. In Oregon in the same year, a total of 105,000 acres was planted to oats, with well over half the total in northwestern counties of the state, particularly Yamhill, Washington, and Polk counties. However, Klamath County in southeast Oregon led the state in acreage in oats although production was considerably below that of Polk County. In Washington 31,000 of the 85,000 acres planted to oats was in western Washington, with Clark County exceeding the rest by a wide margin; while another 27,000 acres, half of it in Spokane County, produced oats in northeastern Washington. A final point should be made regarding oats: Not all the acreage planted is harvested—in 1989 in Idaho only 60,000 of 90,000 acres, in Oregon 70,000 of 105,000 acres, and in Washington 45,000 of 85,000 acres. (Figure 1)

Brief mention should also be made of corn, which is grown for both grain and silage. In 1989 Idaho had 130,000 acres planted to corn, of which about 50,000 acres was for grain and 80,000 for silage. In Oregon, of the 50,000 acres planted, 22,000 acres was harvested for grain, and in Washington, of 130,000 acres planted, 90,000 acres was harvested for grain. Although grown widely across all three states, corn is most successfully cultivated on irrigated land east of the Cascades and along the Snake River valley. Canyon County in Idaho, Umatilla and Malheur counties in Oregon, and Benton, Franklin, and Grant counties in Washington, are the leading producers. (Figure 1)

Other Major Field Crops

Sugar beet: Since the 1890s *sugar beet* has been a mainstay of Idaho agriculture. By contrast, in Oregon and Washington virtually all attempts to establish sugar beet as a viable field crop—in the western counties of each state without irrigation, and in counties east of the Cascades with irrigation—have generally been short-lived. It is safe to say that of all the crops grown in the Pacific Northwest sugar beet is the most *localized*, being heavily concentrated in the Snake River valley. In 1989 a total of 179,000 acres was planted to sugar beet in Idaho, with Minidoka (41,000 acres), Canyon (32,400 acres), and Cassia (25,000) being the highest producing counties. Idaho's production of slightly more than four million tons placed the state third in the nation behind Minnesota with 5.4 million tons, and California with 4.9 million tons. (Figure 1)

Photo 5. Alfalfa hay being gathered by a field chopper and blown into a truck driven alongside the tractor. Near Quincy, Washington. (U.S. Department of the Interior, Bureau of Reclamation, Pacific Northwest Region).

Potatoes: Three-fifths of the nation's potatoes were grown in the Pacific Northwest in 1989, a tribute in large part to the ready availability of water for irrigation in Idaho and Washington. Idaho alone produced close to one-third of the nation's crop while Washington produced approximately one-fifth. Two regions are especially important: the first is the Snake River valley of Idaho, where Bingham (60,000 acres), Bonneville (37,500), Madison (35,000), and Cassia (30,000) counties all produce more than eight million cwt: each, while four other counties produce in excess of 5 million cwt.; Power, Fremont, Madison and Minidoka. The second region is the Middle Columbia valley of eastern Washington, where Grant (18.9 million cwt.), Franklin (15.3 million cwt.), and Adams (8.6 million cwt.) counties account for approximately two-thirds of the state's total production of 64.3 million cwt. Skagit County in western Washington is the only other county to exceed one million cwt. (Figure 1)

Hay: High on the list of agricultural products in all three states and the 16 counties of western Montana is hay, a term that encompasses many grasses such as timothy, ryegrass, fescue, and others, as well as alfalfa. In most areas, the latter crop is heavily dependent on irrigation, especially in the Columbia Basin, most parts of eastern Oregon, and south and southeastern Idaho.

Idaho ranked seventh in the nation in the production of alfalfa hay in 1989, and throughout the 1980s Idaho consistently led in the production of all hay in the Pacific Northwest, with Oregon and Washington alternating for second place. The production of hay in western Montana was approximately 60 percent that of Oregon or Washington, and production for the whole region in 1989 totalled ten million tons.

Much of the hay produced is marketed, and large volumes are regularly transported by truck from eastern Oregon and eastern Washington to dairy farms and feed lots west of the Cascades. In recent years Japan has become a major overseas market for timothy hay and alfalfa from eastern Washington counties such as Kittitas.

Other field crops include dry beans, dry peas, hops, peppermint, spearmint, and grass seeds. *Dry beans* are a major crop in southeastern Washington and the Snake River valley; while *dry edible peas* are more heavily localized in the Palouse region of Washington, with smaller acreages planted in the Snake River valley and a few counties of northeastern Oregon.

Spearmint and *peppermint*, the two varieties of mint grown for their *essential oils*, are among the many *specialty crops* of the Pacific Northwest. In 1989 Washington accounted for 60 percent of the nation's production of spearmint while Oregon accounted for about three percent. In the same year, Oregon led the nation in the production of peppermint, with almost 47 percent while Washington produced another 24%.

With 73 percent of the total, Washington led the nation in 1989 in the production of *hops*. Oregon with about 20 percent was in second place. Particularly important in the growing of hops is the Yakima Valley of Washington, where the tall poles with their network of wire make hops one of the most easily recognizable of crops.

The cultivation of various *grass seeds* is particularly important in Oregon, which leads the nation in the production of ryegrass seed, fescue seed, orchardgrass seed, and Kentucky bluegrass seed, all of which are produced in large quantities in the Willamette Valley. Alfalfa seed, however, is produced mainly in the lower Snake Valley just east of the Oregon border and in the Palouse region of southeastern Washington.

Vegetables and Fruits

A wide variety of vegetables and equally large number of fruits are grown in parts of the Pacific Northwest on a commercial scale. Truck farms, berry farms, vineyards, and various orchards are the major providers of these. Most of the products are heavily dependent on large amounts of labor, despite the introduction in recent years of many labor-saving machines, fast and efficient transportation, sophisticated marketing techniques, suitable packing and storage sheds, and cooling and refrigeration facilities.

Photo 6. A field of Great Northern beans north of Nampa, Idaho (U.S. Department of the Interior, Bureau of Reclamation, Pacific Northwest Region).

Vegetables grown in the region include carrots, green peas, asparagus, sweet corn, onions, cauliflower, snap beans, broccoli, cucumbers, and more. Washington leads the country in the production of carrots and green peas, much of which are packaged and frozen, and is second only to California in the production of asparagus. Washington is also an important producer of sweet corn and Walla Walla sweet onions. Oregon is a major producer of onions, snap beans, cauliflower, broccoli, sweet corn, and green peas, being in the top five state producers of each of these. Idaho's vegetable crops are smaller and less varied, although onions and sweet corn production rank Idaho high among the states.

Both the "wet" west and the "dry" east of the Pacific Northwest are equally involved in these activities, but irrigation is a prerequisite east of the Cascades in the Yakima and Snake River valleys.

Berry farming is largely confined to western areas. Oregon had approximately 20,000 acres devoted to berry production in 1989: strawberries 6,200 acres, red raspberries 4,000, black raspberries 1,400, blackberries 4,900, boysenberries 900, and cranberries 1,400. In western Washington a little less than half Oregon's acreage was used for berry production: red raspberries 4,200 acres, strawberries 1,900, cranberries 1,400, and blueberries 900. The total value of these berries was close to $100 million with Oregon producing $66 million and Washington $33 million. The proximity of many of the berry farms to large urban areas, like Seattle, Portland, Tacoma, Everett, and Bellingham, has resulted in a decline in acreages in recent years, as suburban developments have replaced berry fields in such areas as Oregon's Willamette Valley and the Green River and Nooksack valleys of Washington.

Apple Farming. Since soon after the beginning of the century, apple farming has been a lucrative occupation for many farmers in eastern Washington and parts of Oregon and Idaho.

Washington has long led the country in apple production accounting in 1989 for more than half the nation's total with a production in excess of 2,500,000 tons. The benchlands of the valleys of the Yakima, Okanogan, and Columbia rivers are clustered with thousands of small 20 to 50 acre orchards, which totalled some 150,000 acres in 1989. The principal varieties grown are red delicious and golden delicious, although demand for greater variety in recent years has resulted in the reintroduction of many earlier favorites as well as the introduction of several new varieties. The value of the apple crop is measured in the hundreds of millions of dollars—close to $500 million in 1989—and traditionally apples rank with milk and wheat as Washington's three leading products.

Much smaller, but far from insignificant, quantities of apples are grown in Oregon and Idaho although the sale of these is regional rather than nationwide or worldwide as is Washington's crop.

Other Orchard Crops include pears, for which Washington in first place accounts for close to 40 percent of the nation's crop and Oregon in third place with about 24 percent. Extensive pear orchards are located in Washington in the same valleys as the apple orchards, and in Oregon in the area between Medford and Jacksonville in the southwestern part of the state and the hillslopes on the leeside of Mount Hood.

The Pacific Northwest is also an important producer of sweet cherries. Washington was the country's leading producer with 43 percent of the crop in 1989, and Oregon was second with 27 percent. Idaho came in a very distant fifth with less than two percent. Other stone fruit of importance in the northwest are apricots, peaches, tart cherries, prunes, and plums, and Washington and Oregon consistently place among the top producing states of these crops.

Viticulture. Grapes have long been grown in the Pacific Northwest, but until recently almost all of these have been the *vitis labrusca* Concord grape. Since the 1960s, however, the planting of varieties of *vitis vinifera*, the European grape, has led to the emergence of a major new agricultural industry in the region: the making of first-class table wines.

Concord grapes are still grown in far larger quantity than the various vinifera grapes, but unlike the latter they are seldom used for the making of wine. In 1989 production totalled 186,000 tons with a market value of more than $50 million, whereas the production of vinifera grapes—notably White Riesling, Chardonnay, Chenin Blanc, Cabernet Sauvignon, Sauvignon Blanc, Semillon, and Merlot—totalled 43,000 tons with a market value of about $20 million. In Oregon production is considerably less, and most of it is in vinifera varieties. Oregon's total crop in 1989 was just under 7,500 tons, with a market value of $5.5 million.

Idaho, which does have a small number of wineries, imports most of the grapes needed from Washington, which also exports sizeable quantities of vinifera grapes, notably Riesling and Chardonnay, to Oregon and British Columbia. The major grape-producing districts are the Yakima and Columbia Valleys of Washington and the Willamette Valley of Oregon.

Livestock Farming

Although livestock farming is varied and widespread in all parts of the Pacific Northwest, today's livestock industry is dominated by dairying, especially in western Washington, western Oregon, and some parts of the Snake River valley, and by the raising of beef cattle and calves on ranches, located for the most part in eastern Oregon, eastern Washington, central and southern Idaho, and western Montana. Of somewhat lesser importance are poultry farming and sheep farming. Pig farming, the raising of mink and the breeding of horses for recreation and sport are other types of livestock farming found in the Pacific Northwest.

Sheep, which were more numerous than cattle a hundred years ago, have declined in importance and in numbers as wool has been replaced by other fibers in many products and as imported lamb from Australia and New Zealand has replaced local supplies. In Idaho, for example, sheep outnumbered cattle by close to two to one in 1890; a hundred years later cattle outnumbered sheep by more than five to one. Much the same story can be repeated for Oregon, Washington, and western Montana. Horses and mules, likewise, have declined rapidly in number, as tractors replaced animal power after about 1920. And horses raised today are more likely to be the lighter Arabian or part Arabian breeds required for recreation and horse racing. Hogs and pigs, however, have continued to be of some importance on the family farm, but only a few large-scale commercial pig farming operations have come into existence in the Pacific Northwest. By contrast, poultry farming, once a minor and usually a female activity, has become increasingly a highly capitalized form of agribusiness, with eggs and table birds as the only products of the operation—one that can be conducted on quite small acreages on which huge, climate-controlled buildings, each capable of accommodating tens of thousands of chickens, occupy most of the ground.

	Idaho	Oregon	Washington
Total value of livestock products	$1,172,149,000	$797,451,000	$1,230,978,000
Cattle & Calves	—	57.7%	—
Fattened cattle	40.3%	—-	32.4%
Other cattle & calves	30.4%	—-	17.3%
Dairy products	20.6%	22.5%	36.1%
Sheep & lambs	2.1%	5.3%	—
Hogs & Pigs	—	1.9%	—
Poultry	—	9.3%	10.2%
All other livestock	6.6%	3.3%	4.0%

Table 13. Value of Livestock Sold, 1987.
Source: Census of Agriculture, 1987.

In 1987 the total sales of livestock and livestock products in the three northwestern states totalled more than $3 billion or close to 35 percent of the value of all agricultural sales (Table 13 provides details of the composition of livestock sales).

Dairy Farming: The days when every farm had its own small herd of dairy cows, and when many non-farm households also kept a milch cow, are long past. Since the beginning of the century, the dairy industry has undergone revolutionary changes, and the number of dairies and creameries has declined dramatically. The use of artificial insemination and other genetic aids and the control by antibiotics of a wide range of diseases has resulted in greatly improved dairy herds and phenomenal increases in milk yields per animal. Today one animal might produce the same quantity of milk it took five or more animals to produce some decades ago. Herd composition partly accounts for this, as increasingly a few high-yielding breeds have come to constitute the vast majority of milk cows in the region, as they do in most other parts of the United States. Particularly important breeds are the Holstein-Friesians and to a lesser extent the Jerseys and Guernseys. The higher butterfat content of the milk obtained from the latter two breeds partly explains the current preference for Holstein-Friesian cows. And as electricity became available on virtually every farm, the replacement of hand-milking by machine-milking equipment and the greater ease of pasteurization were among the earliest responses. Faster and more efficient transportation and the introduction of truck refrigeration made possible the replacement of the ubiquitous small farm dairy by regional diaries and the establishment of regional *milksheds*.

	Number	Production (in lbs)	Leading Counties
Idaho	157,665	2,669,000,000	Jerome (20,480)
			Gooding (18,053)
			Canyon (14,288)
Oregon	95,235	1,509,000,000	Tillamook (22,098)
			Marion (13,464)
Washington	220,649	4,097,000,000	Whatcom (55,631)
			Skagit (19,211)
			Pierce (12,078)
W. Montana	16,614		Gallatin (6,675)

Table 14. Numbers of Dairy Cows, Milk Production and Leading Counties, 1987.
Source: Census of Agriculture, 1987.

As a result of these and other developments, the dairy industry has become an increasingly localized one, concentrated in areas most geographically advantageous for the pasturing of milk cows. Table 14 provides details on the numbers of dairy cows, their milk production and the leading counties in each state.

Cattle and Calves: Important as milk and milk products are to the states of the Pacific Northwest, the actual number of milk cows is quite small when compared to the number of cows raised for meat, as Table 15 indicates.

Aberdeen-Angus, Hereford, and Shorthorn are the major beef cattle breeds along with smaller numbers of Charolais, Highland, Brahman, Santa Gertrudis and other breeds. The majority of farms raising beef have herds of fewer than 50 head, but by far the greatest number of cattle are on farms and ranches that maintain herds of a thousand or more. In 1987, for instance, Washington had 39 farms with 2,500 head or more of cattle and another 71 farms with more than 1,000 head, for a grand total of 412,609 animals. This contrasts markedly with the 9,292 farms that had fewer than 50 head each for a total of 160,260. The same is true in Idaho, Oregon, and western Montana, in each of which the greater proportion of beef cattle are raised on farms or ranches that carry more than 1,000 animals.

	All cattle	Cows & heifers that had calved		Heifers & heifer calves	Steers, steer calves, bulls & bull calves
		Milk cows	Beef cows		
Idaho	1,772,756	157,665	558,229	499,252	557,610
Oregon	1,503,625	95,325	616,857	398,119	391,324
Washington	1,304,573	220,849	334,966	398,954	349,904

Table 15. Numbers of Cattle and Calves in the Pacific Northwest, 1987.
Source: Census of Agriculture, 1987.

Extensive acreages of each state are now given over to cattle raising, especially in those areas less suited to arable farming or places where abundant amounts of water for irrigation are not available. More rugged upland areas and some forest land are also used for cattle raising. Ten counties in the Pacific Northwest have herds numbering more than 100,000, but only one of these, Whatcom County, Washington, is west of the Cascades and it is also the leading dairy-producing county of the region, where beef cattle are as numerous as milk cows. The other nine counties are: Canyon (130,677), Twin Falls (109,680), and Owyhee (106,755) in Idaho; Harney (118,202), Lake (107,350), and Baker (105,913) in Oregon; Yakima (163,524) in Washington and Madison (101,176) in western Montana. (Figure 1)

As in all other branches of farming, the cattle industry has been revolutionized with the adoption of a wide range of technological innovations all the way from genetic research and artificial insemination to packaging and marketing. The open range has given way in large measure to managed and improved pastures, and the cowboy of old has been replaced by cattlemen with technical "know-how" and advanced university training.

Hundreds of autobiographical and fictional works have been written about life on the open range, the cowboy, and the various cattle trails. In contrast, the record of the modern cattle industry is found in the more sober technical reports issued by USDA and the agricultural colleges of the Land Grant universities. However, works such as D. W. Meinig's *Great Columbia Plain* and Alec McGregor's *Counting Sheep* provide much solid information on the transformation of livestock farming in the Pacific Northwest, while the novels of Ivan Doig, *This House of Sky, English Creek,* and *Dancing at the Rascal Fair,* and Herman Oliver's *Gold and Cattle Country* include excellent background material on, respectively, western Montana and the John Day Country of Oregon.

Poultry Farming: Poultry farming has steadily increased in importance in the Pacific Northwest during the past half century. There is little doubt that it is in poultry farming that the technological changes have been the most revolutionary. This branch of farming is now so far removed from other farming that it more closely resembles the automated factory. Here is agribusiness on the grand scale: chickens raised by the tens, even hundreds of thousands, in climate-controlled sheds, their feed carefully regulated as to quantity and timing. The small producer still accounts for a slight share of the commercial market, but in the past few decades it is the relatively small number of "giants" working under contracts with major food chains that have come to dominate the industry.

As a result, localization and concentration of poultry farming has taken place, with ultimate success dependent largely on proximity to a large urban market and access to a major highway or interstate. In Oregon two counties, both within short distances of Portland, account for more than 80 percent of the state's poultry industry: Clackamas County with 1,367,036 chickens and Marion County with 1,065,580. In Washington the concentration is not quite so pronounced. Six counties, all close to a major metropolitan area and served by an interstate highway, account for most of the state's poultry farming: Thurston County with 1,347,738 had in 1987 close to one-quarter of the state's total, followed by Skagit County with 772,761, Spokane County with 488,461, Yakima County with 365,406, Lewis County with 240,358, and Whatcom County with 239,584. (Figure 1)

AGRICULTURAL REGIONS OF THE PACIFIC NORTHWEST

In this final section on agriculture, a brief comment is needed about agricultural regions. Any attempt to regionalize is, according to many, an exercise in futility, and to attempt to regionalize such an evolving and ever-changing activity as agriculture is particularly so.

Within each state the USDA, in collaboration with the state's department of agriculture, has established a system of agricultural districts or regions—there are four in Idaho and five in each of Oregon and Washington. And although these might be called *functional regions*, in fact the only function they serve is the regular monthly, quarterly, or annual collection of information pertaining to climate, productivity, market conditions, and the like.

Formal regions, which depend on the careful choice of "scientific" criteria for their delineation, pose much greater problems. The acreages planted to particular crops or used as pasture, the particular crops or mix of crops grown and animals raised, and the productivity per acre or unit, are in constant flux. The best, therefore that can be achieved is the delineation of *generalized farming regions*, within which an accepted range of conditions and the undertaking of specific types of farming—arable or pastoral—will usually be found. (Figure 2)

The map prepared a half century ago by the Pacific Northwest Regional Planning Commission to show the generalized farming regions of the three northwest states and western Montana is still of use today. Despite the loss of agricultural land in such

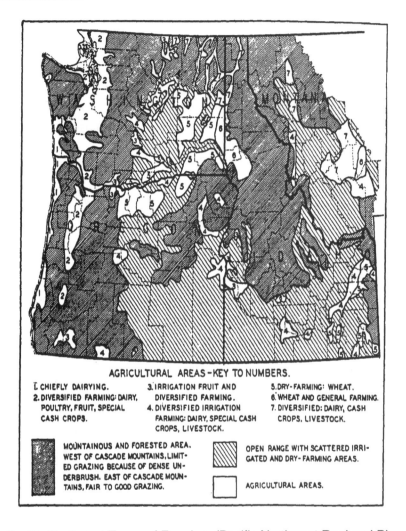

Figure 2. Pacific Northwest Types of Farming. (Pacific Northwest Regional Planning Commission.)

areas as the Puget Sound lowland or the Willamette and Snake valleys, the types of farming carried on in these areas remains largely unchanged. Further, the boundaries of the generalized farming regions shown on the map require only minor changes, as in the Columbia Basin and the Snake River valley, where the extension of irrigation to an additional million or so acres has resulted in the reduction of rangeland and in "scattered" irrigation, as it is called, becoming "extensive" irrigation.

As long as irrigation is available at current levels in the Pacific Northwest—although there are numerous experts and scholars who think that this will not continue to be the situation for very many more decades—it seems likely that agriculture will remain a dominant activity in the region not only in the "wet" west but also in the "dry" east.

REFERENCES

This chapter depends very heavily on the *1987 Census of Agriculture*, including the *Atlas of American Agriculture*, and on the *Agricultural Statistics* volumes issued by the states of Idaho, Montana, Oregon and Washington annually. Generally the 1989 and 1990 volumes were those most often consulted.

Works cited in the chapter other than the above are:

Carson, Rachel. *Silent Spring*. New York, 1962

Doig, Ivan. *Dancing at the Rascal Fair*. New York: Atheneum, 1987

English Creek. New York: Atheneum, 1984

This House of Sky. New York: Atheneum, 1981

Freeman, Otis W. & Howard H. Martin. *The Pacific Northwest: An Overall Appreciation*. New York: Wiley, 1954

Hart, John Fraser. *The Land That Feeds Us*. New York: Norton, 1991

McGregor, Alexander Campbell. *Counting Sheep: From Open Range to Agribusiness on the Columbia Plateau*. Seattle: University of Washington Press, 1982

Meining, D.W. *The Great Columbia Plain*. Seattle: University of Washington Press, 1968

Oliver, Herman. *Gold and Cattle Country*. Portland: Binfords & Mort, 1961

Reisner, Marc. *Cadillac Desert: The American West and Its Disappearing Water*. New York: Viking, 1986

Worster, Donald. *Rivers of Empire: Water, Aridity and the Growth of the American West*. New York: Pantheon Books, 1985

TOURISM AND RECREATION

Chapter 14

Larry King
Portland Community College

INTRODUCTION

The Pacific Northwest offers the visitor a multitude of scenic vistas, recreational opportunities, and tourist attractions. A mix of rugged coast lines, snow enshrouded mountains, rushing rivers, and enchanting settlements are but a few of the Pacific Northwests' tourist enticements.

Tourism is one of the top three economic activities for the Pacific Northwest states. Only in the past two decades, however, have state departments of tourism been created and budgets established to promote tourism in the region. Today all Pacific Northwest states use sophisticated promotional programs designed to attract tourists both nationally and internationally.

For example, Travel Montana, Montana's tourist promotion division, actively and aggressively promotes Montana as a destination to the international travel trade, with special emphasis on Germany, United Kingdom, Japan, and Taiwan. Montana accomplishes this by sending representatives to key international travel trade shows, advertising efforts, brochure distribution, education, and increased tour operator and travel agent awareness of Montana. The state also maintains offices in Tokyo, Japan and Taipei, Taiwan to oversee these activities.

In short, Pacific Northwest state governments perceive tourism and recreation as important to a healthy state economy and have created apparatus and strategies to encourage the continued growth of this economic asset.

This chapter will explore tourism and recreation in the Pacific Northwest. Beside defining what is meant by tourism and recreation, the chapter will examine the importance of tourism and recreation to each of the states and summarize the effect tourism and recreation have on the region. Finally, several examples of tourist attractions and recreational sites located throughout the Pacific Northwest will be discussed.

No definition of tourism is universally accepted. Although a link exists between tourism, travel, recreation, and leisure, the link is fuzzy. All tourism involves travel, yet not all travel is tourism. All tourism involves recreation, yet not all recreation is tourism. All tourism occurs during leisure time, but not all leisure time is given to tourist pursuits.

Though several definitions of tourism and recreation exist, the one chosen for use in this chapter is the following: Tourism includes all activities, commercial and noncommercial, associated with a traveler (resident or nonresident) who traveled from his or her home community for purposes or pleasure and leisure—including outdoor recreation.

A variety of studies report that tourists travel to view scenery, to learn about other cultures, or to visit friends and relatives. In a different context, the reasons why people travel can be classified into five categories.

1. Sophisticated Adventure Seekers: These travelers look to vacations for adventure and excitement. Their ideal vacation consists of a whirlwind of activities and includes everything from skiing and hiking to fine dining and sightseeing. These vacationers are educated, affluent professionals. They are typically married, and about half have children living at home.
2. Upscale Comfort Seekers: The Upscale Comfort Seekers enjoy the good things in life. They like to be pampered and consider vacations a time to relax and indulge, a time to become reacquainted with their families. These travelers are the most affluent of all groups. They are college educated and hold professional managerial positions. They are typically married, though most do not have dependent children living at home.
3. Middle America: This group represents the average person traveling in America. Their ideal vacation offers a variety of experiences, and they choose destinations with a range of activities. They seek first-class treatment, but not necessarily luxury, and are apt to seek out bargains. This group, although less affluent than the previous two groups, is comfortable financially. Most are high school graduates working in a variety of blue collar and white collar professions. Almost three-quarters are married, but fewer than half have dependent children living at home.
4. Restricted Income Travelers: These vacationers steer away from trendy or sophisticated travel destinations; rather, they elect to spend time off with their friends or relatives, often choosing to combine vacations with visits to family. Though they like to experience some new things, they will do so in a framework that is familiar. More than others, this group chooses sedentary activities—relaxing, sightseeing, and dining on local specialties. They have limited incomes, and most have a high school or grade school education. They hold clerical or blue collar jobs, some are retired, and most are married.
5. Fun Seekers on a Budget: This group knows how to have fun on a limited budget. They like to experience excitement and see new places. They are the most social of all the groups and enjoy active nightspots when on vacation. A vacation for this group is a change of pace. Like the previous group, these travelers have

limited incomes and usually no more than high school or vocational school degrees. Just over half are married and work in clerical, service, or unskilled labor positions.

A composite of an average leisure traveler would include the following characteristics: They would be married (67% and average 45 years of age. Thirty-eight percent would have children at home. Seventy percent would have more than a high school education. There would be 1.6 wage earners per household, and the family median income would be $43,000.

What does the average traveler do on vacation? They sightsee (93%), dine on local specialties (90%), shop (89%), visit museums and historical sites (87%), visit national and area parks (83%), relax at the beach (77%), attend special events (73%), participate in water sports (73%), and visit theme parks (71%).

There are generally three components of the tourist experience: motivation, objectives, and satisfaction. The motivation to travel is primarily a psychological phenomenon based on a person's desires and lifestyle. It is often defined as an escape from a monotonous and stressful urban existence. The objective of travel is the explanation given to others, usually presented in terms of the desirable attributes of the destination. The satisfaction of travel is the actual response that results from traveling.

Tourism accounts for a significant part of the economies of various Pacific Northwest communities, counties, and states. Tourists are attracted to these areas because they have certain natural or cultural features. The number of tourists attracted depends partially on accessibility of the attraction and how much advertisement or media attention it receives. However, recent studies show that a high percentage of American tourists receive their travel destination information from personal sources, such as family and friends.

The importance of tourism to the state and local communities, in the form of tax receipts and employment, should not be underestimated. A headline in an April, 1991, Portland, Oregon, newspaper emphasizes the importance of tourism to one small coastal community. "Slide—it buried coast tourism, too." The headline refers to a huge landslide in the Oregon Coast Range that buried a section of the Wilson River Highway linking Tillamook to the Portland area. According to a restaurant owner, whose establishment is located along this highway, "It's a nightmare. We live and die by the Portland clientele that flows over the mountain. Suddenly, that business is gone, and we're in real trouble." The owner expects to lose as much as $20,000 before the highway is reopened, and other businesses in Tillamook and surrounding resort communities also predict staggering losses.

Travel and tourism have been fast growing sectors of the economy, even in periods of economic decline. They employ many marginal workers who would otherwise have a difficult time finding a job, and the increase in outside money coming into the community has a multiplier effect, benefiting firms that do not serve tourist directly.

Nevertheless, negative effects of tourism on an economy include a generally low-wage, seasonal employment, and susceptibility to changes in the taste of visitors. Large scale tourism can also have a social impact as residents and visitors compete for limited space on streets and in other public areas. However, this is seldom a perceived problem except in the most heavily visited destination areas.

Attempts have been made to determine and measure the various factors which contribute to the satisfaction a person receives from visiting a tourist attraction. The influencing factors found to be most important include natural features, cultural and social factors, attitudes of residents, accessibility, prices, recreational and educational activities, and retail outlets and commercial activities.

Furthermore, the principle difference between the tourist and the resident is that the tourist is more interested in the history and culture of a place than the resident is. This is possibly because in tourist settings, traditions and history are more readily accessible to the tourist than are differences in the settings' art, language, and leisure activities.

Natural features are also important tourist attractions. These include environments that tourists can relate to either passively or actively. Scenic sites, such as Oregon's Crater Lake and the Washington-Oregon Coast, are examples of combination passive/active environments. Many tourists are content to spend a rainy weekend in front of a cozy fire watching storm waves pound the coast line. In contrast, other tourists prefer to utilize these scenic environments rather than just view them as pretty landscapes. Instead of viewing Crater Lake, they will hike around it or boat across it.

Besides the attraction of natural features, many tourists are drawn to communities which have been designated as unique by the federal government or have acquired a positive reputation among tourists. These places can be grouped into categories including designated historic districts, theme communities, and resorts. Designated historic districts are registered with the National Registrar of Historic Places. Theme communities, as the term implies, select a theme usually associated with an important current or historic economic activity (gold rush, cannery row) or some unique site feature.

A recent study of 52 Pacific Northwest thematic retail districts indicated that 33 considered tourism as the primary purpose for using a design theme in an older retail area. Twelve respondents considered tourism as a secondary purpose, and only one considered it of no significance to the district. The primary reason tourism is a popular emphasis in older retail districts is because of the money it generates.

In most cases, private developers and property owners introduced theme concepts in districts emphasizing tourism (78 percent), while city agencies were the least likely to take such action (33 percent). The history of the community, the surrounding natural environment, nearby recreational opportunities, and cultural attractions influenced the selection and development of tourism themes.

Based on the study's findings, the typical tourist oriented, thematic retail district is characterized by the following attributes:

Waterfront or western theme
Introduction by the private sector
Importance of community history
Importance of location, especially the surrounding environment
Low public sector involvement
Strong retail associations with group advertising
Design control through peer pressure, rather than ordinance
Some tourism related industry prior to the theme development
Tourists as a major customer group

Sisters, Oregon, provides a prime example of these characteristics.

TOURISM BY STATE

The four states comprising the Pacific Northwest have each conducted studies aimed at assembling a comprehensive data base on tourists, collected information useful to state and regional tourism promotional efforts and identified the extent of the impacts of the travel and tourism industry on the state's economy. Some or all of these themes will be discussed in the following pages.

Idaho

Who Are Idaho's Leisure Travelers?

Sixty-three percent of Idaho's leisure travelers come from outside the state. Of those coming from outside Idaho, the largest percentage originates in Washington (25 percent), followed by Montana (12 percent), California (12 percent), Canada (10 percent), Oregon (8 percent), and Utah (6 percent). Thirty-seven percent of Idaho's leisure travelers begin their trips in Idaho. The greatest number are from southwestern Idaho, including Boise.

The average age of leisure travelers is 44 years; the largest age segment are travelers in their thirties and sixties. Fifty-two percent are men, and 48 percent are women. Three percent have some elementary education, 43 percent attended high school, 39 percent attended college and 16 percent attended graduate school.

Twenty-four percent are professional/technical workers, 24 percent are retired people, and 11 percent are homemakers. Sixty percent have household incomes under $30,000, 33 percent have incomes between $30,000 and $70,000, and seven percent have household incomes of more than $70,000.

Seventy-eight percent of the leisure travelers stay overnight, and 22 percent take day trips. Thirty-six percent of the day users are nonresidents, and 64 percent are residents.

What Do Leisure Travelers Do in Idaho?

Sixty-one percent of travelers in Idaho have the state as a destination, while the other 39 percent are traveling through to some other state. Thirty-six percent of all leisure travelers travel in the summer, 20 percent in the fall, 24 percent in the spring and 20 percent in the winter. Thirty-two percent of leisure travelers stay at inns, 32 percent camp, 29 percent stay with friends and relatives, and four percent stay at second homes. Of all the campers, 43 percent use public campgrounds, 33 percent use private campgrounds, 23 percent use roadside areas, and only five percent use backcountry sites.

Of the more than 75 different activities leisure travelers engage in while in Idaho, fifty-five take place outdoors, and 41 are dependent on the natural resource setting. Primary activities include sightseeing, visiting friends and relatives, fishing, dining and shopping, attending special events, hunting, and downhill skiing. Hiking, camping, and photography were also listed as important activities.

The greatest share of travel expenditures are made for lodging (33 percent), transportation (26 percent), and food and beverages (23 percent); four percent is spent on recreation services and fees, four percent on recreation-related retail items, and ten percent on other retail items. Nonresident groups spend nearly 43 percent more, on the average, than do resident groups.

How Do Idaho's Leisure Travelers Think about and Make Travel Decisions?

The three purposes for leisure travel in Idaho are (1) being with friends and relatives, (2) visiting attractions or natural resources, and (3) participating in recreational and leisure activities. People traveling to an Idaho destination are more likely to participate in recreation and leisure activities than people traveling through to another state.

The most positive Idaho qualities influencing people's travel decisions include its scenery, its natural attractions, and its recreation opportunities; the most negative include its road system, unattractiveness, and its high prices. The top six images (as expressed in one word) with which Idaho leisure travelers characterize Idaho include "attractive," "wonderful," "natural," "family-related," "good," and "friendly." When comparing Idaho's services to those in other western states, the top three rated as better in Idaho include its public campgrounds, its outfitting/guide services, and its friendly people; those three assessed as worse include Idaho's night life and entertainment, its public restrooms, and its roadside services (roadside information, signs, etc.).

Forty-eight percent of leisure travelers spent less than one day planning their trip, 23 percent spent more than one day but less than a week, 14 percent spent more than a week but less than a month, and 12 percent spent more than one month.

Friends and relatives, prior travel, and maps and books are used by more travelers in Idaho than other sources of information. This finding suggests that, along with providing quality services, ensuring that residents and tourism professionals are knowledgeable about these services and Idaho's tourism resources is critical to the state's tourism future.

Summary

Tourists traveling through the state to non-Idaho destinations spend more on the average than do tourists traveling to Idaho destinations. Consequently, extending lengths-of-stay in Idaho to include multiple destinations in the state will be critical for maximizing tourism revenues.

The vast majority of leisure travel groups in Idaho relied upon motor vehicles for transportation, underscoring the importance of roads, roadside information centers, and restrooms as vital components of the state's recreational and tourism system. Unfortunately, Idaho's road system was the most frequently mentioned negative attribute affecting a person's decision to travel in the state, and restrooms and roadside services were listed as two of Idaho's worst services. These findings suggest that investment in Idaho's road infrastructure (signs, roads, restrooms) could pay tourism dividends while, at the same time, enhancing other aspects of Idaho's economy.

States surrounding Idaho are all important geographic markets for maintaining and expanding Idaho's tourism industry. In addition to attracting travelers from the outside, keeping Idahoans in Idaho and meeting their recreational needs is important.

Idaho's primary geographic competitors for existing leisure travelers are Washington, Montana, and Western Canadian Provinces. Distant secondary competitors include California, Oregon, Wyoming and Utah.

Retirees are of major importance since they currently comprise 24 percent of Idaho's leisure travelers and are becoming the largest segment of U.S. travelers. Although these travelers have discretionary time and money, they are currently more likely to have destinations outside of Idaho. Reversing this pattern could be a key strategy.

Finally, natural resources are a big part of Idaho's appeal and image as a tourism destination. Central to this role are the outdoor activities in which Idaho's leisure travelers participate. Federal, state, and local policy should consider carefully the tourism values associated with Idaho's natural resource lands.

Montana

Who Are Montana's Leisure Travelers?

Trip origin data show that most visitors to Montana came from Pacific Coast states (Oregon, Washington, and California) followed by the Rocky Mountain states (Idaho,

Wyoming, Colorado, and Utah), and the North Central states. Approximately eight percent of Montana's visitors came from Canada.

More visitors came to Montana from Washington (13%) than from any other state. California was second, accounting for 11 percent of the visitors. Minnesota, Idaho, Wyoming, Nebraska, and Colorado each accounted for four percent.

The median age of Montana's tourists was 48 years. Sixty percent were men and 40 percent were women. Fifty-one percent had a college degree while another 24 percent had some college education.

Thirty-eight were professionals; 11 percent were managers. Approximately 23 percent were retired persons and 5 percent were homemakers. Thirty-four had household incomes over $50,000.

What Do Leisure Travelers Do in Montana?

Tourists come to Montana for a variety of reasons. Over half of all visitors come for vacations, and 36 percent come to visit family or friends. Forty-three percent reported that they were just passing through Montana on their way to another destination. Of course, many visitors may have multiple purposes in coming to Montana.

Since over half of Montana's visitors come to vacation, it is important to know what types of recreational attractions have drawn these visitors to Montana. Over 80 percent indicated that scenery was an important reason for coming to vacation in Montana. In addition, 23 percent of Montana's tourists said that fishing was a reason they came to the state. About 80 percent of the tourists sampled said that they had visited Montana previously.

A review of the recreational activities in which tourists participate indicates the top three are day hiking/walking, photography, and interest in historic sites.

The greatest share of travel expenditures are made for retail goods, restaurant/bar, and lodging.

How Do Montana's Leisure Travelers Think about and Make Travel Decisions?

About one-fourth of Montana's visitors started thinking about visiting the state one to three months prior to actually coming. Another one-fifth started thinking about the their trip three to six months in advance. About 14 percent of all visitors began thinking about their trip one week to one month in advance of coming to Montana.

Montana tourists indicated that their most important source of information about the state was family and friends. Prior visits ranked second as a source of tourist information. One fourth of Montana summer visitors came to visit relatives and friends.

In 1988, 5.1 million nonresident tourists visited Montana. It is estimated that these out-of-state tourists spent $658 million that resulted in $1.5 billion of total economic impact to Montana, including the generation of 25,000 jobs.

Montana's tourism budget for 1988 was $4.7 million. This funding was provided by a statewide four percent accommodations tax, which has been in affect since 1988. According to state law, about 25 percent of the funding is returned annually to tourism regions and eligible cities for regional and local tourism promotions.

Summary

Proximity to Montana is the most significant variable in attracting visitors. Over half of Montana's out-of-state vacation planners are unfamiliar with the state. As familiarity increases, so does the chance for visitation. To increase visitation, Montana must create a vacation image that in turn creates consumer demand. By building an image for Montana as a vacation destination, the state will also positively influence the media and the travel trade.

Montana's major appeal is in its scenic beauty, clean environment, mountains, national parks, outdoor recreational opportunities, and friendly people. Montana is a good vacation value, but the cost of getting there—both in terms of time and money—is a deterrent. Montana ranks 47th in the nation in distance from the major metropolitan areas. Economic factors (gas prices, airfares, currency exchange fluctuation), obviously, influence visitation.

Oregon

Who Are Oregon's Leisure Travelers?

Eight percent of Oregon visitors are foreigners, the largest percentage from Canada. Oregon is primarily a domestic U.S. travel destination, although growth probably will occur in foreign visitation.

The bulk of visitors to Oregon are from the West, nearly half from the Far West. This area clearly is Oregon's current primary market. The Great Lakes area contributes 11 percent of Oregon's visitors, followed closely by the South at nine percent.

Nearly a third of Oregon visitors live in California and about an eighth in Washington. Other states lag relatively far behind, with the most important being Texas, Arizona, and Florida.

California represents Oregon's primary market for several reasons. First, it has a major concentration of relatively well-off households that tend to travel. Second, Oregon lies astride the route to Washington and British Columbia, other major destinations for California travelers.

Washington also is a significant market for Oregon. However, its population is smaller, and its resource base is similar to that of Oregon, thereby diminishing Oregon's drawing power.

Nearly two thirds of Oregon's visitors have traveled to Oregon previously, indicating a strong repeat visitor market and the importance of satisfying visitors.

Pleasure travelers and those visiting friends and relatives are the most common in Oregon, comprising a total of 70 percent of all travelers. Auto travelers comprise 55 percent of travelers, with another 16 percent arriving in an RV. Over a fifth arrive by air, a large portion of whom rent an automobile. The most common length of stay for Oregon visitors is between three and five nights. Most travel parties consist of two people.

Relatively few Oregon visitors are under 35 years old; the remainder are fairly evenly distributed among age categories above this. Nearly a fifth are 65 or over.

Family income of Oregon visitors is spread fairly evenly between $20,000 and $75,000 per year, with few travelers above this range. Median family income for all travelers is $42,700 per year. The bulk of those earning more than $75,000 are business travelers.

Travelers to Oregon are highly educated, with well over half holding a bachelor's or graduate degree.

What Do Leisure Travelers Do in Oregon?

The most common activities of Oregon travelers consist of relaxing or sightseeing, shopping in a small town or metro area, or visiting friends and relatives. The prevalence of sightseeing among nearly four out of five visitors shows that enjoying coastal, mountain, and other areas in at least a passive manner is a dominant form of pleasure travel in Oregon.

The next most common activities relate to a number of Oregon's historic or natural resources. Almost half of visitors report visiting a historic site or area. About a third hike or picnic, and about a quarter view or study wildlife or camp. Smaller but still substantial proportions run rivers or fish.

Urban or cultural activities are moderately popular, consisting of visiting a restaurant or club primarily for entertainment or attending a cultural or artistic event.

About half of Oregon visitors traveled to a number of locations in the state, not focusing on one or a few in particular. About a quarter focused their visit on one location.

The Coast and the Portland area are the most popular destinations for visitors to Oregon. Over 62 percent of Oregon visitors travel to or through at least one portion of the coast, and over three-fifths travel to the Portland area (which includes the Columbia Gorge and Mt. Hood.)

Approximately 6 million visitors traveled to Oregon in 1987.

How Do Oregon's Leisure Travelers Think about and Make Travel Decisions?

Only about a quarter of Oregon's travelers plan their trip carefully, with the remainder either planning the overall route but not individual stops, or doing little or no planning.

The most common sources of information are friends and relatives and the traveler's own prior experience as an Oregon visitor. The latter is by far the most common information source for repeat visitors. For visitors overall, the next most common information sources are the automobile clubs, local chambers of commerce and visitor bureaus, and commercial guidebooks.

Visitors report that they are highly satisfied with their visit to Oregon. Nearly three quarters say they are very satisfied with their visit, and another one-fourth say they are satisfied. The proportions are similar for first-time visitors as well as repeat visitors.

Travelers score Oregon's natural resources very highly as attractions. Over two-thirds consider Oregon's scenery exceptional, with other high scores for the coast and mountains. The friendliness of Oregonians and Oregon's recreational opportunities also receive high scores.

Scores are lowest for urban activities including restaurants, cultural activities, and directional signs. Even in these cases, however, the proportion of respondents who say these attractions are disappointing or very poor is quite small, no more than 2 percent.

Summary

Overall, Oregon's visitor industry has fared somewhat better than industry in general between 1975 to 1987. During this period when commercial accommodations employment grew by a 32 percent, employment for all industries in Oregon grew more slowly at 26 percent.

Total travel-generated expenditures for Oregon amount to about $1.8 billion (1987). The associated payroll is about $319 million, representing nearly 37,000 jobs. Travel-generated local tax receipts, which derive from local room taxes, amount to $19.2 million. State tax receipts from gasoline taxes and corporate income taxes total $48.5 million.

Travel and tourism is tied with high technology manufacturing for third among export-oriented sectors of the Oregon economy, following forest products and agriculture.

The Tourism Division's budget for the 1987-89 biennium was $4.5 million. Funding increased for the 1989-91 biennium. The Tourism Division's budget, $5.6 million, received $2.6 million from the general fund, while lottery proceeds funded $3 million.

Washington

Who Are Washington's Leisure Travelers?

The characteristics of a Washington resident traveler would include the following. Sixty-eight percent are married and have a mean age of 41 years. Forty-four percent have children at home. Sixty-four percent have more than a high school education, and the 1.6 wage earners per household earn a median income of $41,000.

A profile of Washington's out-of-state visitors in many ways parallels that of the Washington resident. The greatest percentage of these visitors come from California. They travel to Washington by automobile usually in the month of July. Over one-half of these out-of-state visitors are married, adult couples ranging in age from 36 to 45 years. Most have a bachelor's degree and a mean income of $38,000.

For ninety percent of these visitors, the purpose of the visit is a vacation, and the main attraction is Washington's scenery. Over seventy percent list sightseeing as a major activity.

What Do Leisure Travelers Do In Washington?

When in the state, visitors stay primarily in hotels or bed and breakfast accommodations. Residents, more than others, vacation in RV parks.

Locals tend to leave the metropolitan areas and head to the mountains and beaches for vacations. Residents from the east of the Cascades often travel to Seattle. Out-of-state tourists are drawn to the cities, especially Seattle, Spokane, and Olympia. Overall, the Seattle area attracts more people than any other region in Washington. Other areas of interest include Northwest Washington and the Olympic Peninsula. Attractions most frequently visited by out-of-state travelers were Mt. St. Helens, Mt. Rainier, and the Puget Sound ferries.

The activities people enjoyed most while visiting Washington included sightseeing, dining on local specialties, shopping, visiting museums and historical sites, attending special events, and hiking and mountain climbing.

How Do Washington's Leisure Travelers Think About And Make Travel Decisions?

Travelers most often associate Washington with beautiful scenes of grand mountain ranges, lush green forests, and interesting coastlines. This image is especially true among Washington visitors. Washington does not suffer from a negative image among travelers; half of U.S. and almost three-quarters of Canadian travelers could not name anything negative about the state.

If there is a "down-side" to Washington, it is bad weather. It is by far the biggest complaint about vacations here from both in-state and out-of-state U.S. travelers. However, Canadians don't complain about it. Among visitors, especially Washington residents, issues of the state's growth have been seen to affect vacations here in a negative sense. Perceptions of additional crowds and urban sprawl may interfere with the ideal vacation in the state.

Summary

Tourism is Washington's fourth largest industry. In 1989, close to 17.5 million nonresident Washington tourists spent $5 billion on tourist related activities. The greatest expenditure was for food service, which accounted for $1.2 billion. These expenditures created a total payroll of $1.2 billion for a total employment of 92,000 (which ranked 4th among the states).

According to data released by the U.S. Travel Data Center, Washington state dropped from 35th to 40th position based on its 1989-90 tourism budget of $2.6 million. The average state travel budget is $6.8 million, an increase of 6 percent from the year before. A four-fold increase was reported in total state budgets during the last decade. The top five budgets are as follows: Hawaii $22.5 million, New York $22.0 million, Illinois $22.0 million, Texas $18.8 million, Pennsylvania $15.9 million. Oregon ranked 35th with a tourism budget of $3.0 million, Idaho was 41st with $2.3 million, and Montana was 30th with $4.5 million.

Recreation in the Pacific Northwest

The Pacific Northwest offers a wide variety of outdoor recreational opportunities which, as was seen earlier, attract tourists from throughout the U.S. and result in important economic contributions in the region. The diversity of Pacific Northwest landscapes offers opportunities for visitors to climb mountains, ski, hunt, fish, beachcomb, white water raft, hike, and pursue numerous other interests.

The fact that the public has access to much of the Pacific Northwest's outdoor attractions has stimulated the growth of tourism as residents of states without these advantages have migrated to the Pacific Northwest for these outdoor recreational opportunities.

However, if a large share of a state's tourist economy is tied to natural conditions—snow, wind, and other weather related factors—the result can be periodic economic peaks and valleys. For example, the Oregon ski industry suffered an economic downturn in late 1989 as low snowfall delayed the opening of the ski season. Most Oregon ski areas open in late November, yet it was mid-January, 1990, before winter storms began to build the necessary snowpack for operation. The ski season was finally in full swing at the end of January. The late start resulted in Mt. Bachelor, near Bend in the Cascades, suffering a decline of 18 percent in skier visits.

Furthermore, during the summer of 1989, Oregon's natural wind tunnel, the Columbia River Gorge, was calm, leading to cancelled events and to a decrease in windsurfing activities in this area. Windsurfing is considered the backbone of the Gorge's tourism economy.

Several of the most outstanding scenic and historic resources in the Pacific Northwest are administered by the National Park System. These include national parks, national monuments, national historic sites, and national recreational areas. (Figure 1).

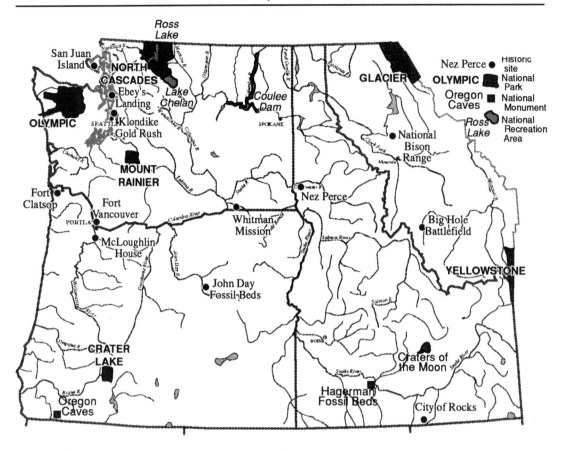

Figure 1. National Parks of the Pacific Northwest.

A Few Lesser Known National Park Attractions

Fort Clatsop National Memorial, Oregon—Fort Clatsop stands on a tree shrouded knoll some 30 feet above the Lewis and Clark River. Trails lead from the fort to a fresh water spring and to the river's edge where members of Lewis and Clark's Corps of Discovery secured their dugout canoes.

Without the modern park service building, picnic tables, parking lots and signs identifying vegetation along the trails, Lewis and Clark would feel right at home today. Reconstructed Fort Clatsop resembles their original fort to the smallest detail.

When Lewis and Clark arrived on the Oregon coast in November, 1805, they spent ten miserable days on the Columbia's north shore. The rain rotted their clothes and dampened their spirits. Clark even complained bitterly about the ocean's noise, "This roaring has continued . . .Since we arrived in Sight of the Great Western: (for I cannot say Pacific) *Ocian* as I have not Seen one pacific day Since my arrival in its vicinity..."

Finding only deer available for hunting, the captains, on the advice of local Indians, relocated to the river's south side where elk were in abundance.

Besides game, Lewis and Clark thought trading ships, from which they could buy Indian trinkets for use on their return trip, would more likely visit the south shore. They also hoped to obtain salt needed for curing and seasoning meat there. Even Sacajawea urged the captains to cross the river because she wanted to go where there were plenty of wappatos, a root resembling potatoes, which she liked.

Reaching the Columbia's south shore on November 29, they made camp on a low and marshy neck of land (probably Tongue Point) while Lewis and five men searched for a more suitable winter campsite. After six days, Captain Lewis returned to the main party to report the discovery of "a tolerable good place for our winter quarters."

On December 8 the entire party moved to the new location on the first high point of land on the west side of the Lewis and Clark River, about three miles from its junction with the Columbia River. Clark said the site, ". . . . is on a rise about 30 feet higher than the high tides leavel and thickly Covered with lofty pine . . . . this is most certainly the most eligable Situation for our purposes. . . . ."

The men, under Lewis' direction, immediately started cutting and shaping trees for the dwellings. By December 14 seven cabins stood, and workers began splitting logs for roofing. One of the roofers wrote, ". . . . the timber splits butifully, and of any width."

By December 24, 1805, the camp was nearly completed and the whole party moved into their quarters. Lewis named the structure, Fort Clatsop, "After the name of the Clatsop nation of Indians who live nearest to us."

They celebrated Christmas Day with a dinner that, according to Clark, ". . . . concisted of pore Elk, so much Spoiled that we eate it thro' mear necessity, Some Spoiled pounded fish and a fiew roots."

Beginning New Year's Day, the captains enacted rules to regulate the operation of the fort. A guard, consisting of one sergeant and three privates, would clear the fort of uninvited Indians at sunset and lock the gates for the night. A guard was to be posted in front of the captains' quarters at all times and would report the arrival of all Indians to Lewis and Clark. Every 24 hours the guard would check the meat room for spoilage and the security of the canoes at the landing.

Much of the winter's activity at the fort revolved around food. Hunters were constantly in the field. During their three months stay, they killed 131 elk and 20 deer. The meat was jerked to prevent spoilage, and the hides were made into clothes and moccasins. Each man made ten pair of moccasins for the return trip. Lewis and Clark traded with the Indians for roots, berries, fish and game. They also learned of the smelt runs, which usually occurred in March, from the Indians.

Life at the fort was far from pleasant. It rained every day but 12 of the 106 days at Fort Clatsop. The dampness gave nearly everyone rheumatism or colds; some suffered from dysentery. Many of the men acquired venereal diseases from the natives who, in turn, had been infected by the first sailors to visit the Oregon coast. Others had dislocated shoulders, injured legs, and back pains. All the men battled fleas, which infested their bedding and made sleep difficult.

Lewis and Clark spent much of their time at Fort Clatsop working on their maps and journals and recording additional information concerning the local flora and fauna, along with observations of the Indians' appearance, dress, customs, and ways of life. Clark went into some detail describing and illustrating the Clatsop's head flattening ritual.

On March 23, 1806, Lewis and Clark left Fort Clatsop for the return trip east. Clark writes, "The rain ceased, and it became fair. . . . at which time we loaded our canoes and at 1 p.m. left Fort Clatsop on our homeward-bound journey." The captains made the local chief, Chief Coboway, a present of Fort Clatsop, including all the furniture.

According to reports by later explorers and settlers, the fort deteriorated quickly after Lewis' and Clark's departure. In 1811, Gabriel Franchere commented that the remains, "were but piles of rough, unhewn logs, overgrown with parasite creepers." The next year Ross Cox said, "Logs of the house were still standing and marked with names of several of their party." By 1853 George Gibbs wrote, "The site of their log hut is still visible, the foundation logs rotting where they lay." Mrs. Victor noted, in 1870, that, "not only have sixty years effaced all traces of their encampment but a house which stood on the same site in 1853, has overgrown with trees now twenty feet in height."

In May, 1900, an old Indian woman named Tsin-is-tum (also known as Jennie Michell), whose father was a Clatsop Indian who had witnessed Lewis and Clark's activities, showed a committee from the Oregon Historical Society the site of Fort Clatsop. The Historical Society acquired the site in that same year. During July and August, 1955, a replica of Fort Clatsop was build in preparation for the August, 1955, Lewis and Clark sesquicentennial celebration. The replica went up in a month, the same amount of time it took Lewis and Clark to erect their fort. It was reconstructed after a sketch on the cover of Clark's elkhide field book.

Fort Clatsop is located five miles southwest of Astoria, Oregon, off U.S. 101. A visitor center contains exhibits, audio-visual programs, and a bookstore. Restrooms, a picnic area, and hiking trails are nearby.

Fort Vancouver National Historic Site, Washington (Figure 2).—In 1825 after seven years at Fort George (Astoria), the Hudson's Bay Company, a British owned firm, moved its trapping and trading headquarters 100 miles upstream to Fort Vancouver. Both the United States and Great Britain had vied for ownership of the northern Pacific Coast, a region rich in furs. By 1818 the two nations had agreed to joint occupancy of the Oregon Country until they could agree upon a final boundary.

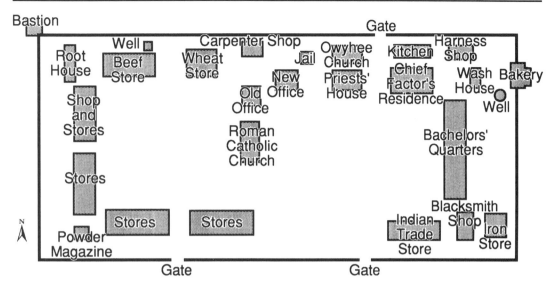

Figure 2. Fort Vancouver, Washington.

The fort's inland location was chosen by Hudson's Bay Company officers Governor George Simpson and Chief Factor Dr. John McLoughlin. They wanted a site for Fort Vancouver that would be on the north bank of the Columbia River, provide opportunities for farming and be at the confluence of the Willamette and Columbia Rivers, for trading purposes. Both men believed that the Columbia River would eventually serve as the boundary separating the United States from British territory, thus the north bank site.

From 1825 to 1846, Fort Vancouver, with a population of clerks, artisans, servants, and traders numbering as many as 200 to 300, served as the nerve center for the Hudson's Bay Company operations west of the Rocky Mountains from California to Alaska. The fort was a busy place, especially from early summer to late fall when the annual fur ship departed. Furs were cleaned, packed, and loaded aboard. Clerks prepared correspondence and account ledgers that would be scrutinized by Company officials in London. In the meantime, trapping expeditions were sent off and returning ones were welcomed with salutes and cheers.

Much of the social and cultural life of the vast region was centered at the fort. Here the Hudson's Bay Company established the first school, the first circulating library, the first theater, and the earliest churches.

When the United States took possession of Fort Vancouver in 1849 and the Hudson's Bay Company relocated to Victoria, British Columbia, the fort slowly decayed from lack of use. Settlers scavenged it for building materials and firewood, and in 1866 the remainder of the fort was destroyed by fire, under mysterious conditions. Eventually, in 1948 the National Park Service obtained the fort site. In

1966, after archaeologists had researched the site, the Park Service began to reconstruct Fort Vancouver's stockade and five major buildings on their original locations. Built originally to prevent theft, the stockade walls stood 15 feet high and by 1845 had been expanded five times to stretch 734 feet by 318 feet. About 35 to 40 people, including wives and children of the factors and clerks, lived within the stockade walls. All laborers lived outside the fort in a village immediately to the south.

Much of the information pertaining to the fort's early appearance and dimensions came from John Fremont, who in 1841 had conducted an inventory for the U.S. Army. He had drawn detailed sketches and noted measurements for Fort Vancouver's buildings, even including window and paneling specifics—size, placement, and types of materials.

Five replicas of the fort's nearly 20 buildings are now in place, held together with wooden pegs and square nails. Others await construction funds. One of the completed structures is the bastion overlooking the fort from its northwest corner. Built in 1845 to defend the fort against possible American aggression, the bastion's three pound cannons were never fired in anger. Their only function was to salute arriving ships.

Craters of the Moon National Monument, Idaho—Southwest of Arco, Idaho, along Highway 20 is the Craters of the Moon National Monument, an eerie domain that early pioneers made a superstitious point to avoid. The sagebrush terrain changes drastically into desolate moonscape, revealing the violent age of volcanos. An astounding variety of caves and craters are hidden among sharp mountains of black cinders and cones. Moon mission astronauts have even trained here.

A seven mile loop road winds through the monument. In places, the lava changes from coal black to rusty red, much of it fiercely sharp. Broad fissures in the volcanic rift where molten lava once spewed forth can be observed.

Along the loop, short walking paths lead to Indian Tunnel and other ice-lined lava tubes which can be explored by flashlight. Also located throughout the area are rock shelters built years ago by Indian hunting parties. Nearby, is a 1 1/2 mile trail leading to the Great Owl Cavern. The visitors' center offers a short guided tour.

Grant-Kohrs Ranch National Historic Site, Montana—This ranch is highly significant in the history of the range cattle industry. John Grant, the original owner, was the son of a Scottish employee of the Hudson's Bay Company and is sometimes credited with being the founder of the cattle industry in Montana. In 1853 he began his career in Deer Lodge Valley. Acquiring rundown cattle in the vicinity of Fort Hall, about 240 miles to the south, he fattened them on his ranch and sold them at a good profit. By 1863 the ranch was running 4,000 head of cattle and nearly as many horses and was providing most of the beef for the miners at Bannack and Virginia City.

In 1866 Conrad Kohrs purchased the ranch. He had been employed in a butcher shop in Bannack, had borrowed money to go into the cattle business, and had risen quickly. As soon as he purchased the ranch, he stocked it with Shorthorn bulls to improve the blood of his stock, the first constructive cattle-breeding effort in Montana. Becoming one of the foremost cattle kings of his era, he figured prominently in the organization of the Montana Stockgrowers Association. Until his death in this century, Kohrs remained a leading Montana cattleman.

The Grant-Kohrs Ranch has been preserved as a working ranch with a bunkhouse and numerous outbuildings. The original frame ranchhouse, erected by Grant in 1862, is still standing, though in the 1890s Kohrs built a large brick addition on the west side. Other old structures include several log cabins, probably built in the 1850s and 1860s, and old corrals.

Ebey's Landing National Historical Reserve, Washington—Ebey's Landing National Historical Reserve is the first national reserve to be established by Congress in 1978. The 17,000 acre reserve is located on Washington's Whidbey Island. Its purpose is to preserve and protect a rural community that provides an unbroken historic record from the nineteenth century exploration and settlement in Puget Sound to the present. It commemorates the first thorough exploration of the Puget Sound area by Captain George Vancouver in 1792; the settlement by Colonel Isaac Neff Ebey, who led the first permanent white settlers to Whidbey Island; the early active years until the Donation Land Law of 1850-55; and the growth since 1883 of the historic town of Coupeville.

There are 48 historic structures in Coupeville. The town and its historic environs were listed on the National Register of Historic Places with the establishment of Central Whidbey Historic District in December, 1955.

The oldest home in Coupeville is the Captain Thomas Coupe residence. Captain Coupe staked his claim here in November, 1852. In September, 1853, he contracted with carpenters to build the house with California redwood lumber brought from a supplier in Bellingham, Washington. He remained a seafaring man all his life, building three ships for ferry service on Puget Sound.

When the Coupeville townsite plat was officially recorded in 1883, there were two hardware stores, a drug store, three hotels, two saloons, a blacksmith shop, a courthouse, a post office, a school house, and a church.

By 1900, central Whidbey Island was known as a place of old settlers and longtime residents. The farmers and sea captains had staked their claims, built their homes, and passed the land on to their children. The area continues to be remarkably stable.

Developed Recreation: Parks and Campgrounds

A map of state campgrounds, as shown in the *Atlas of the Pacific Northwest*, indicates an uneven distribution. State campgrounds tend to be located in particularly scenic areas that offer a variety of recreational activities to the camper. The majority of Pacific Northwest campgrounds are managed by the U.S. Forest Service. Other federal agencies that maintain campgrounds include the Bureau of Land Management, the Army Corps of Engineers, and the National Park Service. States, counties, and cities also provide and maintain parks.

Oregon and Montana have recently completed studies of their park systems. Montana found that approximately 1.3 million nonresidents camped in the state's developed campgrounds in 1988. These campers stayed an average of seven days in Montana. The study found that the median age of the campers was 52 years with some college education, and 30 percent of them had a household income of over $50,000. The study also found that visitors to developed campgrounds relied mainly on prior visits in deciding to travel to Montana. Family, friends, and maps were the second and third most important sources of planning information. The largest proportion of campers started thinking about their trip one to six months in advance.

Finally, the study revealed that visitors to developed campgrounds spend less per day than many other nonresident travelers. Retail and gasoline purchases made up the majority of their expenditures. Average trip (7 days) expenditures per group (2.9 people) are about $414.

When asked what facilities needed improving in Montana State Parks, campers most often mentioned rest rooms, water supply, and picnic tables.

Oregon's 1988 state park survey is based on over 17,000 responses. Of these, 55 percent were from overnight campers, and 45 percent were from day visitors. Slightly over 50 percent of the campers were from out-of-state; nearly 40 percent of the day visitors were from out-of-state.

Nearly one fourth of the visitors were 13 years old or less, and 22 percent were between 35 and 49, indicating that families with young children formed a strong contingent among the visitors.

Visitors' income and education were high, especially among out-of-state visitors, who had an average income of over $37,000 per year compared to $29,000 for Oregon day visitors and $33,000 for Oregon campers. Visitors responding to the survey overall had an average of over 14 years of education.

Visitors traveled an average of 124 miles to reach the park. A study of Idaho residents indicates that the average distance traveled for camping trips is well over 100 miles. For most, the Oregon park where they were staying was a planned destination, but most out-of-state visitors were also visiting other state parks during their stay in Oregon. The common activities engaged in during the park visit were viewing scenery, relaxing, enjoying solitude, and watching wildlife.

Park visitors have a substantial effect on the communities in the vicinity of the park. Overall, each visitor will spend an average of $26 per day within 25 miles of the park. Based on 1987 attendance, these expenditures total about $250 million a year.

Over 90 percent of the visitors rated the parks as excellent or good. When rating specific park qualities, all were rated as satisfactory or better. Highest marks went to employee courtesy, clean grounds, abundant picnic sites, and easy registration.

Increases in camping fees are seen as one way to solve the budgetary difficulties of supporting the Oregon State Parks System. When possible camping fee increases were presented to visitors, about 80 percent said a $1 increase would have no effect on their park camping. However, those with lower incomes were more likely to say they would "stop camping" than those in higher income groups. Yet, it is these lower income groups which are already underutilizing the parks system.

The recent increase in popularity of downhill and cross-country skiing, along with other winter activities such as snowmobiling, snowshoeing, and snowcamping, has extended the recreation-tourism season into the winter months in the region's mountainous areas. Coastal tourism also reflects this same pattern. People now journey to the coast during winter storms to shop at quaint coastal communities like Long Beach in Washington and Cannon Beach, Seaside, and Lincoln City in Oregon.

Wilderness Recreation

In 1964 Congress passed the National Wilderness Act. It defined wilderness as an area where, "the earth and its community of life are untrammeled by man, where man himself is a visitor who does not remain."

To qualify for inclusion in the wilderness system, an area must be large, at least 5,000 acres, unaffected by human activity, including permanent improvements and places of human habitation, and characterized by outstanding opportunities for primitive recreation and solitude.

Wilderness areas must be on federal lands but can be administered by any one of several agencies: the National Park Service, the U.S. Forest Service, and the Bureau of Land Management. Some remote and scenic stretches of free-flowing rivers have been established as Wild and Scenic Rivers in accordance with 1968 legislation. These sections of river and their shorelines are to be maintained in a natural condition.

Unfortunately, the demand for recreation in these wilderness settings has increased to the point that managers are facing serious problems of environmental degradation caused by overuse of fragile ecosystems.

Theme Communities

The history of the Pacific Northwest is also attractive to recreational travelers. Sites that recall Indian cultures, early European settlement, and the colorful days of early logging, mining, and ranching, all have potential visitor interest. Some Pacific Northwest settlements have parlayed these historical traditions into present day tourist attractions called theme communities. Theme communities are a fairly recent landscape phenomenon.

The current revitalization trend in small community retail districts is one of image development. Image enhancement occurs through both promotional activities and the physical design of buildings and street environments. Much of this is done to attract tourists and their dollars.

According to Lew, the process of image development has moved away from the idiosyncrasies that make a place unique to images that appeal to the masses or have mass identities. Mass identities are superficial identities of place, lacking in traditionally accepted values and symbols. Mass identities destroy local uniqueness and authenticity and create a landscape which is identified by a theme name—"Old Town," "Historic District," "Western," and "Bavarian." Furthermore, tourism has helped create these unauthentic, placeless landscapes in the older retail districts of many communities. The degree to which mass images are incorporated into a retail district's design and promotion varies from one place to another.

What is being created is what Grady Clay calls an "epitome district." Epitome districts are special places which contain symbols that have the capacity to generate an emotional response. Tourist oriented epitome districts are unique in that they rely heavily on advertising to create larger-than-life myths out of older retail districts. These newly created epitome districts can be identified by some of the following traits: a name ("western," "historic," "old town"), well defined boundaries, and local history made evident by maps and brochures. Another three resources have a bearing on the type and extent of theme development in the retail district. These include (1) quality of the district's architecture, (2) location in terms of access and site characteristics, and (3) a community's cultural heritage. These resources, along with the concept of authentic versus unauthentic, will be reflected in the following case studies. The two theme retail districts selected have promoted themselves for purposes of tourism.

Sisters, Oregon

In Sisters, buildings look frontier western, with wooden eaves, second-floor balconies, timbered sidewalks out front, and even hitching posts for horses. Except for the paved streets and gas stations with plastic signs, visitors could be entering a time warp into the last century.

It was not always like this. The idea to remake Sisters into a cowboy town originated in about 1970 with the local Jaycees. They organized a community effort to remodel two small buildings, one at each entrance to the community. By 1977, the Western facade theme had become so pervasive throughout the retail district that it was made mandatory by a city ordinance adopted in 1979. Today tourism's enthusiasts hold up Sisters as a small scale model of what all of Oregon should become. "We've used Sisters as an example of a town that turned itself around from nothing to destination tourism," the state's former tourism director, Ed Remington, says.

Sisters first served as temporary home to cattle drivers and sheepherders. As more came and stayed, a permanent community took root. The first school was established in 1885, a post office in 1888. The town was platted in 1901. But in 1923 every building on the north side of the main street except the hotel burned down. In 1924 every building on the south side was destroyed the same way.

Sisters rebuilt. From the 1940s, timber kept the small town alive until the industry's decline in the early 1970s. What saved Sisters was a new, 1,280 acre development ten miles west called Black Butte Ranch. Black Butte's developers intended it to be a quiet mountain retreat, not a tourist center. They didn't want retail shops within the subdivision.

Black Butte's developers listened with interest when Harold Barclay, owner of some of Sisters' best commercial properties, suggested that the town be renovated with a Western motif to serve as Black Butte's commercial center. Some controversy has arisen over the forced remodeling of the few buildings whose authentic historic value reflected periods other than the 1800s. However, because these buildings are so few and the community has become so tourism-dependent, such objections have not been effective. In the case of Sisters, the unauthentic false facade has proven to be both successful and acceptable to tourists and local residents.

Poulsbo, Washington

The Poulsbo community has a strong Norwegian heritage. During the initial half of its first 100 years, Poulsbo was so Norwegian that any non-Norwegian speaking person would be singled out as a curiosity (Figure 3).

The founder of Poulsbo was Jorgen Eliason. Eliason, his sister, and his six-year-old son came to Poulsbo from Fordefjord, Norway, by way of Michigan in 1883. He homesteaded the land on what is today the city of Poulsbo. A month after Eliason's arrival, Iver Moe, his wife and three sons arrived from Paulsbo, Norway, by way of Minnesota.

In 1886, Iver Moe made application for a post office and named it Paulsbo. The Postmaster General, misreading Moe's handwriting, listed the new post office as Poulsbo.

Poulsbo was settled in the late 1880s by fisherman, loggers, and farmers who likened Liberty Bay and its surroundings to the fjords of Norway and the adjoining Scandinavian countries. Fishermen from the Bering Sea brought their catch of codfish here, to one of the largest processing plants in the Northwest, for salting and preserving.

In 1959 after decades of complacency, Poulsbo woke up to realize it must do something drastic to make the town attractive to the outside world. A Bureau of Community Development survey made by the University of Washington in 1961 showed an overwhelming sentiment by the entire North Kitsap community for the adoption of the "Little Norway" theme for Poulsbo.

The World's Fair in Seattle in 1962 was the stimulus for action. In a statewide beautification contest with 43 entries, Poulsbo won second place. Much of the improvement in Poulsbo's appearance and its image can be traced to that competition.

Today Poulsbo's rich Scandinavian heritage is retained and displayed in the unique storefronts; outside murals depicting dancers in Norwegian costume; the annual Viking Fest, Scandia Midsommarfest, Yule Fest; and Norwegian named streets such as Lindvig Way, Moe Street, and Jensen Way.

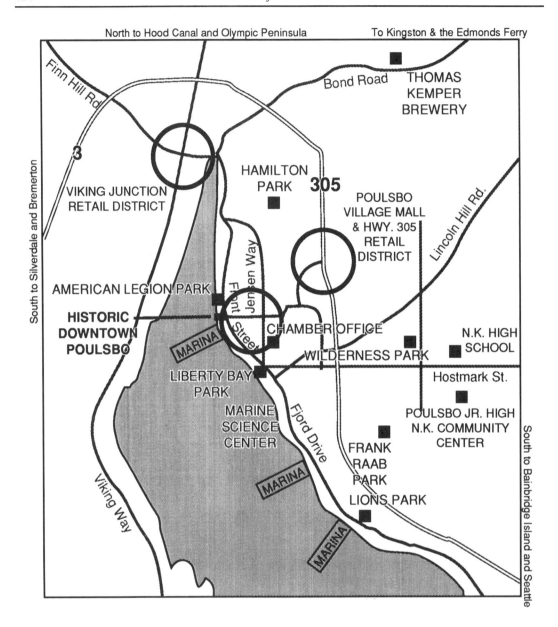

Figure 3.

Though many of Poulsbo's retail businesses have a Scandinavian appearance, there are no municipal architecture standards that specify how building exteriors should look.

Conclusion

Travel destinations have faced stiff competition in the 1980s. State governments in 1989-90 allocated more than $340 million for state tourism agency budgets, a six percent increase over the previous year. At the local level, convention and visitors' bureaus have proliferated, and cities have recognized the importance of revitalized downtowns, cultural attractions, and special events as tourism attractors.

The tourism and travel industry, despite the recession and societal changes (increased mobility of the nation's senior citizens), grew substantially during the 1980s. The U.S. Travel and Tourism Administration projects that international visitors to the U.S. will climb 11 percent to a record of more than 40 million in 1990 and will gain an additional nine percent in 1991. Further, foreign visitors' share of the total U.S. tourism and travel industry will continue to increase. One reason for this prediction is the strong appeal of the United States to foreign visitors; few other countries can match the United States in terms of diverse scenery, geography, cultures, climates, and travel experiences. The travel industry is now the nation's third largest retailer and second largest employer.

This increase in tourism, however, presents a problem for many Pacific Northwest communities, according to Steve Lawton, coordinator of the office of tourism for Oregon. "Even though tourism produces jobs, many are the low paying, entry level kind. And for all its appeal as a clean industry, tourism does have an environmental impact. Tourism can result in an aesthetic blight of wax museums and fast-food restaurants that denigrate the scenic beauty and recreational opportunities for which Oregon is known." Of course, this conclusion could apply to all states in the Pacific Northwest.

REFERENCES

Clay, Grady. *Close-up How to Read the American City*. New York. Praeger, 1973.
Dean Runyan Associates. "Oregon Travel and Tourism: Visitor Profile, Marketing and Economic Impact." Prepared for Tourism Division, Oregon Economic Development Department, 1989.
Ferris, Robert, Editor. *Prospector, Cowhand, and Sodbuster, Volume XI,* The National Survey of Historic Sites and Buildings, United States Department of the Interior, National Park Service, Washington, 1967.
Flexman, Janet and Christensin, Neal. "Montana Travel and Tourism Statistics Sourcebook." Institute for Tourism and Recreational Research. University of Montana, Missoula, Montana, 1991.

Harris, Charles, "The 1987 Idaho Leisure Travel and Recreation Study." Department of Wildland Recreation Management, University of Idaho, Moscow, Idaho, 1988.

Kimerling, A. Jon and Jackson, Philip L. *Atlas of the Pacific Northwest.* Corvallis, Oregon State University Press, 1985.

Lew, Alan August. *Thematic Revitalization of Older Retail Districts in the Pacific Northwest,* Thesis, University of Oregon, 1983.

Loy, William G. *Atlas of Oregon.* Eugene, University of Oregon Books, 1976.

Richard, Debra. *The Preservation of Cultural Resources in Oregon National Register Historic Districts,* Thesis, University of Oregon, 1982.

Robertson, H. *A Geography of Tourism.* London, MacDonald and Evans, 1976.

The Gilmore Research Group. "Washington State Tourism: Our Consumer Profile." Prepared for Tourism Development Division, State of Washington, 1990.

U.S. Travel Data Center. "The Economic Impact of Travel on Oregon Counties, 1985." Prepared for Tourism Division, Oregon Economic Development Department, 1986.

URBAN SYSTEMS — Chapter 15

Ronald R. Boyce and William Woodward
Seattle Pacific University

Two instincts war against each other in the American soul. From the earliest settlements, the peoples of the United States have sought Opportunity. For no less time have they yearned for Community. Opportunity and Community offer polarities of allure, precariously balanced between, on the one hand, enticements to get rich quick or simply get a fresh start, or, on the other, longings for security, stability, and a sense of belonging. In the ambivalence and contradictions of the American city may be found the synthesis of these opposing hungers.

In the mythic imaginings and cold realities of the American West are recorded this metaphorical Tale of Two Cities. That is so because for several centuries, in wave upon wave, the adventurous sought their future in the West. As Ivan Doig (1980:4–5) remarks:

> "America began as West, the direction off the ends of the docks of Europe. Then the firstcomers from the East of this continent to its West, advance parties of the American quest for place . . . imprinted our many contour lines of frontier."

Indeed, the West has long been as much a state of mind as a compass direction. For Europeans, who came voluntarily, as well as for Africans forcibly transported, the direction was literally westward. For latecomers, it was eastward from Asia, northward from Mexico and the Caribbean, southwestward from the Canadian prairies, or north and west from the rural south. From every direction, they came seeking opportunity, and settled in community. And when the West's "empty wilderness" was filled and the future thus appeared in jeopardy (say about 1900), novelist Emerson Hough would feel compelled to mourn that "America is not American" any more (Athearn, 1986).

Ironic, then, is the reality masked by the myth: that the Pacific Northwest for most of the century-and-a-half of its existence has been dominated by its *urban* centers. There the life of the region unfolded, even while products of soil and sea, forest and mine flowed out along with most of the profits. There the people of the region congregated, despite overwhelming dependence on a rural-based extractive economy. There, in fact, the character of the region took shape.

Thus the cities of the Pacific Northwest are burdened with the weight of ancient aspirations. For thousands of emigrés over scores of years, the region's cities have been places of promise, sites where opportunity and community beckoned.

THE QUEST FOR COMMUNITY

Spatial factors and historical experience have intersected with personal dreams to shape the region's urban centers. That is to say, individual newcomers sought their new farms, new jobs, new challenges, new lives in particular places at particular times. In the process, they built cities.

Settlers came in communities and settled in communities. Though they saw themselves as individuals, though they followed the siren call of individual opportunity and used the legal fiction of individual property title to displace the Natives, they also uprooted themselves from extended community networks to cross the continent in "transient communities" (Boorstin, 1965:49–112).

They did as their forebears had done in earlier "wests." Indeed, all the way back to the earliest English colonies one discerns three distinct modes of community-building. This tripartite pattern may be taken as a sociological model and historical precedent to explain why settlers create towns.

England's colonization of the Eastern coastline of North America represented a continuation of a demographic, economic, and strategic out-thrust across the North Atlantic (Meinig, 1986). Differing intentions shaped the new settlements. At least one colony, Georgia, was established as a *Community of Coercion*. The founders of Georgia prescribed a blueprint for utopia: a community whose essential character was to be unitary and disciplined according to the dictates of a ruling elite. Yet in this respect as in many others, Georgia was an anomaly. More typical was Massachusetts, one of several colonies founded to house *Communities of Commitment*. Intended to be no less unitary and disciplined than Georgia, the "City on the Hill" in Puritan New England differed in that the prescribed blueprint was not imposed from above but composed from below, according to the willing choices of participants.

But in the earliest pattern of colonization, the essential character was not predetermined but open, not unitary but diverse, not prescribed but invented. The community at Jamestown offered an open-ended context where individual aspirations could be separately pursued. To be sure it was a community created out of necessity (and at that not a very successful one at first), but a *Community of Convenience* nonetheless.

Once established according to one of these three modes, communities tend to mature in one of two directions. If over time an essential character of stability and tradition emerges, molded by established social and economic patterns, a classic small town develops—a *Community of Continuity*—in which, it may be surmised, a palpable sense of community is purchased at the price of limited opportunity. But in a certain few cases, a combination of natural advantages, historical circumstance, and openness to individual enterprise serves as a catalyst to change. The essential character alters. Now a diverse and changing *Community of Centralization* emerges as a political, economic, and cultural hub that comes to dominate its surrounding hinterland. Here, it seems, is a place of enhanced opportunity but eroding community.

These types reappear in the founding of towns in the Pacific Northwest. The company town is one kind of community of coercion. Dominated by a single firm, usually a timber company, such towns as Port Gamble, Longview, and Everett long felt the domination of a single employer who usually wielded political as well as economic clout. The railroads exercised the same kind of power in several cities. Tacoma, first a mill town, then the designated terminus of the Northern Pacific, could become (as its slogan hopefully announced) a City of Destiny only on terms set by the railroad. Yakima relocated itself several miles northward to conform to the rail line, and when Sprague burned and the railroad moved its operation to Spokane, the latter emerged as the capital of the so-called Inland Empire.

In some respects certain booster towns also fit the model of a community of coercion. Port Townsend's speculative boom in the 1880s rested on flimsy hopes, but its architecture and social character were forced to fit the blueprint of the elite. In a different category altogether, the various tribal reservations suffered under various kinds of coercive, if paternalistic, regimens.

The resident populations of military posts also, at first glance, seem to fit the coercive category. But on closer inspection, they are better considered as communities of commitment. Except during the rare occasions when they were populated by unenthusiastic draftees, the forts constituted enclaves of active duty service folk who had chosen the military life. The communities created in these often remote outposts usually functioned autonomously and cooperatively, whether in the frontier era (Steilacoom, Walla Walla), the coastal defense interlude around the turn of the century (Canby at Cape Disappointment, Worden at Port Townsend, Casey on Whidbey Island), or in the Twentieth Century (Fairchild Air Force Base adjacent to Spokane, Fort Lewis and McChord Air Force Base south of Tacoma).

However, it is the Mormon settlements that provide the premier examples of communities of commitment, of course. Many southern Idaho towns began as deliberate Mormon colonies. Less successful but no less earnest have been an assortment of other religious and utopian communal efforts, ranging from the pietist Aurora community south of Portland to the cooperative colony at Freeland on Whidbey Island to the 1980s spectacle of Rajneeshpuram in the once and future town of Antelope, Oregon. And one might imagine that certain native tribes, in the process of wresting control over their reservations, are creating communities of commitment out of communities of coercion: the Warm Springs reservation in Oregon comes especially to mind.

Despite these significant if scattered examples, the dominant mode of community-founding in the Northwest has been the community of convenience. From the town that grew up around Fort Walla Walla to the dominant regional metropolis that developed far beyond the confines of the Skid Road and Yesler's mill on Elliott Bay, individual convenience, which is to say opportunity, has shaped the urban Northwest.

That communities of convenience have dominated is clear from historic patterns of settlement. But to focus on convenience as primary motivation in part begs the question—of *whose* convenience.

TIDES OF WHITE SETTLEMENT

Three hydrologic metaphors can help to clarify the historic process of town settlement in the Pacific Northwest: tides, channels, and waves. First, like great tidal flows, in-migration occurred in successive surges. Second, settlers followed identifiable avenues of approach to and then within the region. Finally, like swells atop the incoming tide, distinct waves of settlers followed these channels to particular places in quest of particular economic objects.

Across North America, settlement of each new "frontier" has unfolded in roughly congruent patterns. Indeed, the American usage of the word "frontier" reflected an assumption of such common patterns: a concept originally connoting a fixed boundary came to mean a mobile and open process. The term describes (1) a *process of encroachment* by Euro-Americans; (2) for the *purpose of a distinct activity* (e.g. a "mining frontier"); (3) *against particular indigenous peoples and places*. (Although scholars now seek, not altogether successfully, to escape Turner's language, it retains utility.) Thus in each successive western destination there has been, to quote Richard Wade (1959), an "urban frontier."

First explorers and traders initiated a period of *Penetration*. Next, reports of these firstcomers induced others to come as settlers during a period of *Pioneering*. Then came a period of *Permanency*, as waves of settlers, now assured of sufficient safety and opportunity, pushed into what they called the New Country, transforming the area from frontier to settled society. It remained only for the area to become fully integrated into the national culture, in an ensuing period of *Parity*.

Few townsites survive from the period of Penetration, 1770s–1820s, of the Northwest: Astoria, Vancouver, and Boise stand as exceptions. What the earliest arrivals accomplished was rather the opening of future channels of access. Seaborne explorers, followed by the otter traders, identified the Columbia River and Puget Sound as key entrepots; land explorers, followed by British and American enterprises in quest of beaver, opened the overland channel via the Platte and Snake rivers.

It was in the ensuing *Pioneering* epoch, 1830s–1870s, when early settlers founded almost all of the leading metropolitan centers of today. Folk wisdom homogenizes this pioneer tide into an epic trek of hardy farmers crossing the plains to virgin agricultural lands west of the Cascades. Apart from its genial ethnocentrism, this familiar story obscures some important variations. More helpful is to identify five distinct waves: 1) First, to be sure, came the farmers, following the overland trail to the Willamette valley, settled north to south in the 1840s and 1850s. 2) A second wave closely followed: to selected Puget Sound sites, by sea as well as overland, often to establish lumber mills to capitalize on the sudden post-'49er demand in California. 3) As the East convulsed in Civil War, a series of mining rushes initiated a third wave, channeled back up the Columbia via Walla Walla to the Boise basin (and later, via the newly blazed Mullan Road, to Northern Idaho and Western Montana). 4) Then, following the same channel in the 1860s and 1870s came the first farmers to the interior. Like the miners, these farmers generally established their settlements not on the waterways, as in Puget Sound and the Willamette, but on the adjoining rimlands. 5) Finally, throughout the mid-century decades, American military posts augmented the emerging demographic landscape, sometimes in advance of, but often in the wake of the pioneer settlements.

The railroad ushered in the era of *Permanency*, marked by a renewed tidal influx. Channeled by the Northern Pacific (1883), Great Northern (1893) and other routes, waves of settlers engulfed new sectors, especially areas east of the Cascades. But more important was the growth of the existing settlements, which rapidly sorted out into their roles as future urban places. Small towns like Seattle and Portland boasted startlingly rapid growth during the decades surrounding the turn of the century.

It was an age of boosterism, and so an age of boom and bust. Many a crossroads village cherished inflated visions, usually tied to a railroad link, of an expansive future. For example, go to Crockett Lake on central Whidbey Island today, adjacent to the Keystone ferry terminal, and observe the pilings of a misguided booster's railway trestle to nowhere. But in fact few new towns emerged along the rail corridors. A few prospered, most notably Spokane, but most did not. A minor but interesting motif was the founding of a network of communal settlements around the Sound.

In the mid-20th century the Depression and World War II wrought new transformations in the region, and brought another tidal surge of settlers. As these 20th century pioneers sought work on the Grand Coulee Dam or at the Kaiser shipyards, or trained at a military base prior to embarkation, they helped the region achieve cultural and economic *Parity* with the rest of the country.

Waves of displaced farmers and jobless workers poured into the work camps and mills and yards. Hooverville shacks disappeared from Seattle, leaving city officials to figure out how to house all the new arrivals. An instant town, Vanport, sprouted on the Willamette delta. Another rose in the barren desert at the new, supersecret Hanford works. The region's metropolitan centers had now matured. Opportunities and challenges henceforth would mirror the urban scene nationwide.

In this context, Seattle and Portland, as Communities of Centralization, vied for regional leadership. Both had grown proportionately since the first decade of the 20th Century. The Klondike Gold Rush accounted for Seattle's slight population edge, but both cities flourished by several measures. For example, each hosted a World's Fair (Portland in 1905, Seattle in 1909); each gained a federal reserve bank when the system was set up in 1913. For the next several decades, each city functioned as a central metropolis for both immediate and distant hinterlands. Portland claimed dominance over the agricultural regions of Oregon and southwestern Idaho, while Seattle tapped the mineral and timber, as well as fish and farm resources of Washington, the Idaho panhandle, and even beyond to Alaska and western Montana.

In the decades after World War II, the city on the sound outstripped its rival to the south although Portland continued to grow as its regional hinterlands developed. For its part, Seattle turned outward to enter national and international commercial and financial networks. The result, argues Carl Abbott (1992), was the emergence of Seattle as the region's primary city. Portland, "process-centered, cautious, and localized," remained a "Northwest city"; Seattle, "project-centered, entrepreneurial, and expansive," grew into a world-class "Pacific city." The newest wave of migration in the 1970s and 1980s, a combination of the Sunbelt shift and the new immigration from Asia and Central America, reflects this decisive leap forward by the self-styled Emerald City.

The metropolitan mosaic on the eve of the 21st century displays the fruits of a relatively brief but energetic history of in-migration. Repeated tidal flows of settlers, arriving in differential waves along channels of successive transportation corridors, established new communities for various reasons: according to coercive blueprints, or mutual commitments, or (most often) individual convenience. Ultimately, a select few urban hubs—Communities of Centralization—have developed in what was once a remote hinterland.

Although these communities began and grew as Communities of Convenience, a certain ironic ambiguity attaches to the label. For if most cities of the Pacific Northwest were communities of convenience, they often flourished at the convenience of someplace else. Mill towns grew thanks to outside capital and distant markets, first oriented toward San Francisco and later toward the Upper Midwest; only recently have these towns matured into self-sustaining communities. The railroads, controlled by Eastern financiers, picked a terminus at Tacoma, then another at Seattle, and dictated important intermediate junctions (e.g. Pocatello or Spokane). Even today it might be said that culturally and politically it is California, and specifically the Los Angeles area, which exercises outside influence.

This dominance from the outside persists, at least in part, because in so many ways cities of the Pacific Northwest remain in a kind of infant stage. Their economies in many ways display their youth. Although fading fast, agriculture, forestry, fishing, and mining economies are part of their heritage. Conversely, the major growth sectors of the urban economies are recent creations: aircraft, "high-tech," defense industries and installations. Even the most productive farms, the produce of which channels through the major port, are relatively new. Irrigated agriculture in vast tracts of Eastern Washington dates only to the fairly recent completion of the Grand Coulee Dam in 1942, while only since the 1960s has deep-well sprinkler irrigation begun to replace the older small-farm surface irrigation systems of the Snake River Plain.

In many respects, then, Northwest cities remain in the early stages of development. Yet by certain key economic and cultural measures those same cities are pace-setters: in outdoor and leisure wear (REI and Nike), computer software (Microsoft), even wine, gourmet coffee, "grunge rock" music, and an openness to Japanese enterprise (Nintendo and its Seattle Mariners). As historian Carlos Schwantes (1989:382–383) argues, "the juxtaposition of metropolitan trend-setter and hinterland is, in fact, the defining quality of life in the modern Northwest."

Such is the ambiguity that marks the Pacific Northwest: hub and hinterland, juvenile and mature, a field of dreams for both opportunity and community. And perhaps *most* ambiguous is the question of whether it can truly be deemed a region at all.

REGIONAL INTEGRITY OF THE PACIFIC NORTHWEST AND ITS CITY SYSTEMS

The Pacific Northwest is not a strong region geographically. It fails the test as a uniform (homogeneous) region and as a nodal region. It lacks a single city that controls the entire area; instead, the area is subdivided among five or more separate cities representing assorted parts of the three states. Each of the states, in turn, awkwardly spans natural physiographic boundaries. There is no intrinsic unity. Moreover, the region continues to show degrees of dependency on cities and regions outside of the territory. In fact, its regionality is primarily the result of historical happenstance, being one of the last major chunks of the coterminous United States to be incorporated into the nation.

The area could hardly be more diversified—physically, culturally, and economically. Physical geographers find the area almost a textbook case of unusual features. Replete with specialized economies, it embraces a Babel of religious ideologies.

From a religious standpoint, the urban population is extraordinarily dissimilar. Southeastern Idaho is dominated by Salt Lake City and "Mormondom." Northern Idaho has been an important missionary field for Catholics, a group that has had great impact on Seattle as well. (The oldest Catholic mission in the Northwest was at Cataldo in the Idaho Panhandle, but other Catholic settlements previously existed). Idaho, in fact, has been dubbed a state with three capitals: Boise, Salt Lake City, and Spokane. Walla Walla all the way to Oregon was an early Protestant mission ground. In short, the region has long felt the influence of assorted religious groups, ranging from the conventionally orthodox to various Jesus-sects and New Age experiments. Paradoxically, however, today the area today is flooded with the generally unchurched, perhaps residuals from earlier mines, timber camps, and ranches. Church attendance here ranks among the lowest in the United States.

Physically, the Pacific Northwest ranges from a rain forest on the windward slopes of the Olympic Mountains to a veritable desert in Eastern Washington and Oregon. It contains assorted mountains, plains, high lava plateaus, and great river gorges. The slice of Idaho between the Boise-Twin Falls axis northward to Coeur d'Alene and Missoula is among the most isolated in the United States.

Indeed, Idaho contains some of the strangest geology on earth, evoking a variety of infernal metaphors. The Craters of the Moon National Park is only one of many tortured lava flow areas; Hell's Half Acre near Shelley is yet another example. Located between Mud Lake and Arco, the Birch, Little Lost, and Big Lost Rivers simply disappear into the sands. In and about Yellowstone National Park is a place described by explorer John Fremont in 1843 as the nearest thing to Hell he had ever seen, an area all pioneers should avoid. The Snake River in its great bend to the Columbia passes through Hells Canyon, a raging white-water terror to explorer and pioneer alike. (The Oregon Trail crossed the Blue Mountains between La Grande and Pendleton, rather than continue on the Snake River route.)

And then there are the famous scab lands and potholes—the result of a great Columbia River flood in geologic times—of Eastern Washington, and the massive volcanic mountain peaks—Baker, Rainier, Adams, St. Helens, Hood, and Crater Lake—along the Cascade range. Finally, Puget Sound is the result of a great depression caused by a massive Ice Age glacier.

Such cultural and physical diversity has shaped divergent locations and patterns for particular urban centers. As Edward Ullman rightfully noted in 1980, cities are bonded together along certain rivers, but the diversity of physiography and climates creates separate urban realms and decidedly different economic bases. Thus, the physical variation is linked with the historical and religious diversity, dotting the Pacific Northwest with various combinations of separate city domains, instead of an integrated regional system of cities.

URBAN ECONOMIC BASE THEORY

Cities attain a certain population level and growth rate in accordance with their ability to provide employment for their inhabitants. Money from outside the city must be generated by activities within the city. Such earnings can occur from any source: manufacturing, retailing, services, retirement income of its inhabitants, or tourism. Persons in the community who receive such funds are called "basic workers." For each worker there is an associated family coefficient. If a basic worker is one of a family of three, and 500 basic workers (or equivalent) are added to a community, 1,500 new persons will also be immediately added. In addition, there is also a relationship between the number of basic workers and the number of "service" workers. The larger the city, generally the larger the multiplier effect—i.e. the basic-to service-worker ratio. If the ratio is one basic worker for every two service workers (about what it is for a metropolis) and 500 basic workers are added, the service worker population would be twice as large, or some 3,000 persons (assuming the family coefficient is the same between basic and service workers). In total, therefore, the addition of 500 basic workers would cause a net new population increase of 4,500.

On this basis it is easy to understand why the population of the Greater Seattle Metropolitan Area (SMA) is about 2,000,000. Boeing alone employs about 100,000 people. Given the family coefficients and multiplier effects as above, these 100,000 basic workers create a population of 900,000 (100,000 basic workers, plus 200,000 associated family members, 200,000 service workers, plus 400,000 of their associated family members). Other basic workers include those in the military, those in transportation, such as are associated with the port, and assorted manufacturing and service workers, such as many of those employed at the University of Washington, including the University Medical Center. In fact, various specialized hospitals in the Seattle area such as the Fred Hutchinson Cancer Center and the Children's Orthopedic Hospital generate considerable dollars in revenue for the metropolis. As regional capitals, Seattle and Portland generate many service jobs in local, state, and federal government, and as headquarter offices for major businesses.

Cities grow in size and number because of their ability to generate money from outside their boundaries. Cities can, of course, "make their living" in various ways. The old fashioned, traditional way has been to rely on nearby natural resources and simply modify and tranship these materials—products of farms, mines, seas, and forests—as they may be needed to other areas. These extractive and processing enterprises have constituted the primary economic base of most Northwest cities.

Within this pattern, many second-order cities in the region have become highly specialized. Lewiston, Idaho, and Everett in Washington have been dominated by the pulp and paper industry. Wallace, Kellogg, and Smelterville in Idaho, and Butte, Anaconda and other nearby towns in Western Montana have made their living by mining. In Washington, Bremerton owes its existence to the U.S. Navy, as do other burgeoning communities on the Kitsap Peninsula. Meanwhile, Oak Harbor fights to keep its Naval Air Station open while Mountain Home, Idaho, worries about the closure of its Air Force base. Washington's Tri-Cities (Richland, Kennewick, and Pasco) are largely the creation of the federal government's Atomic Energy Commission. Pendleton, Ontario, Twin Falls, Moses Lake, Dillon, and dozens of other communities owe their existence to farming and ranching. Others, like Pullman, Washington, and nearby Moscow, Idaho, gain their living from state universities. Tourism sustains many small towns—towns along the Oregon coast; in the Idaho Panhandle; in "Olde West" mockups or relics like Winthrop, Washington, and Virginia City, Montana; in ski destinations like Sun Valley and Bend. Many such towns and small cities that are highly specialized with respect to economic base are highly vulnerable to changing conditions.

Another kind of Northwest single-industry town attracted national attention in early 1993 thanks to President Clinton's "timber summit." Along the windward foothills of the Olympics, Cascades, and Rockies are multitudes of communities that have long earned their living from the woods. Such communities include Forks, Port Angeles, Darrington, Aberdeen-Hoquiam, Shelton, and Longview in Western Washington and Spokane and Clarkston in Eastern Washington; Coos Bay in Oregon; Lewiston, Boise, and Sandpoint in Idaho, to name a few.

The larger cities, as befits communities of centralization, are generally much more diversified. Idaho Falls relies on a mix of agriculture and nuclear energy functions; Pocatello on agriculture, forestry, mining, and transportation; Spokane on agriculture, mining, defense, and assorted transportation; Boise on forestry, agriculture, and state government.

Portland and Seattle stand out as significant regional capitals and manufacturing cities and ports that have long contended with one another for economic control of much of the interior, especially for the massive wheat shipments out of the Palouse. Today, containerized shipping dominates the Puget Sound ports. Increasingly the commerce of the Pacific—automobiles, electronic items, and agricultural produce—passes through the Sound.

SYSTEMS OF CITIES IN THE PACIFIC NORTHWEST

No single city or urban complex controls the Pacific Northwest. Conurbations, or clusters of somewhat separate city systems, occur in each state, but none are dominated by a single city in the region.

The communities about the lowland of the Puget Sound contain about two-thirds of Washington's population and dominate most of the state; Spokane and Portland control the remaining sections. Portland and the Willamette lowland, likewise, contain about two-thirds of Oregon's population and dominate southwestern Washington and much of eastern Oregon. The Snake River communities, spreading in curved fashion from Rexburg to Ontario, Oregon, contain well over 65 percent of Idaho's population. The region's other major metropolis, Spokane, is somewhat of an anomaly: a small urban complex situated in the midst of its self-styled Inland Empire. Although the cities in these conurbations occupy only a small percent of the land, they control a vast, rural, and largely uninhabited area.

But in a larger context, the entire Pacific Northwest is itself economically dominated by Los Angeles. Neither Portland nor Seattle dominates the Pacific Northwest fully; both are controlled in various ways by California, continuing the historical domination of the region from the South as well as the East.

City Systems in the Snake River Plain

The Snake River Plain, now settled in linear fashion, was initially passed by. Historian Ross Peterson expressed it well when he declared:

> *"Pioneer after pioneer, oxen after oxen, and wagon after wagon passed across southern Idaho, finding not the slightest temptation to stay." (Boyce: 1987: 51)*

Explorer John Fremont erroneously reported that the area was unsuitable for agriculture, a dictum believed by Brigham Young. Mormons were discouraged from settlement north of 43° degrees north latitude (the Utah-Idaho border), especially after the disastrous Lemhi settlement experience. It was not until the 1880s that Mormons ventured into the Upper Snake River valley in Southeastern Idaho and established villages and canal systems for flood irrigation.

The earliest impetus for settlement was the discovery of gold above the Boise basin. Miners poured in; mining camps sprouted and spread north and east. Ignoring the Civil War back east, the prospectors tracked the tantalizing color all the way to Virginia City and Helena in Montana. What once had been all Washington Territory was quickly subdivided into the present political configurations.

The cradle-shaped Snake River Plain now contains a necklace-like beading of cities. Culturally, this seeming system of river-oriented places is bifurcated between Boise and the Twin Falls-Pocatello-Idaho Falls group of cities. From Twin Falls eastward, Southeastern Idaho is simply an extension of Mormonism. Salt Lake City might well be considered the capital of the area. Twin Falls is the territorial capital of such farm towns as Gooding, Buhl, Rupert, and Burley. In the Upper Snake River Plain of Southeastern Idaho, jurisdiction of control was first achieved by Pocatello, but now is increasingly dominated by Idaho Falls, thanks both to the nuclear energy complex near Arco and the conversion of former dry farming wheat areas to deep well sprinkler irrigation for potatoes. Lesser cities in Southeastern Idaho include American Falls and Blackfoot, generally dominated by Pocatello, and Shelley, Rigby, Roberts, Rexburg, and St. Anthony, which are dominated by Idaho Falls. (The extreme southeastern Idaho communities such as Montpelier generally fall into the domain of Logan, Utah.)

The Snake River cities about Boise are oriented northwest-southeast, are generally linear in shape, but are wider than the narrow strip of cities in southern and southeastern Idaho. The Boise basin includes over a half dozen agricultural service centers including Weiser, Nyssa, Payette, Caldwell, and Nampa. Nearby communities in Oregon, such as Ontario, and as far west as Baker and La Grande are in the general domination of Boise. In fact, with a population of almost 200,000, Boise, is headquarters for several important regional and national firms and has now achieved genuine metropolitan area status.

Culturally, this region contains a great diversity of religions. Mormonism is much evident, but there are many assorted Protestant denominations, including smaller groups such as Seventh Day Adventists and Nazarenes. Catholicism is also well represented. The politics of the region is extremely individualistic and conservative. The Mormon settlements in Southeastern Idaho, however, although politically conservative, constitute a cohesive religious-based society that is often isolated and at odds with others in the area. Through the stake and the local ward the Mormon Church in its hierarchical wisdom shapes social activity and political thinking. Thus the old Mormon culture of community (with an independent Republican twist) permeates all aspects of life. Yet big farms, big houses, and big stakes are built into the bones of this kind of Idaho Mormon—unbowed and casual, content on balance to subordinate community to the open road of economic opportunity.

Puget Sound Lowland Cities

The Puget Sound lowland is rimmed with cities and towns, most of which are ports situated near river estuaries. On the western shores of Puget Sound are Port Townsend, Port Orchard, Bremerton, and many smaller, has-been lumber towns. The eastern shores of the Sound contain a half dozen or so thriving centers including Bellingham, Anacortes, Everett, Seattle, Tacoma, and Olympia. All are ports, but only Seattle and Tacoma have become diversified in commerce beyond primary-based commodities.

Destinies of Puget Sound cities and towns have changed considerably. Under the British, Fort Steilacoom near Tacoma was the prominent center. By the 1850s, several small communities had been founded in the more fertile agricultural areas (i.e. about major rivers which contained an alluvial soil, suitable means of transportation, and generally a more cleared and flatter area than otherwise.)

A second generation of towns came with the development of the timber industry. Egress from the region was through the Strait of Juan de Fuca; communities on the northern Sound and especially on its western shore prospered. Among these were Port Townsend, Port Ludlow on Hood Canal, Port Madison at Agate Pass, and Port Blakely on Bainbridge Island. Accessible by steamer, Coupeville on Whidbey Island, of similar vintage, enjoyed a brief heyday as a tourist retreat.

By contrast, east Sound communities such as Everett, Seattle, and Tacoma were able to eclipse the west Sound communities only after the coming of the railroads. The initial railroad link was the Northern Pacific connection from Portland. Soon thereafter, Tacoma was chosen as a direct Puget Sound terminus. The Great Northern threaded in from the north via Everett, terminating at Seattle. The Chicago, Milwaukee, St. Paul and Pacific Railroad took the central route over the Cascades to Auburn in the Green River Valley, intermediately situated between Seattle and Tacoma.

Historically, the primary contest for control on the Sound pitted Seattle against Tacoma (Boyce: 1986, 66–81). Before the rails arrived, both were mere villages with approximately equal populations. The historical record thereafter is mixed. At times, Tacoma looked like the victor, at other times Seattle. The first major contest occurred with the NP's decision to locate its terminal at Tacoma, perhaps a blessing in disguise for Seattle. Their rival to the north thereby avoided domination by the railroads. Never a "railroad town," Seattle eventually ended up as the terminus for all the railroads coming into the Puget Sound Lowland—whether from south, east, or north.

A second coup for Seattle was the Klondike Gold Rush in the 1890s. This is ironic because Tacoma had served Alaska, whereas Seattle had focused on its famous "Mosquito Fleet" to serve various communities about the Sound, particularly the prosperous mill towns on the west side of Puget Sound. However, thanks to the promotional huff and puff of the Seattle Chamber of Commerce, under the tutelage of Erastus Brainerd, Seattle convinced those with "Klondike Gold Fever" that Seattle was the only reasonable outfitting and departure point to the Alaska gold fields.

Another major benefit to Seattle was its selection as the major transhipment point for the silk from China to the East Coast. From Seattle, great "silk trains" left for the East. This precedent has been revived in the past several decades by the massive container cargo trade. However, thanks to some effective management and promotion the Port of Tacoma has taken from Seattle at least one half of the container trade industry during the past decade. Nonetheless, the Port of Seattle controls the monstrous Seattle-Tacoma Airport—a great money maker but a colossal headache because of its adverse effects on the surrounding community. (Tacoma paid to have its name included, but is otherwise uninvolved).

Perhaps Tacoma's great coup was obtaining Fort Lewis for its backyard in 1917. This massive division headquarters facility was largely the result of successful lobbying by the Tacoma community (Boyce: 1986, 93–95). McChord Air Force Base is also a substantial economic asset to Tacoma. By contrast, Seattle contained only very small military bases such as Fort Lawton and Sand Point Naval Station, both of which have now virtually been closed.

While not a contender in the rivalry, Everett lately has had some major additions to its economic base. Characteristically a log exporter, manufacturer of tissue paper, and major Weyerhaeuser pulp and paper town, it has been able to weather the decline in these primary-sector activities by providing a more modern economic base. In the late 1960s, Boeing built the largest building in the world (in cubic feet) at Everett as the production center for its famed 747. Although the project is still in the early stages, the Navy is constructing a major aircraft carrier home base in the port of Everett.

Seattle is unique among the cities of the Sound in that it contains one deepwater lake and adjoins another. The Lake Washington Ship Canal, with the U.S. Army Corps of Engineer locks at Ballard, provides a corridor of access through Lake Union to Lake Washington. The interior fresh waters of the canal and lakes are very valuable for the fishing fleet and for other shipping purposes. Even before the locks and canals, the lakes provided opportunities for transporting coal and other resources from the Renton area; coal was Seattle's most valuable export during the 1880s. In fact, there were even serious plans about that time to build an iron and steel complex on the eastern shores of Lake Washington at Kirkland. Today, Lake Washington is more of a barrier than a corridor, thereby necessitating the building of very expensive floating bridges across the deep waters.

Since World War II, thanks to some very effective civic leaders, Seattle has accomplished important projects that have catapulted Seattle ahead not only of other communities on the Sound, but also of its historic rival, Portland. The first major publicity event was the Seattle World's Fair, that placed Seattle in the national consciousness. It greatly bolstered tourism to the region and provided facilities that have proven effective in making Seattle of regional, even national repute—e.g. the Pacific Science Center, the Opera House (Seattle Symphony and Opera), and the

Coliseum (Seattle Supersonics). The initiation of METRO (now absorbed into King County) led to a county-wide sewer project to clean up Lake Washington, Green Lake, and other water bodies, setting the standard for intergovernmental cooperation in cleaning up the environment. The construction of the Kingdome, although ugly and poorly located, nonetheless allowed professional baseball and football, as well as other mass entertainment events, to occur. (Portland was handicapped by too small facilities. Likewise, the Tacoma Dome, although more attractive, is smaller).

Downtown Seattle is also one of the great success stories in America with respect to office space. The Downtown Development Association, for decades headed by John Gilmore, has proven very successful in obtaining many projects for the downtown area: the Freeway park covering Interstate 5, the Washington State Convention Center, and the elaborate and expensive Bus Tunnel. Another very effective leader has been Jim Ellis, originator of Forward Thrust, a major civic improvement campaign in the 1970s.

Cities in the Willamette-Columbia Confluence Zone

The Willamette Valley was considered the jewel of the western United States by pioneers of the 1840s. The Oregon Trail made a beeline from Missouri to the promised land from Portland southward. So favored was the valley that all other areas were overlooked in the trek westward. In fact, many of the new migrants were former New Englanders, people who even by-passed the Midwest in favor of the far west. They poured into the Willamette lowland, naming communities after those back home: e.g. Portland and Salem. For a season, this region was the most impressive part of the Pacific Northwest, and indeed of the whole continent.

It was so successful, in fact, that latecomers, thinking that all the good lands and opportunities in Portland and the Willamette plain had been taken, chose other areas. Their disappointment with Portland and the Willamette caused the Denny party to trek north into the Puget Sound lowland to found Seattle.

Except for the mid-1840s when Oregon City was the more prominent settlement in Oregon Territory, Portland has been the dominant center. Linear strips of cities range southward up the Willamette River, eastward to the Cascades, and northward beyond Vancouver to Longview and Kelso—all subservient to the "city that gravity built" at the confluence of the two rivers.

As a regional capital, Portland controls portions of three states: southeastern Washington, portions of western Idaho, and, of course, Oregon. The Columbia Gorge provides barge, rail, and road access to the east. Grains from the productive Palouse exit mainly through Portland, while produce from the fertile Columbia Basin, watered from the Grand Coulee Dam, flows to Seattle.

Seattle versus Portland

Until the 1950s, the Portland metropolitan complex held the advantage over Greater Seattle. Its port was solid and growing, it contained many headquarters companies that dominated Washington, and, dubbed "The City of Roses," it had a reputation as a very pleasant place to live and work. However, it took a conservative stance—this is what had worked in the past—on its port, not moving toward container cargo as did Seattle and Tacoma, and failing to achieve consensus on several of its major public projects. By contrast, Seattle with its Forward Thrust was jubilantly successful on almost every count. The rivalry between Portland and Seattle ended with Seattle as the dominant center (Abbott: 1992).

Yet despite the regional preeminence of Seattle, and despite the local dominance of each urban cluster, the Pacific Northwest is as a whole dominated by places elsewhere much as it was in the past. It is, of course, controlled in its economy by Eastern demands. It also is dominated from the south: San Francisco determined the initial economic base of Puget Sound as it sought timber for the mines and lumber to build the city. With the Gold Rush, San Francisco skyrocketed to become the seventh largest city in the United States by 1870. Such rapid growth dictated the economy throughout the Pacific Northwest littoral. The Willamette Valley suddenly found a commercial outlet for its agricultural produce. Workers for the Writers' Program of the Work Projects Administration stated San Francisco's importance well:

> "It 'never was a village'—this had been its proud boast. Where barren sand dunes, marshes, and brackish lagoons had surrounded an abandoned mission and a decaying fort with rusty cannon, San Francisco sprang into life overnight—a lusty, brawling he-man town of tents and deserted ships. Business, mushroom-like, flourished in mud-deep streets. Almost before it had achieved a corporate identity, San Francisco was a metropolis—to be named in the same breath with Boston or Buenos Aires, Stockholm or Shanghai. When other cities of the Coast were still hamlets in forest clearings or desert cow-towns, San Francisco was 'The City.'" (1940: 3)

Today, Los Angeles has eclipsed San Francisco to become the supreme center for the entire West. Even so, San Francisco remains an important influence over the Pacific Northwest and is in many respects a rival and equal of Los Angeles. Following the precedent set by Chicago, Los Angeles overcame the handicaps of nature—scarcity of water, lack of a natural port, and distance from Eastern markets—to become the undisputed regional capital of the Western United States.

Despite being closer to the Orient, Seattle and Portland remain in the shadow of the Port of Los Angeles (at Long Beach). Los Angeles has become the great manufacturing and wholesaling colossus of the West Coast, the possessor of the greatest port traffic, and the primary determinant of trends for much of the United States for sportswear, outdoor living, and general lifestyle. But the future still holds interesting possibilities for the two premier cities of the Northwest.

THE PUGET SOUND-SALEM CORRIDOR: A FUTURE MEGALOPOLIS?

Some analysts project an eventual merging of the Puget Sound lowland and Williamette-Columbia confluence complexes .The East Coast of the United States provides a clear precedent. The Eastern Seaboard from about Boston to Washington has been called by various names: Main Street, Bos-Wash, and even Megalopolis by Jean Gottmann in a famous book of the same name. This linear pattern of almost adjacent conurbations, connected by rail and highway routes, is somewhat spatially analogous—albeit antipodal in terms of coasts—to the urban strip that runs from Vancouver, Canada, to Eugene, Oregon.

Indeed, the grouping of a half dozen cities rimming Puget Sound has been dubbed "Pugetopolis" for several decades. And southward from Olympia, despite the largely rural character of Lewis and Cowlitz counties, Interstate 5 provides an open invitation for future "megalopolitan" urbanization extending to Portland and onward, perhaps even all the way to the southern Willamette Valley as far as Eugene-Springfield. Of course, this is far from the reality now, but it is distinguishable on a map of the United States as a light strip of urbanization in a general sea of openness. What could bring it closer to reality is the ever-continuing sprawl of suburbanization.

CENTRAL CITIES AND SUBURBIA: NORTHWEST STYLE

Throughout America, central cities have been on a downward skid demographically, economically, and politically since the 1950s. They have lost upwards of one-half of their pre-World War II populations; their remaining inhabitants have become older, poorer, and more minority-ridden; retail sales have plummeted such that many outlying shopping centers record more sales than their central business district (CBD) counterparts; and jobs have decentralized to districts outside the central city. Since 1970, over one-half the urban population has resided outside the central cities.

Massive Federal projects such as urban freeway development and urban renewal have failed miserably to restore the central cities and their CBDs to their former prosperity. In many metropolises the CBDs are ghost towns of their former selves. Department stores, apparel stores, and furniture stores—almost all functions in the classic downtown—have departed for suburban locations.

Such has been the destiny of even many moderate-sized cities in the Pacific Northwest. Certainly, downtown Idaho Falls is but a shell of its former self, as is downtown Pocatello, Twin Falls, and Boise. Even downtown Spokane, despite its massive central city measures, reflects decay and abandonment.

Tacoma contains one of the most devastated downtowns in the nation. It has lost all its department stores, despite the construction of an attractive, but largely vacant, downtown mall. It has also constructed several public garages, but these also lie largely empty. Escalators designed to connect one steep downtown street with another were misplaced and a total failure. Even public offices have moved out of the downtown area. Weyerhaeuser relocated several decades ago from its downtown Tacoma complex to Federal Way in King County. Initially, downtown Tacoma was by-passed by Interstate 5, causing the CBD to be severely off-center. The final blow to retailing in the CBD occurred when the Tacoma Mall was constructed on Interstate 5 about four miles south of downtown. Likewise, the Tacoma Dome sits on Interstate 5 a mile or so south of downtown. Most recently, Tacomans learned that the University of Puget Sound had sold its law school down the Sound—to Seattle University.

By contrast, the downtowns of both Seattle and Portland are powerful exceptions. Of course, both have suffered severe losses in retail sales because of outlying shopping centers, but they have compensated for this loss by their skyscraper-laden office districts and by various civic projects that have aided their downtowns. Given the great amenity-laden lures of the suburban communities, this accomplishment is nothing short of astounding.

Both the Seattle and Portland downtowns are important tourist attractions. Seattle and Portland both contain sports facilities adjacent to the downtown and both have promoted historical districts designed to attract tourists. Seattle has captured the Washington State Convention Center and has recently completed a massive underground bus tunnel and the upscale Westlake Mall. The tunnel also provides an underground shopping access corridor. Portland has been successful in its light rail system between Gresham and downtown and has many of the same convention and tourist attractions as Seattle. Both downtowns are well-kept, generally prosperous, and unusually attractive.

Unfortunately, the central cities adjacent to the central business districts of Portland and Seattle have suffered much the same fate as central cities elsewhere. Both have lost population, have suffered severe decentralization of jobs and sales, and have a less prosperous, older, and more minority population than earlier. In these respects, they are little different from other central cities. The middle and upper class population, the jobs, and the sales have either moved or been newly created in what is loosely called suburbia.

In fact, both cities are highly multi-nucleated. Both have massive employment centers outside the city limits, for example Boeing and Microsoft located in Everett, Renton, and Redmond. In fact, neighboring downtown Bellevue has its own impressive CBD, skyscrapers, and a growing regional importance. Beaverton with Nike and Tektronix, Parkrose, Gresham, and Milwaukie all contain considerable shopping, office, and industrial development in Greater Portland. Both cities have massive planned shopping centers, burgeoning modern office parks, industrial parks, and considerable employment and moderately dense housing nuclei on the periphery of their metropolises.

Nonetheless, the most recent model of great "Edge Cities" (Garreau, 1991) is only now just becoming a possibility in the Pacific Northwest. There are as yet no Tyson II's, nothing approaching the Mall of America near Minneapolis, nor the great retailing-office complexes found at the outer margins of some large cities in other parts of the nation.

The real test for the future in Washington is the Growth Management Act. This legislation requires suburban communities and central cities to work out a growth plan for the future, one that subdivides the expected growth and allocates it to various communities. The Act also is designed to curtail further spread and sprawl of the suburban areas into rural areas. And, of course, it calls for a carefully coordinated and integrated transportation and land use plan.

It is too early to tell whether the Growth Management Act will make any appreciable difference in the general areal extent and morphology of Pugetopolis. But it will not stem the continuing tide of new arrivals. The waves of settlers will keep coming.

CONCLUSION: GROWING TOWARD COMMUNITY?

The 1990s pioneer, rolling in on the interstate in a rented U-Haul, observes an unusual characteristic of Northwest urban neighborhoods: back east, front porches still adorn the older homes, and expansive front lawns and flowerbeds function as a kind of private modern agora. In the East or Midwest, the view from the street suggests, in short, an environment that fosters interaction—community. But Northwest homes and yards seem suffused with an architectural modesty befitting a Muslim woman: high fences enclose decks and patios with an almost pugnacious privacy. Perhaps it is the youthfulness of the Northwest metropolis, or perhaps it is a legacy of the escapist, individualistic frontier mentality. Whatever the reason, the Northwest residential pattern has clearly made the choice for privatized opportunity over civic community.

Of course, once again the generalization must be qualified into ambiguity. Community evidently gains the upper hand in the Mormon-dominated places in the Upper Snake River Valley. Yet the effect is not thorough-going. In the economic sphere, the yeoman spirit of the farmer, mirrored in the towns, insists on the primacy of opportunity, expressed in an unmanicured free-style economic ethos.

In sum, then, denizens of the cities and surrounding metropolitan environs in the Pacific Northwest have flourished in their efforts to build new communities by prioritizing individual fulfillment. They have thereby created, recently and swiftly, a thoroughly urbanized culture. Concentrated geographically into two strips flanking Puget Sound and the lower Willamette, it is a culture marked by a profoundly self-oriented character and remarkably high population density.

Moreover, there has been a dark side to their quest for opportunity and community. Individuals have often pursued their personal ambitions at the expense of the land, the native people, and even their fellow immigrants. Community has been reduced at times to mere conformity, or a nostalgic and obsessive grasping for a risk-free life.

In consequence, a certain adolescent character marks the region's urban culture. Urbanism has not necessarily fostered urbanity. Ethnic groups multiply in number and visibility and enriching contribution, but racial tensions flare. The arts flourish, but public education teeters on the edge of mediocrity. A Northwest style and cuisine attract attention in New York, but inordinate interest in the latest list of "most livable cities" betrays a kind of touchy parochialism. Nordstrom and Nike, Starbucks and Group Health Cooperative, Boeing and Microsoft provide pace-setting innovation, yet Northwesterners remain afflicted by an inferiority complex about Californians.

Thus the Northwest city, trendsetter and hinterland, checkered by ambiguities of all sorts, continues to hold prospects of both fulfillment and frustration. Whether bright or bleak, however, the region's future clearly will unfold in the metropolitan Northwest, for in its urban places will be found the greatest challenges, opportunities—and yes, at least potentially—its most vibrant communities.

REFERENCES

Abbott, Carl. "The Metropolitan Region: Western Cities in the New Urban Era," in Gerald D. Nash and Richard W. Etulain, eds., *The Twentieth Century West* (Albuquerque: Univ. of New Mexico Press, 1989).

Abbott, Carl. *Portland: Planning, Politics and Growth in a Twentieth-Century City* (Lincoln: Univ. of Nebraska Press, 1983)

Abbott, Carl. "Regional City and Network City: Portland and Seattle in the Twentieth Century," *Western Historical Quarterly*, August 1992, pp. 293–322.

Alwin, John A. *Western Montana: A Portrait of the Land and Its People* (Helena, Montana: Montana Magazine, 1983).

Alwin, John A. *Between the Mountains: A Portrait of Eastern Washington* (Bozeman, Montana: Northwest Panorama Publishing, 1986).

Athearn, Robert. *The Mythic West* (Lawrence: The University Press of Kansas, 1986)

Boorstin, Daniel. *The Americans: The National Experience* (N.Y.: Random House, 1965).

Boyce, Ronald R. *Seattle-Tacoma and The Southern Sound* (Bozeman, Montana: Northwest Panorama Publishing, Inc., Northwest Geographer Series, Number 3, 1986)

Boyce, Ronald R. "The Mormon Invasion and Settlement of the Upper Snake River Plain in the 1880's: The Case of Lewisville, Idaho," *Pacific Northwest Quarterly*, January–April 1987, pp. 50–58.

Doig, Ivan. *Winter Brothers* (N.Y.: Harcourt Brace Jovanovich, 1980).

Fahey, John. *The Inland Empire: Unfolding Years, 1879–1929* (Seattle: Univ. of Washington Press, 1986).

Fahey, John. "The Million Dollar Corner: The Development of Downtown Spokane," *Pacific Northwest Quarterly*, April 1971, pp. 77–85.

Faragher, John Mack. *Women and Men on the Oregon Trail* (New Haven: Yale Univ. Press, 1979).

Findlay, John M. "Far Western Cityscapes and American Culture since 1940," *Western Historical Quarterly*, February 1991, pp. 19–43.

Findlay, John M. *Magic Lands: Western Cityscapes and American Culture after 1940* (Berkeley: Univ. of California Press, 1992).

Fishman, Robert. "America's New City: Megalopolis Unbound," *The Wilson Quarterly*, Winter 1990, pp. 24–45.

Garreau, Joel. *Edge City: Life on the New Frontier* (New York: Doubleday, 1991)

LeWarne, Charles. *Utopias on Puget Sound, 1885–1915* (Seattle: Univ. of Washington Press, 1975)

Limerick, Patricia Nelson et al. *Trails: Toward a New Western History* (Lawrence: Univ. Press of Kansas, 1991).

Meinig, Donald W. *The Great Columbia Plain: A Historical Geography, 1805–1910* (Seattle: Univ. of Washington Press, 1968)

Meinig, Donald W. "The Mormon Culture Region: Strategies and Patterns in the Geography of the American West, 1847–64," *Annals of the Association of the American Geographers*, June 1965, pp. 191–220.

Meinig, Donald W. *The Shaping of America*: Vol 1: *Atlantic America* (New Haven: Yale Univ. Press, 1986)

Morgan, Murray. *The Mill on the Boot: The Story of the St. Paul and Tacoma Lumber Company* (Seattle: Univ. of Washington Press, 1982)

Morgan, Murray. *Puget's Sound: A Narrative of Early Tacoma and the Southern Sound* (Seattle: Univ. of Washington Press, 1979)

Morgan, Murray. *Skid Road: An Informal Portrait of Seattle* ([3d ed.] Seattle: Univ. of Washington Press, 1982)

Sale, Roger. *Seattle, Past to Present* (Seattle: Univ. of Washington Press, 1976)

Schwantes, Carlos. *The Pacific Northwest: An Interpretive History* (Lincoln: Univ. of Nebraska Press, 1989)

Ullman, Edward L. "Rivers as Regional Bonds: The Columbia-Snake Example," in *Geography as Spatial Interaction*, ed. Ronald R. Boyce (Seattle: Univ. of Washington Press, 1980), pp. 107–121.

Wade, Richard C. *The Urban Frontier: The Rise of Western Cities, 1790–1830* (Cambridge MA: Harvard Univ. Press, 1959).

Workers of the Writers' Program of the Work Projects Administration. *San Francisco: The Bay and Its Cities* (New York: Hastings House, 1940).

Ronald R. Boyce is Professor of Urban and Regional Studies and former Dean of the School of Social and Behavioral Sciences at Seattle Pacific University. William Woodward is Professor of History at Seattle Pacific, and also serves as the Command Historian of the Washington National Guard.

The authors acknowledge ideas initially offered in different venues by G. Thomas Edwards, Roger Sale and David Nicandri as seed concepts for the analytical model developed here.

External Relations

THE NORTHWEST AND THE PACIFIC RIM

Chapter 16

John A. Alwin
Central Washington University

New spatial forces are dramatically influencing today's Pacific Northwest. Actual geographical locations have not changed, but relative locations are experiencing profound reorientations. In America's northwest corner the strongest among new spatial influences stem from a Pacific-facing location and ascendance of the burgeoning Pacific Rim, those lands fronting the Pacific Ocean. Northwest links with lands of the Pacific littoral are nothing new, but during the last three decades the intensity of these links and their relative importance have grown with the shift of the world's geoeconomic center of trade and commerce to the Pacific Basin. Pacific Rim ascendance has exciting and profound implications for the Northwest and its dynamic geography. Those ramifications are the subject of this chapter (Figure 1).

Unlike most earlier chapters, which focus on discrete and clearly defined topics, this final chapter deals with a multi-faceted and wide-ranging theme that permeates many aspects of the region's geography. To make the broad topic of the Northwest's Pacific Rim geography more manageable, this brief essay is presented chronologically and concludes with speculation on possible future Northwest Rim geographies. Readers should develop an ability to look at America's Northwest corner from a new, and increasingly more relevant, Pacific Rim perspective.

THE NORTHWEST AND THE PACIFIC RIM: YESTERDAY

The Northwest's pre-1900 Pacific Rim heritage dates back deep into prehistory, when North America's first resident *Homo sapiens* diffused out of Asia, crossed the Bering Land Bridge and headed south to our Northwest via both a coastal and inland Cordilleran route. The earliest well-documented archaeological sites show that the area's first pioneers arrived at least 11,000 to 12,000 years ago. Waning of Pleistocene glaciation and rising sea levels eventually flooded Beringia, severing the direct land link between eastern and western Pacific Rim. For thousands of years limited technology and the Pacific Ocean's vast expanse must have prohibited regular trans-Pa-

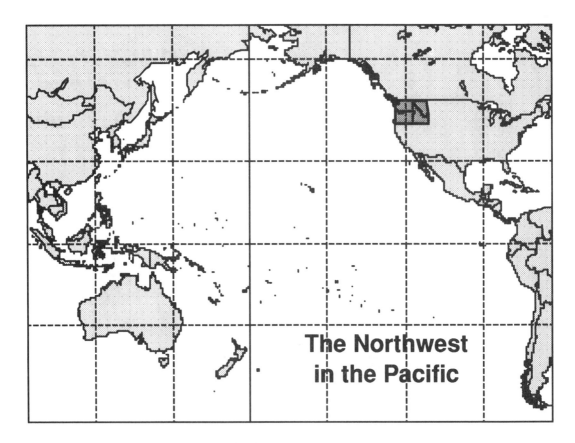

Figure 1. The Northwest in the Pacific Rim.

cific interaction, but initial transoceanic contacts between Asian cultures and Amerindians may have occurred surprisingly early.

The prospect of pre-Columbian, trans-Pacific contacts has intrigued cultural geographers for generations. Citing evidence as diverse as potatoes, pan pipes, anchors, jade, pyramids, pottery, races of chickens and knots-on-string record keeping, some researchers are convinced there were early trans-Pacific contacts. Much of the discussion centers on western Pac Rim ties with Latin America, although some also involves trans-Pacific interaction with the Pacific Northwest.

The same clockwise, north Pacific Ocean current that transports glass fish-net floats from the East Asian coastal areas and, to the delight of beachcombers, deposits them on the Washington and Oregon coasts, also carried early visitors from the East. Oral tradition among tribes of Washington's Olympic Peninsula tells of emaciated and strange-looking individuals whom the Indians rescued in offshore waters and held as slaves. Historical records document the arrival of Japanese fishermen blown off course and carried by currents to waters off the Northwest. One of the most referenced cases involved three Japanese castaways who were disabled off Japan and drifted across the North Pacific for more than a year before they made landfall near Cape Flattery in the far northwest tip of Washington's Olympic Peninsula in 1833 or 1834. Enslaved by the local Indians, they were rescued by the Hudson's Bay Company and taken to Fort Vancouver before traveling on to London, England and eventually China.

Pre-Columbian similarities in such cultural features as languages, house construction, ceramics, fishing techniques and metals in the Northwest and sections of the western Rim suggest that erratic, accidental cross-Pacific drift voyages, perhaps in both directions, may have influenced pre- and protohistoric Northwest culture. The well-preserved and well-documented Olympic Peninsula Ozette Village site has yielded some of the most irrefutable evidence of trans-Pacific linkage. Less sporadic and more intentional interchange had to await arrival of Europeans and their wooden sailing ships in the Pacific Basin.

The Maritime Fur Trade as Pacific Rim Commerce

Although other seagoing European explorers had preceded him in Northwest waters, it was Britisher Captain James Cook's arrival in 1778 that ushered in a new phase of Northwest trans-Pacific interchange. On his third Pacific voyage Cook's primary objective was discovery of a presumed and elusive Northwest Passage. He failed to locate a western entrance to an all-water route across North America, but the expedition proved to be the spark that ignited regular trans-Pacific trade and the region's initial participation in globe-girdling commerce.

Sea otter pelts were the Northwest's first commodity of "international" trade. Although Cook was killed in 1779 while wintering in Hawaii, what he called the Sandwich Islands, his comrades returned to Britain via a westerly passage. En route they stopped at Canton, China and were amazed to find that even the worn-out sea otter pelts they had traded from natives on Vancouver Island were highly prized by Chinese Mandarins and fetched unheard-of prices. Word of the guaranteed Oriental market and the great potential for profit in fur spread quickly. British traders in the aptly christened *Sea Otter* arrived off the Northwest Coast in 1785, ushering in the Pacific Northwest's maritime fur trade and the region's first trans-Pacific commerce. In the 1790s Northwest sea otter became the cornerstone of a China Trade dominated by American traders who sailed out of their Boston, Massachusetts home port.

Among the Oregon Historical Society's extensive Northwest collection at their Portland museum is an elaborate Chinese willow ware dinner service Captain Robert Gray carried from Canton to Boston. This 1787–90 voyage out of Boston was one of American firsts. Not only were the *Columbia* and *Lady Washington* the first American ships to attempt the Cape Horn route around the southern tip of South America, Gray's return to Boston Harbor aboard the *Columbia* on August 10, 1790 after a voyage of

41,899 miles, was the first American circumnavigation of the globe. The passage inspired other Boston merchants and set the standard for the transport geography in the nascent American Northwest maritime fur trade.

Russian fur traders already had depleted the northern sea otter population in Alaskan waters, but the more southerly Northwest otter numbers remained substantial. By 1801, 15 American ships conveyed 18,000 skins to Canton worth $500,000. It was customary for "Boston Men," as the Northwest natives called these American traders, to spend one or two seasons sailing slowly up and down the coast, anchoring in protected bays and coves to trade with natives for their sea otter pelts. Iron chisels, blankets, beads, clothing, muskets, rum and an array of trade goods were highly prized by Northwest Indians. If fur returns were not adequate after a single trading season, it became routine to winter-over on the Hawaiian Islands, as was the case with Gray's voyage, and return to the Northwest Coast for a second round of trading. Even if a single season proved adequate, a stopover at the Hawaiian Islands on the Northwest Coast:Canton, China leg to take on supplies, especially fresh water, fruit and meat, became standard operating procedure.

Competition in the Northwest's coastwise sea otter trade was intense and quickly decimated sea otter populations. Incredible as it sounds, after traversing 3,000 miles of wilderness Lewis and Clark arrived at the mouth of the Columbia River in November 1805 to find a depleted sea otter population and discriminating Indians demanding specific trade goods and highly favorable terms. The onslaught continued despite reduced returns. Between 1812 and 1834 annual take averaged only 2,000 skins and emphasis shifted to other fur-bearers. A greatly reduced sea otter trade continued into the 1830s and 1840s, but by that time most pelts were taken north of the 50th parallel. By the early 1850s Hudson's Bay Company (HBC) personnel thought sea otter from the Washington and Oregon coasts must have migrated to waters off Russia and Japan.

The Northwest maritime fur trade, the region's oldest antecedent of today's trans-Pacific commerce, showed the young American nation the great geoeconomic potential of a Pacific face. In the prolonged discussion of the Oregon Question during much of the nineteenth century's first half, the territory was not viewed simply as a promising frontier home for prospective settlers. Equally important, it was seen as a strategic beachhead on the Pacific Rim, a jumping off point to the rich Orient and its vast potential markets.

As this earliest Northwest commodity of Pacific Rim commerce neared extinction, another more abundant resource was about to find a larger and more long-lived Pacific Basin market.

The Northwest's Forests and Pacific Rim Linkage

The moist and moderated maritime clime of the Northwest's Pacific slope makes it one of the world's most prolific forested regions. Early white explorers and settlers

Photo 1. Late nineteenth century view of the mill complex at Port Gamble, Washington. (Washington State Historical Society Photo).

were awed by the scale of these lush west-side forests where Douglas fir standing 150 feet tall and 12 feet across at their base were commonplace.

Hudson's Bay Company employees at Fort Vancouver, so often a factor in Pacific Northwest "firsts," are credited with developing the region's initial regular foreign trade in timber and forest products. In 1824–25 newly built Fort Vancouver, located at the confluence of the Columbia and Willamette rivers, became the headquarters for the reorganized HBC Columbia Department. Soon the post was operating two saw mills, one on Wappatoo Island at the mouth of the Willamette River and another six miles upstream from the post. Production at the mills not only satisfied HBC requirements, it also served an export market. Hawaii was the largest buyer and so persistent an offshore market for spars and planks that by the early 1840s the company was sending two or three cargoes a year to the islands. Hawaiians, or Kanakas, dominated mill employment. The company's annual London supply ship routinely returned to the British Isles loaded with lumber as well as furs, stopping at west coast Latin American ports where lumber was in short supply.

Other non-HBC mills, some geared to the export market, were opened in the upper Willamette Valley/Lower Columbia River reach in the 1840s. Discovery of gold in

California in the late '40s and the resultant rush created new, domestic markets in the gold fields and the burgeoning gateway urban center of San Francisco. Although timberlands in the lower Columbia drainage were nearly ideal, nature had been even more accommodating farther north around the Puget Sound re-entrant. The Sound provided direct access to the prodigious waterside stands and afforded deep and sheltered harbors, which doubled as handy mill sites, protected from the storms of the open sea. Puget Sound was a timber baron's nirvana and, in short order, San Francisco investors began tapping its green gold.

Big timber money had moved into the Puget Sound area by the early 1850s, when tidewater steam sawmills were in operation at Port Gamble, Port Ludlow, Steilacoom, Alki and diminutive Seattle—others followed. San Francisco was initially the dominant market for Northwest and Puget Sound forests, but as the luster of the California gold rush waned, producers scrambled for new markets, many within the Pacific Rim. Hawaii, foreign territory until annexed by the United States in 1898, remained the most persistent offshore market, but trade straddled the Pacific basin. For example, in the fall of 1854 the first shipment bound for Australia left the A. J. Pope and F. C. Talbot owned Puget Mill Company's Port Gamble mill. The next year the company sent lumber to Manila in the Philippines, spars and ship plank to Hong Kong, spars to Valparaiso, Chile and other shipments to Sydney and Melbourne in the British colonies that became Australia. In subsequent years the company delivered fine lumber, spars and ship planking to Shanghai and other Chinese ports. Callao, Peru and Guayaquil, Ecuador also received occasional company cargoes.

Northwest mill companies maintained their own lumber fleets, which collectively spanned the Pacific. By the late 1880s Pope & Talbot alone had 14 ships. This fleet included six large vessels capable of carrying one million or more board feet. Such large capacity craft routinely were used in long-distance trade to South America, Australia and the Orient.

Unique Pacific trade patterns developed by the late 1800s. A triangular ocean-spanning trade emerged between the West Coast, Australia and the Orient. In this trading system forest products from the West Coast were conveyed to Australia, where lumber was off-loaded and coal taken on board. Ships then sailed north to such ports as Hong Kong and Shanghai, where coal was exchanged for cargoes of Oriental goods prior to returning to the West Coast. A trade variant for mill company lumber ships was to carry Puget Sound lumber to Australia where they on-loaded coal for Hawaii. At the islands coal was exchanged for sugar before sailing on to San Francisco. On the final leg home to Puget Sound mill docks, company ships carried supplies from California.

Another early trans-Pacific linkage began tying Portland, Oregon with Hong Kong in the 1870s. Pioneer Portland entrepreneur John Ainsworth purchased three ships specifically for this trade to carry lumber, scrap iron, foodstuffs and passengers west to Hong Kong and return to home port with a diverse cargo that included Chinese laborers, rice, granite and small art objects. Lumber and Chinese laborers were the key cargoes in this early Pacific Rim interchange that continued into the 1890s.

The Northwest's Pacific Rim Grain Trade

Pac Rim grain trade has been a feature of Northwest economic geography for more than 150 years. Ever industrious and innovative, it was HBC's Fort Vancouver that pioneered regular Northwest grain exports. In the decade following the first wheat harvest at the post in 1825, wheat production grew to perhaps 5,000 bushels, enough to allow shipment of flour from London, England, to be discontinued. The first outside shipment of record dates to 1836 when the company's newly acquired steamship, *Beaver*, carried a cargo of wheat to Russian America.

The Hawaiian Islands were an early and reliable market for HBC flour, where it was exchanged for sugar, molasses, nut oil, whalebone and salt. Starting in 1840 the HBC began meeting the requirements of a long-term agreement to supply the Russian-American Company in Sitka with land otter pelts and various provisions at set prices. The Russian contract stipulated that the HBC would annually supply 8,400 bushels of wheat and eight tons of wheat flour. Farmland at Fort Vancouver alone could not meet this quota so a new agricultural operation was begun by the HBC's Puget Sound Agricultural Company at Cowlitz Farm (located midway between today's Centralia and Longview in western Washington). Fort Vancouver and Cowlitz production, combined with grain purchased from the growing contingent of Willamette Valley farmers, allowed the HBC to meet its own needs, as well as Russian American obligations and other export demands.

It was the HBC's Rim export market that provided early Willamette homesteaders with the first export market for their surplus wheat. The HBC maintained two receiving stations and two flour mills in the valley to facilitate this Pacific Rim trade. Not long after the Willamette's first wheat was planted in the 1830s, the valley emerged as a major agricultural mecca, becoming the Northwest's undisputed granary in the decade of the 1840s. The California gold rush of the late 1840s created new domestic demands on the Willamette granary.

The valley's cereal grain prowess made it the focus of an expanding international grain trade. Direct overseas flour shipments occurred early and, for example, flour was shipped from Oregon to Japan in 1857 aboard the *Herman C. Leonard*. Prior to 1868 most Pacific Northwest trade was in the form of flour, shipped in barrels. For two decades the standard procedure was for most Willamette production to be carried to San Francisco in coastwise vessels where it was consumed regionally or blended with California wheat before export.

In 1868 an established trade pattern was broken and the Willamette began plugging directly into the international wheat trade. That year the British bark, *Helen Angier*, with its load of Oregon wheat was dispatched directly to Liverpool, the pivotal port for the European grain trade. By the early 1870s a distinctive Portland

NORTHWEST FLOUR SHIPMENTS, 1868–1874

Destination	Barrels
British Columbia	97,475
Great Britain & Europe	91,636
Australia	1,185
Hawaii	8,674
China	46,050
Brazil	5,436
San Francisco	779,249

Source: John B. Watkins, "Wheat Exporting from the Pacific Northwest," State College of Washington, Agricultural Experiment Station, Bulletin No. 201 (May 1926), p. 11.

wheat fleet had developed, the tall-masted ships alongside Portland's mist- and fog-shrouded wheat wharfs becoming standard fare each fall and winter.

Between 1868 and 1874 the lower Columbia/Willamette wheat region exported just over one million barrels of flour. San Francisco remained the major customer. British Columbia, Europe (mostly Great Britain) and China also were sizable markets, with smaller amounts going to the Pacific destinations of Hawaii and Australia.

Following the initial 1867 downriver shipment of Walla Walla flour to Portland, and eventually on to San Francisco, the proportion of Northwest wheat derived from newly opened, fertile and extensive Columbia Basin farming districts increased dramatically, and soon came to dominate. By the 1870s the prodigious Palouse region had begun adding its bounty of wheat sacks to this downriver flow to Portland. The previously relied upon inland transport system of interconnecting wagons, portage railroads, river boats and frequent transhipment along the Columbia River corridor was no longer adequate. In 1882 the Oregon Railroad & Navigation Company responded with completion of its rail line from Portland to Wallula, downstream from the mouth of the Snake River, and also obtained control of, and improved, the rail linking Wallula with Walla Walla. Other rail service followed, as did improvements in river navigation, giving a growing contingent of inland grain growers more direct access to Portland's wheat wharfs, the Pacific Rim and global economies.

Beginning in the 1880s Puget Sound ports joined Portland as wheat and, especially, flour export centers when rail lines linked them with dryland farming districts east of the Cascades. Portland, however, remained the Northwest's undisputed center of international grain trade. One of that city's most interesting early Pacific Rim ties was a Siberian flour trade begun by the Centennial Milling Company in 1891. The trade reached as high as 200,000 barrels annually until the Russo-Japanese War caused a precipitous decline.

The Northwest as National Pacific Rim Portal

The 1880s arrival of transcontinental rail service ushered in a new phase in the Northwest's evolving Pacific Rim geography. Now the region was able not only to export its own staples as it had through the initial period of white occupancy (especially furs, timber and grain), it also became a transit region and portal through which a rapidly developing American economy articulated with the circum-Pacific. American linkage with the western Pacific Rim, especially the Orient, held great appeal as it had since the time of Lewis and Clark.

Although generally not thought of as Pacific Rim explorers, that is precisely what Lewis and Clark were on their historic 1804–06 expedition that took them from St. Louis to the Northwest Coast and return. Among their many charges from Thomas Jefferson was examination of the portage between the Upper Missouri and the Columbia River headwaters. Jefferson, a proponent of American expansion to the Pacific, was interested in developing an efficient water-based, transcontinental transport corridor that would span the American West and permit two-way trade with the Orient. The Corps of Discovery did not find a short and easy portage across the Northern Rockies, but the quest for a transport tie to the American West Coast and the Pacific Rim threshold persisted.

If nature had failed to provide the transcontinental highway to the Pacific, then certainly railroads could. By the 1840s the topic of a Pacific railroad and the importance of Asian markets was a common one in the U.S. Congress. Pacific Railroad promoters and agitators of the mid-century claimed that rails would reduce the travel time from New York City to China from 150 days to just 30. Advocates claimed that rail service not only would link the breadbasket of the American Midwest with the teaming population of Asia, it also could function as what is today called a landbridge across America over which Asian:European trade could pass.

A transcontinental railroad with a Pacific Northwest terminus became a reality in 1883 when the final section of the Northern Pacific's track was laid in western Montana. Four years later the company extended its line to the shores of Puget Sound, terminating at Tacoma. In 1893 competing transcontinental rail service became a reality in Puget Sound when James Hill's Great Northern Railway entered Seattle. In his vision of a transportation and economic empire spanning the Pacific he saw Seattle as Gateway to the Orient.

Hill realized that low income levels in the Orient ruled out some types of trade. Nevertheless the almost 500 million residents of East Asia represented a huge potential market for such American staples as grain, cotton, iron ore and coal. Hill spoke about such things as the Great Northern's Cotton Route to Asia and even suggested his line might carry Europe's Oriental trade.

Photo 2. The June 17, 1892 arrival of the *Phra Nang*, the first steamship to arrive at Tacoma from the Orient. It was one of the fleet which the Northern Pacific Railroad had employed in the Oriental trade. (Washington State Historical Society Photo, Tacoma, Washington).

Visitors walking the pier-side of Alaska Way in Seattle's central waterfront district of today pass by a brass historical plaque adjacent to Pier 59 that is, indirectly, a tribute to James Hill and his Pacific Rim vision. The plaque reads:

> *At this site on August 31, 1896, the Nippon Yusen Kaisha steamer "Mike Maru" arrived with her cargo of tea. This was the first regularly scheduled steamer service between the Orient and Seattle, and marked the birth of Seattle as an international port.*

Hill previously had signed a traffic agreement with Nippon Yusen Kaisha (NYK, the General Mail Steamship Company) to provide regular steamship service to the Orient in conjunction with his railroad. This was seen as an interim solution only, and within a decade Hill's Great Northern Steamship Company had christened the world's two largest capacity cargo ships to service the Seattle:Orient trans-Pacific route.

With much fanfare the *Minnesota* set sail from Seattle and the Great Northern's docks at today's Pier 90 and 91 on its maiden Pacific crossing on January 23, 1905 loaded with rails, copper ingots, cotton, flour and general merchandise bound for Yokohama, Kobe, Hong Kong and Manila. The sister ship *Dakota* departed on her initial Pacific voyage later that year loaded with 20,000 tons of flour. Problems plagued the ships and the service. The *Dakota* sank in Tokyo Bay in March 1907 and the *Minnesota* left Pacific service in 1915 after 40 round trips to the Orient. Despite their mishaps, Hill's two great ships clearly showed the advantages of a coordinated, intermodal approach to Pacific trade.

Speed and efficiency of both sea and land transport were the hallmark of the colorful silk trade that became the dominant component in Seattle's early twentieth century international trade. Initiation of a regular interchange with the Orient in 1896 meant a range of new and often exotic trade items, none more unique than silk. Raw silk threads, a perishable commodity at extreme temperatures, arrived in Seattle in large burlap-wrapped bales worth about $1,000 each. They were quickly off-loaded and carefully stowed aboard awaiting and heavily guarded Silk Trains for shipment to East Coast weaving mills. Since time was of the essence, only the fastest locomotives and best crews worked these high-speed runs. Other trains were required to yield right of way to the silk express, which stopped only to take on coal and water and to change crews about every 400 miles.

The first silk train left Seattle in 1909. By the 1920s peak, when Seattle handled more raw silk than all other West Coast ports combined, silk accounted for almost half the value of all imports off-loaded. Seattle's silk imports declined dramatically by the 1930s as Japan shifted to exports of finished garments and the all-water route to Eastern mills became more popular.

Many residents of the early twentieth century Northwest felt their region was at the beginning of a new Pacific era and they wanted the world to know of its promising commercial future. This sentiment was perhaps nowhere more blatant than at Portland's 1905 fair, The Lewis and Clark Centennial and American Pacific Exposition and Oriental Fair (The Lewis and Clark Fair), and Seattle's ambitious first world's fair, the 1909 Alaska-Yukon-Pacific Exposition.

At the Lewis and Clark Fair the imposing Oriental Exhibits Building, more than half occupied by Japan, symbolized the growing relevance of the Far East in Portland's international economics. Four years later, on the site of today's University of Washington in Seattle, the Alaska-Yukon-Pacific Exposition was intended to involve all Pacific Rim countries. That goal wasn't achieved, but this even more ambitious fair did celebrate the city's key commercial water links with Alaska and other sections of the Pacific world. Four foreign countries, including Japan, Philippines and Canada erected their own buildings, and 32 additional participating nations housed their exhibits in state buildings. Among the fair's Pac Rim potpourri were an impressive Oriental Building, a Chinese Village and exhibits of native peoples of Siberia and the North. Both international fairs were eerily prophetic in their recognition of an increasingly interconnecting Pacific Rim, a trend that has accelerated in recent decades.

Photo 3. Scene at Seattle's 1909 opulent and visionary Alaska-Yukon-Pacific Exposition. This site now is part of The University of Washington campus. (Museum of History and Industry Photo, Seattle, Washington).

THE NORTHWEST AND THE PACIFIC RIM: TODAY

In late 1991 signs of Asian Gypsy moths were discovered in close proximity to ports at Tacoma and Portland. These first reports of this voracious insect in the United States made members of the Northwest timber industry shudder. The Asian variant makes its immigrant cousin, the European Gypsy moth, look almost benign. Its resident kin, which defoliates millions of acres of forests annually in the eastern part of the United States, is spread by crawling or by wind. In contrast, the female Asian Gypsy moth can fly up to 20 miles, triggering a fast-spreading infestation through both conifers and hardwoods. So serious was the threat of this alien to millions of acres of prime Pacific Northwest timber, that one Washington state official felt its presence had "the potential to be the most serious exotic insect ever to enter the U.S."

Quick eradication efforts, including controversial aerial spraying, were judged effective the following spring when none of the hundreds of moths collected were of the Asian variety. At the time, it was assumed that the Asian moth was inadvertently carried to the Northwest aboard ships that either originated in, or passed through, the Russian Far East. That theory seems to have been verified in March 1992, when four wheat-laden Russian vessels were ordered out of the Port of Vancouver, British Columbia, after Asian Gypsy moths were discovered on board. Fortunately, this biogeographical consequence of the Northwest's Pacific ties is not typical; most contemporary linkages are viewed more positively than the Asian Gypsy moth.

Northwest Ports as Pacific Rim Portals

The Northwest's economic ties to the Pacific Rim, especially those relating to international trade, are the most pervasive and visible. There are few places in the Pacific Northwest where Pac Rim commerce is more apparent than at the region's major ports. Windjammers may no longer tie up dockside along Portland's Willamette banks to take on wheat sacks, and multi-masted lumber ships are gone from Port Gamble and the sites of other once-booming Puget Sound cargo mills, but the Northwest's major ports remain the region's key gateways to the Pacific Rim. Wooden sailing ships have been replaced by swift containerships and custom bulk carriers, but it still is at these portals that Pacific Northwest and Pacific Rim space most inexorably overlap. If one had only a few hours to gain an appreciation for the impact of Rim linkages on Northwest geography, those hours would best be spent visiting one of the large port complexes. In these high energy landscapes of commerce, the scale and geographical consequences of the Northwest's Pacific Rim linkages are most evident.

Since the 1960s oceanic shipping has experienced a transportation revolution. One of the most dramatic changes has involved the widespread adoption of containerized shipping. Containers are large and usually colorful steel boxes commonly 40 feet long, about the size of a standard semi-truck trailer. Modern container shipping may have begun elsewhere in the nation in the 1950s, but the Pacific Northwest may be credited with developing the early forerunner of this innovation. As early as 1949, the Alaska Steamship Company began experimenting with six-foot wooden containers on its passage between Seattle and Alaska and, in the 1950s, shifted to eight-foot steel boxes.

The advantages of containerized shipping are obvious and numerous. With items packed in containers at their point of origin and not unpacked until arrival at their destination, considerable labor is saved in handling cargoes. Containers also are ideally suited to intermodal transport, or being carried by several interconnecting modes of transportation—barge, truck, railroad and ships. Ease of transference from one mode to another means considerable savings in time and money. As well, self-contained "boxes" or "vans" as they sometimes are called, provide dry, secure and sometimes even refrigerated shipping environments.

Photo 4. *Hyundai Commodore* at the Port of Seattle's Terminal 18. The Seattle CBD rises in the distance. (Don Wilson/Port of Seattle Photo).

Today port size and cargo volume prowess often are measured in terms of TEUs, or 20-foot equivalent units—equal to one 20-foot container. A standard 40-foot container thus equals two TEUs. The number of TEUs handled by a port in a year is a good general indicator of the amount of high value cargo passing across its docks. Both Seattle and Tacoma are among the nation's largest container ports, each handling more than one million TEUs annually. Combined, these Washington ports make the Puget Sound region the nation's second largest container import/export region, ranking behind Los Angeles/Long Beach, and ahead of New York/New Jersey. In this category, both Seattle and Tacoma rank among the world's 20 largest container ports. The Port of Portland, where emphasis is less on container cargoes, handles about 200,000 TEUs per year.

At each of the Northwest's three largest ports, trade with Pacific Rim countries dominates traffic. In 1991 the top ten trading partners for both the Port of Seattle and the Port of Tacoma were Pacific Rim nations, and at the Port of Portland and the remainder of its tributary Columbia-Snake River Customs District, seven of the ten largest trade partners were located around the Pacific Rim.

The combined imports/exports for the Pacific Northwest's three largest ports are made up of cargoes produced within the Northwest or headed for destinations in the region, as well as more long distance, pass-through trade that utilizes Seattle, Tacoma or Portland as a portal into or out of the Pacific Rim. Cargoes originating within the Northwest are of greatest interest in this chapter and have the most obvious links to area geography. It is impossible to itemize and describe the associated geography of the hundreds of products and commodities that are produced in the Northwest states of Washington, Oregon, Idaho and Montana and then move to Pac Rim markets through the region's largest ports. But even a sampling of Northwest exports handled by these ports suggests the profound impact Pacific Rim markets have on geography within the ports' Northwest hinterlands. Non-Northwest, pass-through trade moving in both directions has a more restricted impact spatially, with most economic benefits accruing port communities.

The Northwest's Pacific Rim Agricultural Exports

Much of the bounty of the Northwest's agricultural lands, long synonymous with the area's international trade, moves to Pac Rim destinations through the region's three major ports. Soft white wheat, the premier crop in the inland Northwest's extensive and productive dryland grain farming districts, is overwhelmingly an export commodity. It finds markets on the Asian side of the Pacific Rim where its low gluten content makes it especially desirable for noodles, crackers, flat breads and pastries. Western Pac Rim Japan, Taiwan, South Korea and the Philippines are the largest and most consistent markets. Potentially huge and lucrative Northwest wheat sales to the People's Republic of China have been stalled by Chinese fears of a fungus spore known as TCK smut, but in the early 1990s experimentation with post harvest sanitation methods looks promising.

NORTHWEST PORTS/MAJOR TRADING PARTNERS 1991

Port of Seattle

Rank	Country	Total Declared Value ($ Millions)
1	Japan	$11,960
2	South Korea	3,729
3	Taiwan	2,807
4	China	1,854
5	Hong Kong	1,133
6	Canada	1,111
7	Australia	378
8	Thailand	357
9	Singapore	320
10	Philippines	223

Source: Port of Seattle.

Port of Tacoma

Rank	Country	Total Declared Value ($ Millions)
1	Japan	$10,760
2	Taiwan	3,104
3	Hong Kong	2,021
4	China	1,719
5	Canada	1,330
6	South Korea	1,083
7	Singapore	766
8	Indonesia	595
9	Malaysia	527
10	Philippines	430

Source: Port of Tacoma.

Columbia-Snake River Customs District
(Port of Portland and Columbia/Snake River Ports)

Rank	Country	Total Declared Value ($ Millions)
1	Japan	$6,425
2	South Korea	1,155
3	Taiwan	765
4	Australia	574
5	China	294
6	Philippines	230
7	Hong Kong	192
8	Germany	180
9	Italy	160
10	Egypt	152

Source: Port of Portland.

Photo 5. Tug *Lewiston* moving a typical tow of four barges of commodities consisting of wheat, wood chips and break bulk paper on the Columbia-Snake Waterway. (Ackroyd Photography Photo).

Portland is the undisputed epicenter of the Pacific Northwest's wheat export market; prices regionwide routinely are quoted FOB Portland. The city is home to the Wheat Marketing Center, appropriately housed in the adapted and historic Albers Flour Mill, overlooking the Willamette River. Developing new uses for wheat and opening additional markets worldwide are the facility's primary missions. Nearby, the harbor, the nation's largest wheat port in 1992, handles about 22 percent of all U.S. wheat exports, and Columbia downriver ports combine to account for another 15 percent.

About 40 percent of all grain exported from Portland arrives at the port by tug and barges via the Columbia-Snake river waterway. This inland water-highway is maintained at a channel depth of 14 feet along the 355 river miles between the mouth of the Willamette River at Portland and Lewiston, Idaho. The chutes and tramways once used to carry grain down the deep and steep Snake River canyon are gone, but more contemporary modes now convey the small-grain bounty to riverside elevators at such inland ports as Lewiston, Wilma, Almota and Central Ferry. A tug and a four-barge tow are able to convey 14,000 tons of wheat (400,000 bushels), equal to 121 grain hopper rail cars or 470 grain trucks. This cost-effective transportation system is made possible by locks around the eight large mainstem dams blocking the Columbia and Snake rivers. The 1993 completion of a new and larger lock at Bonneville Dam, farthest downstream on the river system, removed the most serious bottleneck to barge traffic on the Columbia/Snake system, and also may open the way for ocean-going barges on the river.

Apples, another product synonymous with Northwest agriculture, also have an important marketing link with the Pacific Rim. Apples are Washington's number one cash crop and the state is the nation's premier grower, supplying more than half of all U.S. fresh apples. Annually, millions of 42-pound cartons filled with apples grown on the 160,000 acres of Yakima and Wenatchee valley orchards along the Cascade's east flank, and in the western part of the Columbia Basin, also head to Asian export markets, primarily via ports at Seattle and Tacoma. Over the five-year span of the late 1980s and early 1990s, seven of Washington's top ten export countries were Pacific Rim, with Taiwan the leading offshore customer. Targeting young Asian consumers in the 15 to 39 age bracket, the Washington Apple Commission has plans to greatly expand overseas markets where demographics suggest significant growth potential. Already, in-store promotions, television campaigns and even street vendors displaying Washington apple logos are not uncommon in Thailand, Indonesia, Philippines, Japan and other western Pac Rim countries. In 1991/92 removal of trade barriers allowed the Mexican market to increase sevenfold and to jump from ninth to number one foreign market for that crop year. In anticipation of still more spectacular growth in Mexican demand, the way was cleared in 1993 for apples and pears to travel directly from the Northwest to Mexico in refrigerated railroad cars, crossing the international border without a time-delaying customs inspection.

The Pacific Northwest's Pacific Rim grain exports affect the agricultural geography of literally thousands of square miles of inland grain growing districts. And Rim apple marketing helps explain the economic geography of fruit-growing areas from the Okanogan and Wenatchee Valley areas south to the Yakima Valley and eastward into the newer orchards in the adjacent Columbia Basin. But not all geographical implications of Northwest:Pacific Rim farming linkages have such diffuse impacts. Scattered about the container yards at the ports of Seattle and Tacoma are 40-foot vans filled with Japan-bound timothy hay from the Kittitas Valley. The timothy trade link between this central Washington valley and Japan provides a classic example of more localized geographical consequences of Pacific Rim agricultural trade.

Set on the east slopes of the Cascade Mountains at about 1,600 feet elevation, the linear Kittitas Valley has a reputation for growing the nation's highest quality timothy hay. Timothy is the common name for 10 species of perennial grasses native to cool and temperate sections of Eurasia; it generally is considered a horse hay. The valley's often cool summer evenings coupled with favorable soil and dry, windy conditions at harvest that quickly cures hay, provide just the right combination of factors.

Early in the 20th century Kittitas hay was shipped to nearby Cascade mining and lumbering communities to feed draft horses and over the mountains where it was fed to horses in pre-automobile Seattle and Tacoma. Beginning in the 1930s dealers began shipping Kittitas timothy to horse racing tracks in California and soon to horse breeders and tracks nationwide.

In the late 1960s Japanese horse breeders visiting Kentucky's horse country spotted bright green Kittitas timothy being fed to thoroughbreds and decided they wanted the same feed for their racehorses back home. A modest trade in timothy was opened immediately with Japan, but was closed for several years beginning in 1976 pending development of a fumigation method to allay Japanese fears of Hessian fly infestation. Exports resumed in 1979 and soon kicked into high gear with the development of a market for the hay among Japanese dairy farmers. Between 1979 and 1989 valley acreage in timothy expanded from 7,500 to 25,000 acres and annual production ballooned from 40,000 to 155,000 tons. In most years 70 to 80 percent of production is loaded into 40-foot containers for export via the ports of Seattle and Tacoma, almost all bound for Japan where the majority has been used by dairy producers.

Back in the Kittitas Valley, the timothy boom rippled through the local economy. Not only were the Japanese buying large amounts of timothy, they were paying high prices. The promise of quick profits convinced many valley operators to shift their farming/ranching operation to accommodate this lucrative new Pacific Rim market. One result has been a modified agricultural landscape, as well as increased demands for limited irrigation water by this water-intensive crop. Timothy acreage was expanded at the expense of other crops and cattle; even more marginal farm ground in the north side of the valley, previously used only for pasture, was planted in timothy. Large hay barns became a highly visible element in the valley's new cultural landscape as many growers erected the structures to protect their valuable crop from moisture damage—Japanese hay brokers primarily are interested in only the highest quality hay.

In 1993 approximately 220 growers produce hay on 25,000 acres, making it the valley's largest cash crop. The area's $18 million timothy hay industry has saved many Kittitas farmers and made others considerably more well-to-do. In Ellensburg the millions of timothy-generated dollars annually have helped buffet the community in otherwise difficult economic times. Yet the newfound prosperity has not been without insecurities. Shipments to Japan in the early 1990s have been down as Japanese dairy operators opt for cheaper feeds, such as Sudan grass. High prices also have invited competition both domestically and from growers in other sections of the Pacific Rim. But valley hay producers are optimistic that double compressed hay, producing bales half the size of a standard bale, soon will cut shipping and warehousing costs in half and make Kittitas timothy more competitive in Japan. To date, a fumigation method for compressed bales has not been approved by the Japanese government.

From specialty vegetable crops in the Willamette Valley to potatoes from the irrigated Snake River Plain of south Idaho and cattle and beef from western Montana valleys, examples of Pacific Northwest agricultural exports to the Pacific Rim are legion. Often the western Pac Rim is the primary overseas market for Northwest food commodities. In East and Southeast Asia, large and often fast growing populations, a limited agricultural land base and rising income levels should mean a continued demand for Northwest food exports and stronger trans-Pacific trade ties.

The Northwest's Pacific Rim Forest Exports

The bounty of the Northwest's extensive coniferous forests continue to be important to the region's trade links with other sections of the Pacific Rim. Demand for tall masts and ship's planking are a thing of the past, but the region's softwood trees and products of their manufacture still find an eager market, especially on the Rim's western edge. In 1991 among export commodities at the Port of Seattle, paper ranked third, wood pulp ninth and lumber twelfth in value. To the south at the Port of Tacoma that same year, logs ranked third in value, lumber fifth and paper eighth. At the Port of Portland and the Columbia River ports forest-related exports are relatively more important; logs ranked third, woodchips fifth and lumber seventh in terms of tonnage exported. Many of these cargoes are produced in the Northwest.

In the early 1990s raw logs were the most controversial aspect of Northwest forest exports. Overseas shipment of logs began in large numbers following the historic Columbus Day storm of 1962, which blew down billions of board feet of lumber in western Washington and western Oregon. Since then, Japan and, to a lesser extent, South Korea and China, have been the chief overseas markets. Exports of logs from national forests have been banned for years, and by 1992 overseas shipment from all Northwest area state-owned forests also was prohibited. Such concerns as preservation of old-growth forests, protection of northern spotted owl habitat, wilderness study/designation and other factors translated into reduced supplies. Logs cut from the maturing plantations of radiata pines in Chile and New Zealand also have meant price competition. In the Northwest, opponents of log exports had lobbied for the bans, arguing that sending logs overseas both further tightened supplies and effectively shipped relatively high-paying wood products manufacturing jobs offshore with the logs.

Between 1989 and 1991 Northwest log exports dropped by about 25 percent; further declines are forecast. Port employment has been impacted at Everett, Port Angeles, Grays Harbor and Longview in Washington and Coos Bay in Oregon, which have been primarily log shipping ports. Export of logs harvested on private lands is permitted, so a port with access to extensive private timberland, such as Everett, is less affected.

Woodchips, only one step removed from raw logs, are a Northwest staple in the export trade. The most visible evidence of this commerce undoubtedly is the Weyerhaeuser wood chip export facility (Puget Sound Chip Center) on the Port of Tacoma's Blair Waterway. Its flat-topped mounds of segregated Douglas fir and red alder chips have been part of the local port geography for two decades. The facility's entire export of about a million tons per year of western Washington woodchips is Japan bound, where chips are consumed at pulp mills and paper-making plants. Stiff competition in the export chip market from other Pacific Rim countries, notably Canada, Chile and Australia, keeps prices in Japan competitive.

Scattered about the forested Pacific Northwest from Montana's Bitterroot Valley westward through the region are log home manufacturers. Many of the homes are custom built, shipped overseas and assembled like Lincoln Log sets. Japan is the primary offshore market for these relatively expensive homes. Wealthy Japanese often erect their log homes in rural areas and use them as weekend retreats. There is considerable prestige associated with log homes in a nation where wood is so highly prized. Japanese fascination with the idealized American Old West and frontier era make this American icon all the more popular. Real Log Homes of Missoula, Montana, services its important Japanese market with dealers for its log home packages in five cities including Tokyo, Nagoya and near Kobe. Farther up the Bitterroot Valley, Rocky Mountain Log Homes of Hamilton ships about 13 percent of its homes to western Pacific Rim customers, with most going to Japan, Taiwan, Thailand and South Korea. They also recently signed a multi-million dollar agreement to assist a Russian-Canadian joint venture in setting up a log home manufacturing plant in Siberia.

Meeting offshore Pacific Rim lumber requirements is another matter. Construction methods in East Asia differ from those in the United States making lumber requirements equally unique. Northwest mills can supply the large squares (beams that are cut into dimension lumber at a building site in Asia), which have been a part of exports since before World War II. But more finished, dimension lumber commonly is provided by local mills in Asia—thus the demand for Northwest logs and a willingness to pay higher than domestic prices.

Some Pacific Northwest sawmills have retooled to serve at least an element of the Asian dimension lumber market. Perhaps no operator has received as much publicity as Vanport Manufacturing in Boring, Oregon, 30 miles outside Portland. In the mid-1970s an abandoned mill in that community was rebuilt to accommodate metric measurements. Foremen were instructed in the fine art of grading wood Japanese style, in which such factors as wood grain and toning are of paramount importance. Most symbolic in the firm's Japanese link is the company-built, traditional Japanese guest house available for overnight visits from prospective Japanese customers. By the late 1980s the mill employed 170, working double shifts to meet the demands of their Japanese customers who purchased 90 percent of production.

Paper is another Northwest forest product that finds a growing market in the Pacific Rim, especially on the Asian edge. Potlatch Corporation's massive mill complex in Lewiston, Idaho, illustrates just how important trans-Pacific linkages can be to Northwest communities. Potlatch's sprawling factory on the Clearwater River near its junction with the Snake dominates city employment, but also has been a point source of air and water pollution. Nine hundred of Potlatch's 2,400 high-paying Lewiston jobs are tied to producing bleached paperboard, a paper used for liquid-tight containers and other packaging. According to company spokesman Mike Sullivan, about 30 percent of paperboard production is exported, with 90 percent going to overseas Pacific Rim markets. Japan dominates that international trade, but South Korea and Taiwan are additional important customers. Bleached paperboard is shipped in the form of large rolls that are loaded in containers and sent to the nearby Port of Lewiston for a downriver barge trip to the Port of Portland. Potlatch-filled containers make up the majority of the approximately 8,000 TEU containers this upriver port sends out annually.

Across the Bitterroot Range in western Montana, the 900-employee Stone Container Corporation paper mill west of Missoula also plugs into the Pacific Rim market. About 13 percent of this mill's production of linerboard (paper used to make the smooth inside and outside walls of corrugated paper) is loaded into containers and shipped to Pacific markets through the ports of Seattle and Tacoma. Japan, Malaysia, Hong Kong and Taiwan are the largest markets, but some production is destined for Rim markets as dispersed as China, Mexico and Ecuador.

Other Northwest Pacific Rim Exports

Although countries of East Asia tend to buy raw materials, products other than those of the Northwest's fields and forests also enter Pacific Rim commerce. Regional exports are many and varied and reflect the Northwest's economy in items as diverse as fish and fish products, electronic equipment, software, biotechnology products, sporting goods, clothing, yachts, machinery and aluminum ingots. Paradoxically, the Northwest's number one export does not pass into world markets via the region's ports. The Boeing Company is the giant of Northwest exporters, and in the early 1990s exports more than any other American manufacturer.

Commercial aircraft sales dominate Boeing's foreign sales, but assembled planes bypass ports, with customers often taking possession outside of Washington. It is not uncommon for buyers to officially take delivery high over the international waters of the North Pacific, where overseas customers can save approximately $12 million Washington state sales tax on a $150 million Boeing 747.

According to Boeing spokesman, Craig Martin, Boeing planners expect the Pacific region to account for 30–plus percent of the world market for new commercial passenger planes between 1993 and 2010. Boeing plans that more than half of those planes will be of its manufacture, with Japan remaining the number one market, and

China equalling the United Kingdom and perhaps even challenging Germany for second rank.

Northwest Ports and National Hinterlands

The Northwest's major ports have been blessed not only with spectacularly beautiful sites, but with a most favored situation relative to trans-Pacific commerce—they are America's closest major mainland ports to the dynamic economies of East Asia. This helps explain why Seattle's foreign waterborne commerce is dominated by links with that region. Japan, South Korea, Taiwan, China and Hong Kong accounted for 83 percent of the Port of Seattle's foreign trade value in 1991. At the Port of Tacoma the trade with the same five overseas trading partners is only a slightly less impressive 79 percent.

Speedy and modern oceanic vessels can make the trans-Pacific voyage between Puget Sound and Japan in 9 or 10 days via the great circle route that arcs northward toward Alaska. Couple such speedy water transport with recent innovations in connecting inland rail transportation and Northwest ports understandably have increasingly functioned as Pac Rim gateways for areas as distant as the American Midwest and East Coast. For example, an early 1993 NYK Line shipping schedule shows that a container leaving the Port of Yokohama, Japan, on March 13 would be at the Port of Seattle on March 21 and could be available for pick-up in Chicago, Illinois, on March 26.

At Portland emphasis is on exports (of both regional and national origins) and serving Northwest markets for inbound cargoes. The ports of Seattle and Tacoma compete in these same areas, but they also are locked in direct competition, each attempting to capture inbound transit, or pass-through, trade that moves on to distant North American locations. At stake are incomes, economies of scale and jobs; containerized cargo generates about one job per 200 TEUs. One key component is an efficient interfacing of water and land transport. Much of this interconnection takes place at each port's on-dock intermodal yards where containers can be swiftly transferred between trains and ships. Railroads serving the major ports have pushed transport innovation and modification to facilitate speed and economy of scale not unlike earlier silk trains. Unit trains (entire trains dedicated to one type of cargo) and double-stacked railcars, in which containers are stacked two high, add to economies on regular runs between such distant cities as Chicago, New York and even Jacksonville, Florida. Up to 80 percent of inbound containers arriving at Seattle and Tacoma are shipped to inland destinations, with 30 percent to 40 percent of those bound for the East Coast.

Automobile imports are an especially visible component of Northwest ports' national Pacific Rim gateway role. The Port of Portland has long-term agreements with Toyota, Honda and Hyundai, and is Subaru's West Coast port of entry. It is the Northwest's largest auto importer and the nation's fifth largest in terms of vehicles, with most leaving the port by rail to more than 40 states. In a recent favorable balance

Photo 6. Port of Portland, Oregon. (Ackroyd Photography Photo).

of trade turn-around, Honda has been shipping vehicles built at its Marysville, Ohio, plant out of the Port of Portland to markets in Japan, Taiwan, South Korea and Australia. In 1992 those exports totaled about 35,000 cars, only a small fraction of the approximately 300,000 vehicles handled annually by Portland, more than any other West Coast port. Recent economic research suggests each car imported through Portland accounts for an additional $350 income to the local economy.

The long-distance origin of the Northwest ports' export cargoes is reflected in the fact that in 1991 cigarettes/tobacco, products of Southeast states, were the Port of Tacoma's most valuable export and the same year cotton ranked sixth at the Port of Seattle.

Not only containerized cargoes use the long-distant landbridge connection with inland locations. For example, in 1991 when 95 huge D-10N Cats needed to be shipped from the Caterpillar factory in Peoria, Illinois, to Russian Far East ports, they were sent via Burlington Northern Railroad to one of the Port of Portland's breakbulk terminals. And recently, when the Union Pacific Railroad imported steel rails used to rebuild their national railnet, they conducted business through the Port of Portland.

Distantly produced bulk cargoes headed for Rim markets also commonly find their way to the Pacific through the Northwest. Almost 1-1/2 million tons of Wyoming soda ash, second only to wheat in tonnage, is shipped annually to the Port of Portland by the Union Pacific Railroad. It is loaded onto ships at Terminal 4, bound for Pacific Rim glass manufacturing plants.

Ports other than the region's three largest full-service ports also handle long-distance cargoes. The Port of Kalama, Washington, downriver from Portland on the Columbia River, is a major grain shipper. Nebraska farmers grow most of the corn shipped out of this port, headed primarily for western Rim markets. The 80,000 rail cars that arrive at Kalama annually, many part of unit trains, make it one of the nation's busiest single rail destinations.

Beyond the Ports—Other Rim Linkages

Many of the Pacific Northwest's economic connections become realities only after considerable effort by states and other regional entities. State trade delegations, commonly including the governor, frequently have crossed the Pacific in search of expanded trade ties. Northwest states also maintain trade offices in such key countries as Japan, Taiwan and Canada. Major ports, with their vested interest in expansion of international trade, have East and Southeast Asia offices/representatives in Japan, South Korea, Hong Kong, Taiwan and Singapore. Sister ports are another vehicle for expanded trade. Most recent were the Port of Tacoma's twinning with the Russian Far Eastern Port of Vladivostok in November 1991 and with the Port of Tiajin, China in April 1993.

Worldwide city twinships for more broad-based objectives are facilitated by Sister Cities International, a program popularized by President Dwight D. Eisenhower in the mid-1950s. According to the 1992 Sister Cities Directory, more than 900 U.S. communities (most commonly cities, but also some counties and states) had a formal sister relationship with more than 1,400 communities in 100 countries around the world. The Pacific Northwest is well represented with 58 communities twinned with 108 overseas sisters. Seventy-three percent of those alliances were with other Pacific Rim counterparts. Twinnings commonly lead to cultural and educational exchanges, tourism and increased commerce.

Direct investment in the Pacific Northwest by individuals and corporations from other sections of the Pacific Rim adds yet another thread in the thickening fabric of Pac Rim spatial interaction. The Japanese have been major investors in the region, often with high profile ventures. One example was the 1992 purchase of the Seattle Mariners baseball team by a consortium funded in large part by Minoru Arakawa, president of Nintendo of America.

Although not likely to make NBC Nightly News, many other Japanese investments are of no less local interest, and sometimes concern. West of Portland, the clustering of Japanese investment in the silicon chip industry around Hillsboro has some area residents concerned that a "little Japan" is emerging. Farther to the east in western Montana, the 1988 acquisition of a large cattle ranch and start-up of the Zenchiku Land and Livestock Company was big news in the Big Sky livestock world. Trade liberalization and improved access for American beef imports into Japan also was behind the 1988 purchase of Washington Beef by Farmland Trading, Ltd., a Japanese group. Purchase of that company's packing plants in Toppenish and Union Gap and the 20 percent of revenues now derived from export of beef to Japan saved the 470 jobs at those facilities according to John Kincaid, Washington Beef President and CEO.

In some cases, even locals may be unaware of a Japanese investment connection with next-door developments. This is the case with the upscale and innovative, 1,000-acre-plus planned community of Mill Creek 20 miles north of Seattle being developed by United Development Corporation, a Japanese consortium. A 116-acre nature preserve, 137-acre championship golf course, country club, tennis complex, swim club and miles of cycling, jogging and walking trails make this south Snohomish County community a residential refuge in the busy and crowding east Puget Sound Corridor. The Japanese corporate approach of investing with a longterm time horizon, as opposed to a more typically American requisite quick return, made this venture possible—investors weathered an initial 11 years of losses.

Not all overseas Pacific Rim investment comes from Japan. A recent study showed in the late 1980s Japan was the first ranking foreign owner of Oregon assets with 21.75 percent of foreign ownership, but the United Kingdom was a close second with 19.18 percent and Canada checked in at 18.68 percent. Although Pac Rim and other foreign investment may be high profile and may profoundly impact some local areas, it accounts for less than 2 percent of Northwest employment.

One of the most interesting and potentially most profound people-to-people linkages tying Northwest residents to others in the Pacific Rim is in higher education. The Pac Rim era of international education has arrived at Northwest colleges and universities. Foreign students from Rim countries dominate international enrollment at area institutions of higher education. At the University of Washington in Seattle, 8 Pac Rim countries rank in the top 10 for student numbers; the number is 9 of the top 10 at Washington State University in Pullman; and at Oregon State University in Corvallis the figure is 7 of the top 10. There is some campus-to-campus variation, but on those three campuses combined, students from the People's Republic of China constitute the largest foreign student contingent.

The geography of Northwest higher education also is being impacted by foreign-run institutions within the region. Mukogawa Fort Wright Institute in Spokane is an example. In 1990 this closed, 72-acre residential campus with spacious brick Victorian buildings was purchased by Japan's Mukogawa Women's University as an

American branch campus. The Japanese school is located in Nishinomiya, a Spokane sister city. University and junior college English majors from the Japanese school spend 14 weeks and 8 weeks respectively studying English language skills and American studies at the Spokane branch. As part of the Institute's community outreach, Mukogawa has established the Japanese Cultural Center, which provides materials on Japan, presents lectures and exhibits, and offers a variety of classes, including Japanese language, ink painting and cooking to the entire community.

Many Northwest communities with Pac Rim sister cities, and others with more informal relationships, exchange public school students and faculty on a regular basis. Opportunities for more in-depth Pacific Rim educational links also are available. At Eugene, Oregon, Yujin Gakuen Japanese Immersion School conducts half of all classes in Japanese. In 1993 fifth-grade students spent three weeks in Japan testing their knowledge of language and culture acquired over the preceding five years.

Pacific Rim immigration, tourism, cultural exchanges, artistic performances, architectural styles, gardens, foods and a wide range of other links strengthen Northwest:Pacific Rim ties and point to a continuing Rim-wide integration.

THE NORTHWEST AND THE PACIFIC RIM: TOMORROW

Prior to the arrival of transcontinental railroads in the 1880s, the Pacific Northwest was, in many ways, closer to other sections of the Pacific Rim than to the American East Coast. From the fur trade era through the early lumber and grain export days, the fortunes of the region were linked more to markets in Canton and Sydney than to those in New York City or Philadelphia. Today the Northwest is on the threshold of a nascent Pacific Century in which not only the region, but the entire nation, is focusing increasingly on the Pacific world. On a continental scale, the American West Coast traditionally has been viewed as America's back door for international trade. Today that position has changed. On an annual basis in the 1980s the value of our nation's trade across the Pacific exceeded that across the Atlantic and the trend is accelerating. The Pacific Northwest now has a front door location, on the threshold of a burgeoning Pacific region. This exciting development means new spatial forces will continue to impact Northwest geography as economic ties strengthen and linkages spill over into other aspects of Northwest life.

New geographies will be related to both inter- and intra-regional forces. Key to changes will be continued improvements in transportation. Increased interaction and demand for accessibility should mean continued transport innovation and modification. It may already be possible to differentiate the Pacific Rim Northwest into several interconnected regions. On the west is the I-5 Corridor, where the Northwest and much of the nation articulates with the Pacific world. Its ports and large urban centers serve as Rim gateways. To the east is a vast hinterland that stretches from the western flank of the Cascades eastward to the Rockies. This is a region of sparse population and is dominantly a producer of raw materials. Along the eastern edge and just beyond the Northwest is an evolving Rocky Mountain Corridor reaching from Edmonton and Calgary, Canada southward to the Mexican border. In Montana and southeastern Idaho north:south Interstate 15 and essentially parallel Burlington Northern/Union Pacific rail service may be helping to define this new international corridor of commerce and trade. Increased cross-border interaction is already a reality and the North American Free Trade Agreement (NAFTA) promises even more north:south commerce.

Linking these north:south regions are major, but few in number, high-order east:west transportation corridors. The Cascade Range is crossed at only three locations by large capacity transport axes; 1) Stevens Pass where the Burlington Northern crosses; 2) Snoqualmie Pass where Interstate 90 crests the divide and 3) the Columbia River Gorge where the Burlington Northern and Union Pacific, Interstate 84 and the Columbia River pass through this deep defile. On the east within and just beyond the Northwest, major transcontinental transportation routes cross the Continental Divide in the form of the Burlington Northern at Marias Pass, Montana Rail Link at MacDonald Pass, Interstate 90 at Homestake Pass and the Union Pacific through the Great Divide Basin in Wyoming Such a sparse transportation system may no longer be adequate for a Northwest that now is on the front line of America's international trade geography. One obvious option is to improve speed and efficiency along existing trunk routes. Already there is discussion of possible funding for high-speed rail service within the I-5 Corridor and even talk of extending service between Seattle and Spokane via Moses Lake and Snoqualmie Pass.

The Columbia/Snake River system also may offer opportunities for continued transport innovation. Joel Kotkin, insightful author of *The Third Century* and other publications, already has called this waterway the "New Northwest Passage." Its future as a low-cost transportation mainline to and from the Pacific Rim, serving an expanding hinterland area that interfaces with the Rocky Mountain Corridor, seems assured if river drawdowns in conjunction with restoration of migrating salmon do not severely impact navigation. This development should bode especially well for Lewiston, Idaho, the river system's most inland port.

Increasingly, global investors will be looking at the Pacific Rim as a unit, deciding, for example, whether they want to locate a new facility in the inland Northwest or in a Special Economic Zone in coastal China. Continued improvements should allow the Northwest to remain an attractive alternative and an active participant in the evolving Pacific Rim mega-region.

A concentration of new Pacific Rim development in the Puget-Willamette Axis, astride Interstate 5, already has contributed to congestion, sprawl and rapidly rising real estate prices. One consequence has been spillover of population and investment east of the Cascades. Some commuters, for example, have found that drive time is shorter to their Seattle area jobs from the Cascades east side community of Cle Elum than from traditional, closer suburbs. Boeing also has felt the squeeze in metro Seattle and has made major investments in Spokane. Such hinterland developments should continue as the entire Northwest adjusts to new Pacific Rim spatial forces.

Major Northwest port cites, especially Seattle/Tacoma and Portland, should continue to evolve as Pac Rim gateways serving much of the nation. Expect Seattle to remain as the undisputed regional capital and the Northwest's leading Pacific Rim world city in the unfolding international cities era. It routinely is classed as one of the nation's most livable and residentially desirable cities. In the November 2, 1992, issue of *Fortune* magazine, it was selected as the nation's number one city for business—world linkages and ready access to the Pacific world were important factors considered in this designation. The city's growing international prowess is reflected in its more than 20 foreign consulates. In 1992 Seattle joined New York, San Francisco and Washington, D.C. as a center of Russian diplomatic activity when Russia selected it as the site for its new consulate, a reflection of the importance of Northwest trade with the Russian Far East.

Residents of Washington, Oregon, Idaho and western Montana commonly think of themselves as citizens of the Northwest. If current trends continue it is not impossible that people in this region may someday refer to themselves as residents of the Northeast—the northeast of a Pan-Pacific community.

New geographies are inevitable in the evolving Pacific Rim Northwest. Unfortunately, most considerations of the Pacific Rim phenomenon are focused almost exclusively on its economic and/or political dimensions. Only rarely are the environmental, resources and quality of life implications of continued region-wide growth and interaction assessed.

From the decimation of the region's sea otter population in the late eighteenth and early nineteenth centuries, to such contemporary concerns as export of raw logs and sprawl of Rim gateway metropolitan areas, the Northwest has paid an environmental price for its Pacific Rim interchange. If the past and present are even general indicators of possible future geographies, there will be major implications for the Northwest's environmental quality and its natural resources in the evolving Pacific Century.

SELECTED READINGS

John A. Alwin. "North American Geographers and the Pacific Rim: Leaders or Laggards," *The Professional Geographer*, Vol. 44 (November, 1992), pp. 369–376.

Gary Dean Best. "James J. Hill's Lost Opportunity on the Pacific," *Pacific Northwest Quarterly*, Vol. 64, No. 1 (January, 1973), pp. 8–11.

Theodore H. Cohn, *et. al.* "North American Cities in an International World: Vancouver and Seattle as International Cities" in *The New International Cities Era: The Global Activities of North American Municipal Governments*, ed. Earl H. Fry, *et. al* (Provo, UT: Brigham Young University, 1989), pp. 73–118.

Thomas R. Cox. "The Passage to India Revisited: Asian Trade in the Development of the Far West, 1850–1950," in *Reflections of Western Historians*, ed. John Alexander Carroll, Western History Studies (Tucson: University of Arizona Press, 1969), pp. 85–103.

_____. *Mills and Markets: A History of the Pacific Coast Lumber Industry* (Seattle: University of Washington Press, 1974).

Mike Douglas. "The Future of Cities on the Pacific Rim," in *Pacific Rim Cities in the World Economy*, ed. Michael Peter Smith, Comparative Urban and Community Research, Vol. 2 (New Brunswick, NJ: Transaction Publishers, 1989), pp. 9–67.

Robert Ficken. *The Forested Land: A History of Lumbering in Western Washington* (Seattle: University of Washington Press, 1987).

D. K. Fleming. "The Port Community: An American View," *Maritime Studies and Management*, Vol. 14, No. 4, 1987, pp. 321–336.

James R. Gibson. *Farming the Frontier: The Agricultural Opening of the Oregon Country, 1786–1846* (Vancouver: University of British Columbia Press, 1985).

_____. *Otter Skins, Boston Ships, and China Goods: The Maritime Fur Trade of the Northwest Coast, 1785–1841* (Seattle: University of Washington Press, 1992).

Gil Latz. "Portland's East Asian Connection," in *Portland's Changing Landscape*, ed. Larry W. Price, Occasional Paper No. 4, prepared for the annual meeting of the Association of American Geographers, April 22–26, 1987, Portland, Oregon (Portland: Department of Geography, Portland State University, 1987), pp. 121–135.

Pacific Gateway, the Port of Tacoma's quarterly publication, is an excellent source for up-to-date information on this port.

Portside, the Port of Portland's quarterly publication, is an excellent source for up-to-date information on this port.

William Thomas. "The Pacific Coast of North America, As Viewed from the Pacific," *Yearbook of the Association of Pacific Coast Geographers*, Vol. 41, 1979, pp. 7–27.

Tradelines, the Port of Seattle's quarterly publication, is an excellent source for up-to-date information on this port.

John B. Watkins. *Wheat Exporting from the Pacific Northwest*, State College of Washington, Agricultural Experiment Station, Bulletin No. 201 (May 1926).

Bert Webber. *Wrecked Japanese Junks Adrift in the North Pacific Ocean* (Fairfield, WA: Ye Galleon Press, 1984).

Simon Winchester. *Pacific Rising: The Emergence of a New World Culture* (New York: Prentice Hall Press, 1991).

Index

Abbott, Carl, 432, 441
Aberdeen, WA, 436
Advection fog, 90
Agate Pass, 438
Aggregates, 292–296, 298, 301, 303–304
Agricultural products, 386
Agricultural region, 397–398
Agriculture, 363–399
Air conditioning, 261
Air drainage, 96
Air Force bases, see Military Posts
Air stagnation, 96
Alaska, 432, 438–439
Alaska current, 79
Alaska Packers Company, 31
Alaska Pollock, 178
Alaskan-Yukon-Pacific Exposition, 459–460
Aleutian Low, 82
Alki, WA, 454
Alpine, 113–114
Aluminium, 263
American Board of Commissioners for Foreign Missions, 12
American Falls, 437
American Fur Company, 10, 11
Anaconda, 304–306
Anaconda Company, 33
Anaconda, MT, 435
Anacortes, 438
Anadromous, 141, 166, 169, 174
Ancestral floras, 100, 102, 104
Animal and crop diseases, 366–367
Annual range in temperature, 78
Antecedent River, 48, 52
Antelope, OR, 429
Apple farming, 392
Aquaculture, 171, 178
Arco, ID, 434, 437
Arctic air, 80
Aspect, 121
Aspen (Populus tremuloides), 108–109
Astor, John Jacob, 3, 7, 10
Astoria, OR, 86, 430
Auburn, WA, 438
Aurora, OR, 429
Average annual temperature, 78

Baby's breath (Gypsophila paniculata), 205
Bainbridge Island, 438
Baja, CA, 81, 82
Baker, Dr. Dorsey, 22
Baker, OR, 437
Ballard (in Seattle), WA, 439
Banks Lake, 378–379
Basic worker, 434–435
Basin and range, 39, 47–49, 5-4, 59–61
Basin and Range geomorphic province, 119–120, 129–131
Bear grass (Xerophyllum tenax), 205
Beaver, 166
Beaverton, OR, 443
Bell-Nelson Lumber Company, 30
Bellingham and Northern Railroad, 24
Bellingham, WA, 78, 83, 438
Belt supergroup (belt rocks), 44, 46–47
Bend, OR, 434–435
Berry farming, 391–392
Berthusen Park, 383
Big Lost River, 434
Biodiversity, 114–115
Biomass, 259, 271
Birch River, 434
Bitterroot Range, 43, 44
Bitterroot Valley, 42, 43
Black Butte, OR, 423
Blackfoot, ID, 437
Blanchet, Father Norbert Francis, 13
Blue Mountains, 434
Bodega y Quadra, Juan Francisco de, 5
Boeing Aircraft and Aerospace, 345, 359–360
Boeing Company, 34, 435, 439, 443, 445, 470, 476
Boise Basin, 297
Boise, ID, 86, 430–431, 433–434, 436–437, 442
Bonanza towns, 19
Bonneville Dam, 34, 466

Bonneville Power Administration (BPA), 228, 240, 376
Boulder batholith, 303–304
Boyce, Ronald, 427, 436, 438–439
Brainerd, Erastus, 439
Bremerton, WA, 435, 438
Bridger, Jim, 9
British thermal unit, 260, 271
"Brookings effect," 79
Brookings, OR, 78–79, 89
Broughton, William, 6
Buhl, ID, 437
Bureau of Land Management, 187, 194, 197, 203–204, 208
Bureau of Reclamation (Bu Rec), 224–225, 227, 250–251
Burley, ID, 437
Burlington Northern Railroad, 473, 476
Butte, MT, 435
Cabrillo, Juan, 4
Calcification, 124–125, 129, 131
Caldwell, ID, 437
California, 431–432, 436, 445
California current, 79
California Gold Rush, 1849, 15, 16
California Trail, 2
Canada (see U.S. Canadian Treaty)
Canadian Pacific Railroad, 7
Canyon convergence, 95
Canyon live oak (Quercus chrysollpis), 191
Capacity, 265, 275–276, 284
Cape Disappointment, 429
Carey Act, 1894, 26, 379
Carnation Company, 34
Carson, Rachel, 367
Cascade Mountains, 39, 48–49, 60–62, 64, 101, 104, 106–109, 113–114, 434, 456, 466–467, 475–476
Cascara (Cascara segrada), 206
Cash-grain farming, 385, 387–388
Cataldo Mission, ID, 13, 433
Catfish, 141, 177
Catholics, 433, 437
Cattle and calves, 395–396
Cayuse Indians, 3, 13
Celilo Falls, 3

479

Central business district (CBD), 442–443
Central Highlands, 49–51, 57
Central Pacific Railroad, 24
Champoeg, 12
Channeled Scabland, 52, 57
Chanterelle (Cantharellus cibarius), 206–207
Chaparral, 111
Chemical Manufacture, 355
Chicago, IL, 441
Chicago, Milwaukee, St. Paul and Pacific Railroad (Milwaukee Road), 438
Chief Coboway, 416
Children's Orthopedic Hospital (Seattle, WA), 435
China, 439
Chinook winds, 96
Christmas greens, 205–206
Clams, 141, 151–153
Clark, Captain William, 7, 11
Clarkston, WA, 436
Clear-cut logging, 135
Clearcutting, 195–196
Climax species, 110
Clinton, William, U.S. President, 435
Clutch, 151
Coal, 259, 279, 439
Coast Range, 102, 104, 106
Cobalt, 295–296
Coeur d'Alene, 296
Coeur d'Alene, ID, 434
Coeur d'Alene Mining War, 1892, 33
Cogeneration, 271
Coliseum (Seattle, WA), 440
Colombia Plateau, 102
Colonies, English, 428
Colonies, Mormon, 429
Columbia Basin, 49–51, 55, 57–58, 440, 456, 466
Columbia Basin Project, 379–380
Columbia Gorge, 80
Columbia Intermontane (Prov.), 39, 48–49, 51, 55, 59–60, 62
Columbia Intermontane geomorphic province, 119–120, 129, 132
Columbia Mountains, 43, 44
Columbia Plateau, 48
Columbia River, 39, 48–49, 56, 61–63, 66, 430, 434, 452–454, 456–457, 465–466, 468, 473, 476
Columbia River Basalt, 51–53, 60, 62–63, 68
Columbia River Gorge, 61, 65, 440, 476
Columbia River Intertribal Fish Commission (CRIFC), 166
Columbia-Snake River Customs District, 463–464
Colville, 303
Combustion turbines, 278
Commercial farming, 25–26
Communal settlements, 429, 431
Communities, types of, 428–430, 432, 435–436, 444–445
Community, quest for, 427–430, 444–445
Composite of leisure traveler, 403
Conservation, 284
Consumptive Use of Water, 247–249
Container cargo trade, 439
Controls of climate, 77
Conurbations, 436, 442
Convectional precipitation, 94
Cook, Captain James, 5, 451
Cooke, Jay, 22
Coos Bay, OR, 436
Copper, 292, 294, 296, 301, 303–306
Cordilleran ice, 48, 52–53, 57, 65, 69
Corporate farms, 372–373
Corps of Engineers (Corps), 225–227, 235–236, 240
Coupe, Thomas, 419
Coupeville, WA, 419, 438
Cowlitz County, WA, 442
Cowlitz Farm, 11, 13
Crab, 141, 147, 151–153, 178
Crater Lake, 108
Crater Lake National Park, 434
Craters of the Moon National Monument, 418, 434
Crockett Lake, 431
Crown-Zellerbach Company, 31
Cyclogenesis, 82
Dairy farming, 393–394
Dalles, the, 3
Dam, 259, 266–268
Darrington, WA, 436
Decentralization of urban centers, 443
Demers, Father Modeste, 13
Denny party, 440
Depression of 1893, 28
Desert Land Act, 1877, 14, 26
Diatomite, 298, 301
Dillon, MT, 435
Disturbance, 100, 110, 189–190
Diversification, 310, 334–337
Doig, Ivan, 396, 427
Dolomite, 303
Donation claims, 15
Douglas, David, 9
Douglas-fir (Pseudotsuga menziesii), 106–108, 110, 115, 191, 206
Downtown Development Association (Seattle), 440
Drainage, 48, 52
Drake, Sir Francis, 4
Dry farming, 365
Dry Shadow, 251
Dwarf Oregon grape (Berberis nervosa), 206
Earthquakes, 47–48, 68
Ebey's Landing National Historical Reserve, 419
Economic base theory, 434
Economic Base and Manufactures, 347–349
Economic development, 309, 334, 337, 339–340, 342–344
Economic diversity, 309, 334, 337
Ecosystems, 188–190, 198–199
Edge Cities, 444
Eldorado Ditch, 298
Electrical Equipment, Instruments, and Components, 358
Electricity, 259, 283–287
Eliza, Francisco de, 5
Ellensburg, WA, 467
Elliott Bay, 430
Ellis, James, 440
Employment, 314–324, 330–332, 335, 338, 340–343
Endangered species, 203, 333
Endangered Species Act, 157, 174
Energy, 259–260, 265
Enlarged Homestead Act, 1909, 14
Entrepots to the Pacific Northwest, historic, 430
Environment(al), 337, 339–340, 343
Environmental change, 4
Environmental Protection Agency, 194
Eugene, OR, 442, 475

Index

Euro-American exploration, 4–10
Everett, WA, 429, 435, 438–439, 443, 468
Evergreen huckleberry (Vaccinium ovatum), 205
Exclusive Economic Zone, 144
Exotic Rivers, 217
Fabricated Metals, 358
Family coefficient, 434
Family farms, 367, 372–373
Farms, 367–368, 370, 376
    average value, 371–372
    labor, 381
    land, 367–368
    machinery, 374–376
    mechanization, 29, 382
    operations, 374–376
    size, 368, 370–372
    types and organization, 372–373
Fault-Block Mountains Section, 43–44
Federal, 333, 339–340
Federal Agencies (Actors), 224–228
Federal Ownership of Land, 224
Federal Way, WA, 443
Ferralo, Bartolome, 4
Fertilizers, 365, 374, 376–377
Ficken, Robert E., 30
Field crops, 388–390
Fire, 100, 108, 110–113, 189, 192, 199–201
    suppression, 100, 114
Fish and Wildlife, 236–239
Fish processing, 31–32
Fisheries, 31–32
Fishery Conservation and Management Act (also known as Magnuson Act), 144, 180
Flatfish, 141, 143–144, 150
Floating Bridges, Lake Washington, 439
Flood Hazard Mitigation, 242–247
Floral greens, 205, 208–209
Food and Kindred Products, 351
Forest Conference, 333, 339
Forest management, 192–194
Forest ownership, 186–187, 192
Forest types, 190–191
Forests, 104
    coastal, 106
    coniferous, 105–106
        grand fir-Douglas-fir, 105, 108
        mixed-conifer, 105–107

        mixed-conifer and broadleaf, 105, 109
        juniper woodland, 109
        ponderosa pine, 105, 108–109
        subalpine, 105, 107
Forks, WA, 435
Fort Astoria, 10–11
Fort Boise, 11
Fort Clatsop, 7
Fort Clatsop National Memorial, 414
Fort Colvile, 11
Fort George, 11
Fort Langley, 11, 17
Fort Nez Perce, 11
Fort Nisqually, 11
Fort Vancouver, 11, 12, 13, 17, 451, 453, 455
Fort Vancouver National Historical Site, 416
Fort Victoria, 12
Forts, *see* Military posts
Forward Thrust, 440, 441
Fossil fuels, 259
Fraser, Simon, 7
Fred Hutchinson Cancer Center, 435
Free trade agreement(s), 333
Freeland, WA, 429
Freeman, Otis, 384
Freeway Park (in Seattle, WA), 440
Freezing rain, 81
Frémont, John C., 9, 434, 437
Frontier, defined, 430
Fronts, 95
Frost, 92
Frost free periods, 85
Fruit Farming, 390–393
Fuca, Juan de, 4
Fuel, 260, 262, 270, 282
Fur trade, 10–12, 451–452
Furnace, 260
Furniture, Lumber, and Wood Pproducts, 355
Galiano, Dionisio Alcala, 5
Garreau, Joel, 444
Gemstones, 294–295, 298, 301, 304
General farming, 384
Genetic enhancement, 377
Geoduck, 153
Georgia colony, 428
Georgia Pacific Corporation, 31
Geothermal, 259, 273
Gilmore, John, 440
Glaciation, 48, 57, 66, 69

Gold, 292, 294–298, 301–306
Gold rushes, 15–17, 21, 431, 438, 441
Gooding, ID, 437
Gottmann, Jean, 442
Grand Coulee, 58
Grand Coulee Dam, 34, 378–379, 431–432, 440
Grand Coulee Project, 26–27
Grant, John, 418
Grant-Kohrs Ranch National Historical Site, 418–419
Graphite, 306
Gray, Captain Robert, 6, 451
Grazing Land Act, 1916, 14
Great Basin, 59, 101–102, 104, 109, 113
Great Northern Railway, 23, 431, 457
Great Northern Steamship Company, 458
Greater Seattle Metropolitan Area (SMA), 435
Green Lake (in Seattle, WA), 440
Green River Valley, 438
Gresham, OR, 443
Ground Water, 222–224, 248–250, 252–253
Groundfish, 141, 143–144
Group Health Cooperative, 445
Growth Management Act, Washington, 444
Guggenheim Corporation, 33
Gulf of Alaska, 79
Hake, 141
Halibut, 141, 147–148, 150
Halophytes, 113
Hanford Reservation, 431
Hanna, Captain James, 5
Harney-High Desert, 49, 54–55
Hart, John Fraser, 367
Hawaiian High, 81–82
Heating cooling, 259
Heceta, Bruno de, 5–6
Helena, 305–306, 437
Hell's Half Acre, ID, 434
Hells Canyon, 55, 60, 434
Hessian fly, 467
High Lava Plains, 49–50, 54, 59
High sun period, 77
Hill, James J., 23, 457–459
Historic Districts, 404
HMS "Chatham," 6
Holloday, Ben, 22
Homestake Pass, 476

Homestead Act, 1862, 14–15
Hood Canal, 438
Hoquiam, WA, 436
Hough, Emerson, 427
Hudson's Bay Company (HBC), 9, 11–12, 18–20, 416–417, 451–453, 455
Hume Brothers Company, 17
Hunt, William Price, 10
Hybrid seeds, 377
Hydro, 259, 263, 269
Hydropower, 222, 232–235
Ice Age, 434
Idaho, 119, 129, 131, 134–135
Idaho Batholith, 43–44, 46–47
Idaho Falls, ID, 436–437, 442
Idaho Panhandle, 78
Idaho Tourism, 405
Importance of tourism, 403
In-migration, 430
Income, 315–316, 324–329, 340–342
Indian prairies, 4
Industrial water use, 253–254
Inland Empire, 23, 429, 436
Inland water transport, 235–236, 465–466
Insects, 199, 201–203
Insolation, 77
Instream use, 232–247
Intermountain Seismic Belt, 47–48, 60
International North Pacific Fishery Commission, 144
International Pacific Halibut Commission, 148
International Workers of the World (IWW), 34
Interstate 15, 476
Interstate 5 (Corridor), 475
Interstate 5, 440
Interstate 84, 476
Interstate 90, 476
Irrigation, 16, 26–28, 228–231, 365, 377–381, 398
Isobars, 81
ITT Rayonier Company, 31
Jamestown colony, 428
Jantzen, Inc., 345, 351, 354, 361
Jefferson, President Thomas, 7, 15, 456
Jet streams, 83
"June Hogs," 157
Kaiser Shipyards, 431
Kamiksu batholith, 44, 46

Kelley, Hall J., 13
Kellogg, ID, 78, 435
Kelso, WA, 440
Kennewick, WA, 435
Kettle Falls, 3
Keystone ferry terminal, 431
King boletus (Boletus edulis), 206–207
Kingdome, 440
Kirkland, WA, 439
Kitsap Peninsula, 435
Kittitas Valley, 466–468
Klamath Mountains, 66, 67, 104, 109
Klondike Gold Rush, 431, 438
Krummholtz, 114
Kumamoto oyster, 152
La Grande, OR, 434, 437
Lake Bonneville, 58, 61
Lake Columbia, 52
Lake Missoula, 48, 52, 57–58, 63, 69
Lake Union, 439
Lake Washington, 439
Lake Washington Ship Canal, 439
Land grants, 187
Landforms as a climate control, 80
Landslides, 56–57, 63
Lapwai Mission, 13, 26
Laramide age, 47
Laramide Orogeny, 47, 52
Latitudinal position, 77
Lee, Reverend Daniel, 12
Lee, Reverend Jason, 12–13
Lemhi Pass, 7
Lemhi settlement (ID), 437
Lewis and Clark, 414–416, 452, 456
Lewis and Clark Centennial and American Pacific Exposition and Oriental Fair (The Lewis and Clark Fair), 459
Lewis and Clark Expedition, 7–8, 10
Lewis and Clark Lineament, 42–43
Lewis, Captain Meriwether, 7, 10
Lewis County, WA, 442
Lewiston, ID, 435–436, 465–466, 469, 476
Lighting, 262
Ling cod, 143–144
Little Lost River, 434
Livestock farming, 393–397
Location quotient, 348, 351

Lodgepole pine (Pinus contorta), 106, 108
Loess, 119, 134
Logan, UT, 437
Logging, 187, 194–198
Lolo Pass, 7–8
Longview, WA, 429, 436, 440, 455, 468
Los Angeles, CA, 432, 436, 441
Louisiana Pacific Corporation, 31
Lovejoy, Asa, 20
Low sun period, 77
LULUs, 352
Lumbering, 29–31
MacDonald Pass, 476
Machinery, Including Computers, 358
Mackenzie, Sir Alexander, 7
Magma, 291
Mall of America, 444
Mammals, 189
Mandan Indians, 7
Manis mastodon site, 3
Manufacturing Location Theory, 345–346
Manzanita (Arctostaphylos spp.), 111
Marble, 293, 306
Marbled murrelet, 115, 333
Marias Pass, 476
Marine West Coast climate, 88
Mass movement, 135
Massachusetts, 428
Matsutake (Tricholoma magniverlare), 206–207
McGregor, Alexander, 373, 396
McGregor Brothers Company, 28, 373
McLoughlin, Dr. John, 11–12, 20, 417
Mean annual temperature, 85
Meares, Captain John, 5
Medford, OR, 82–83, 89
Medicinals, 205–206
Mediterranean climate, 88
Megalopolis, 442
Meinig, D.W., 396, 428
Metaline District, 302
Methane, 272
Methodist missions and missionaries, 12–13, 19
Metro, 317–319, 323, 328, 332, 341–343, 440
Microsoft Corporation, 433, 443, 445

Middle Cascade Section, 62, 63
Middle latitude cyclonic storm, 94
Middle Rocky Mountain geomorphic province, 119
Middle Rocky Mountains, 39, 48, 59, 60
Military posts, 429, 431–473
    Everett naval home port, 439
    Fairchild Air Force Base, 429
    Fort Canby, 429
    Fort Casey, 429
    Fort Lawton, 439
    Fort Lewis, 429, 438
    Fort Steilacoom, 429, 438
    Fort Walla Walla, 429
    Fort Worden, 429
    McChord Air Force Base, 435
    Mountain Home Air Force Base, 435
    Sand Point naval station, 439
    Whidbey Island naval air station, 435
Mill Creek, WA, 474
Mill towns, 431–432, 435–436
Milwaukee Railroad, 24, 438
Milwaukie, OR, 443
Mineral exploitation and mining, 17, 33–34
Mining rushes, 431–432, 440
Minneapolis, MN, 444
Missionary era, 12–14
Missoula, MT, 434, 469, 470
Mist Gas Field, 300
Mitchell Act, 174
Moisture, 87
Molybdenum, 295–296, 303–304
Montana, 119, 129, 131
Montana Central Railroad, 23
Montana Rail Link, 476
Montana Tourism, 407
Montmorillonite, 132
Montpelier, ID, 437
Morel (Morchella esculenta), 206–207
Mormons, 26, 377, 429, 433, 437–438, 444
Moscow, ID, 435
Moses Lake, WA, 435, 476
Mosquito fleet, 438
Moss, 205
Mount Adams, 434
Mount Baker, 434
Mount Hood, 434
Mount Rainier, 434
Mount St. Helens, 434

Mountain Home, ID, 435
Mountain men, 9
Mt. Mazama, 119
Mud Lake, ID, 434
Mukogawa Fort Wright Institute, 474
Mukogawa Women's University, 474
Mullan Road, 431
Multiplier effect, 434
Municipality of Metropolitan Seattle (METRO), 440
Mushrooms, 206–207, 209
Mussels, 154
Nampa, ID, 437
Nash, Gerald, 34
National Park Features, 413
National Reclamation (Newlands) Act, 1902, 26
National Registrar of Historic Places, 404
Native American marts, 3–4
Native American Migration, 1–3
Native American settlement, 3
Native American tribes, 8, 14
Native Americans, 1–4, 187, 192
Natural gas, 259, 270
Naval stations, *see* "military posts"
Nazarenes, 437
New Age religion, 433
New England immigrants to Pacific Northwest, 440
Newlands Act, 1902, 27
Newsom, David, 20
Nez Perce Indians, 3, 13
Nickel, 292, 300
Nike Corporation, 433, 443, 445
NIMBYs, 337
Nintendo Corporation, 433
Nippon Yusen Kaisha (NYK), 458
Nitrogen Super Saturation, 169
Noble fir (Abies procera), 107, 205
Non-timber forest products, See Special Forest Products
Nonmetro, 317–326, 329–332, 335, 340–343
Nonmetro adjacent, 317, 321–324, 341
Nonmetro nonadjacent, 317, 323–324
Nootka Sound Convention, 5
Nordstrom Company, 445
Nori (Porphyra), 178
North American Free Trade Agreement (NAFTA), 333, 476

North Cascades, 51–53
    section, 62
North Pacific current, 79
North West Company, 7, 11
Northern Pacific Railroad, 22–24, 26, 28, 30, 378, 429, 431, 438, 457
Northern Rocky Mountain Province, 39, 42, 44, 48, 51, 57
Northern Rocky Mountain geomorphic province, 118, 129
Northern spotted owl, 115, 333, 468, 483
Northwest Economic Adjustment Initiative, 339
Northwest Manufacturing Concentrations, 348–349, 354–355, 357–359, 361
Northwest Manufacturing Employment Shifts, 348–350
Northwest Manufacturing Shift/Share, 348–350
Northwest Passage, 5
Northwest Power Planning Council (NPPC), 231, 237–239
Nuclear, 259, 282
Nuttall, Thomas, 9
Nyssa, ID, 437
Oak Harbor, WA, 435
Offstream water use, 247–255
Ogden, Peter Skene, 8
Oil, 270
Okanogan City, 302
Old-growth forests, 468
Oliver, Herman, 396
Olivine, 303
Olympia oyster, 151–152, 171
Olympia, WA, 86, 438, 443
Olympic Mountains, 66–68, 106–107, 113–114, 433, 435
Olympic National Park, 67
Ontario, OR, 435–437
Open Range, 28
Opera House (in Seattle, WA), 440
Oregon, 118–119, 126, 129–135
Oregon and California Lands, 187
Oregon and California Railroad, 22
Oregon Central Railroad, 22
Oregon City, OR, 440
Oregon Coast Range, 53, 66–67
Oregon County, 6, 11–12, 14
Oregon Donation Land Act, 1850, 14, 16
Oregon Historical Society, 416

Oregon Oak (Quercus garryana), 108, 110
Oregon Provisional Government, 14
Oregon Short Line, 24
Oregon Steam Navigation Company, 22
Oregon Territory, 14, 15, 430–431, 440
Oregon Tourism, 409
Oregon Trail, 2, 13, 434, 440
Organic farming, 367
Orogenesis, 101, 102
Orographic precipitation, 217
Overland Express, 22
Overland flow, 120, 129
Owyhee Upland, 49–50, 59
Oysters, 141
Pacific American Fisheries Company, 31
Pacific Border, 39, 42
Pacific Fishery Management Council, 144, 147, 172, 180
Pacific Fur Company, 3, 7, 10–11, 19, 21
Pacific madrone (Arbutus menziesii), 109–111, 191
Pacific Northwest-California Intertie, 232
Pacific Northwest Parks and Campgrounds, 419
Pacific Northwest Wilderness Recreation, 421
Pacific NorthWest Economic Region, 333
Pacific Ocean perch, 143–144
Pacific Science Center (in Seattle, WA), 440
Pacific yew, 115
Paleosol, 122, 133–134
Palouse, 90, 112, 436, 440, 456,
Parker, Reverend Samuel, 9, 12–13
Parkrose, OR, 443
Pasco, WA, 435
Payette, ID, 437
Pedogenic processes (internal), 117, 119, 122, 124–125, 128–129, 133–134
Pend Oreille County, 302
Pendleton, OR, 435
Perez, Juan, 5
Perlite, 301
Pests and pesticides, 365–366
Peterson, Ross, 436
Pettygrove, Francis, 20–21
Phosphate, 292–297

Photovoltaic, 276
Pink or Humpback, 132, 135
Pioneer economy, 15–19
Pioneer settlement, 12–15
Pioneer town-building, 19–22
Placer gold, 292, 296–299, 302, 305–306
Platte River, 430
Pleistocene, 101, 102, 114
Pocatello, ID, 86, 432, 436–437, 442
Pogonip, 91
Point Grenville, 5
Poison oak (Rhus diversiloba), 109, 110
Ponderosa pine (Pinus ponderosa), 107–108, 191
Pope and Talbot Lumber Mill, 16, 31
Pope and Talbot, 355, 357
Port Angeles, WA, 435
Port Blakely, WA, 438
Port Gamble, WA, 429, 453–454
Port Ludlow, WA, 438, 454
Port Madison, WA, 438
Port of Kalama, WA, 473
Port of Lewiston, 470
Port of Portland, 463–465, 468, 470–473
Port of Seattle, 439, 462–464, 466–468, 470–472
Port of Tacoma, 463, 466–473
Port Orchard, WA, 438
Port Townsend, WA, 429, 438
Portland, OR, 80, 429, 431–432, 435–436, 438–443, 451, 453, 455–456, 459, 461, 469, 474, 477
Potholes, 434
Potlatch Corporation, 470
Poulsbo, WA, 423
Poultry farming, 396–397
Pressure and wind, 92
Primary Metal Manufactures, 357
Prince's pine (Chimaphila umbellate), 206, 209
Printing and Publishing, 354
Prosser, WA, 89
Protestant missions, 12–13
Protestants, 437
Public supply, 251–253
Public utility districts, 376
Puget Mill Company, 454
Puget Sound (and Puget Sound lowland), 430–431, 434, 436,

438–442, 444, 454, 456–457, 461, 471, 474
Puget Sound Agricultural Company, 11, 455
Puget Sound lowlands, 119, 126, 132, 134–135
Puget Trough, 66, 69
Puget-Willamette trough, 85
Pugetopolis, 442
Pullman, WA, 435
Pumice, 292, 298, 300
Pumice Plateau, 130
Purcell Trench, 42–44, 48
Puritans, 428
Quarternary, 102, 104
Quimper, Manuel, 5
Radiation fog, 90, 91
Radioactive waste, 283
Railroad(s), 22–25, 312, 337
  Great Northern, 23, 431
  Milwaukee Road, 438
  Northern Pacific, 429, 431, 438
Rain shadow, 88
Rajneeshpuram, OR, 429
Range wars, 16
Rawhide Railroad, 22
Razor clam, 152–153
Reasons for tourist travel, 402
Recession, 313–314, 323, 328, 330
Recreation, 239–242
Recreation in the Pacific Northwest, 413
Recreational Equipment Inc. (REI), 433
Red alder (Alnus rubra), 106, 109
Red Sea urchins, 178
Redmond, WA, 443
Refrigeration, 28–29
Regional Actors, 229–231
Regional integrity of Pacific Northwest, 433–434
Reisner, Marc, 379
Renton, WA, 439, 443
Resource dependency, 309
Restructuring, 309, 328, 330
Rexburg, ID, 436, 437
Richland, WA, 435
Rigby, ID, 437
Rivers, major, 220–222
Roberts, ID, 437
Rockfish, 141, 143–144
Rocky Mountain Corridor, 476
Rocky Mountain Fur Company, 11
Rocky Mountain Log Homes, 469

Rocky Mountain Trench, 42–43, 48
Rocky Mountains, 101, 104, 106–108, 113, 435
Rogue River Valley, OR, 87
Roman Catholic Missions, 12–13
Roosevelt era, 379
Roseburg, OR, 87
Ross, Alexander, 3, 4
Roundfish, 141, 143–144
Ruby-Conconully District, 302
Rupert, ID, 437
Rural electrification, 376
Rural Electrification Act, 1936, 376
Sablefish, 143–144
Sagebrush (Artemisia spp.), 109, 111–112
Sagebrush Rebellion, 338
Salal (Gaultheria shallon), 205
Salem, OR, 380, 442
Salinization, 124–125, 129
Salmon, 237–239, 312, 333–334, 343
    Atlantic, 173, 175
    Chinook or King, 155, 157–158, 165, 171, 174
    Chum or Dog, 132–133, 135, 155, 158, 165, 171
    Coho or Silver, 133, 135, 155, 158, 165, 171–175
    Pink or Humpback, 155, 165, 171
    Sockeye, Red or Blueback, 133, 135, 155, 157–158, 165, 174
Salmon farming, 119, 145–146, 173
Salmon ranching, 119, 145–146, 171
Salt Lake City, UT, 433
Saltbush, 113
San Francisco, CA, 432, 441
Sand dunes, 49, 69
Sandpoint, ID, 436
Sapphire, 306
Sashimi, 152, 178
Scab lands, 434
Schwantes, Carlos, 433
Sea cucumber, 152, 178
Sea otter, 451–452, 477
Sea Otter Trade, 9
Seasonal precipitation maxima, 89
Seattle and Walla Walla Railroad, 23
Seattle Lakeshore and Eastern Railroad, 23
Seattle Mariners baseball club, 433

Seattle Symphony orchestra, 440
Seattle-Tacoma International Airport, 439
Seattle, WA, 427, 430–433, 435–436, 438–441, 443, 454, 457–462, 467, 471, 474, 476–477
Semi-permanent high and low pressure cells and wind, 81
Sequim, WA, 88
Serpentine soils, 111
Service worker, 434
Settlement patterns, historical periods of, 430–433
Seventh Day Adventists, 437
Shad, 119, 150, 141, 174
Shelley, ID, 434, 437
Shellfish, 119, 127, 141, 151, 171, 180
Shelton, WA, 436
Sherfey, Florence, 18
Shift/Share Index, 348–350
Ships
    Beaver, 455
    Columbia, 451
    Dakota, 458–459
    Helen Angier, 455
    Herman C. Leonard, 455
    Lady Washington, 451
    Lewiston, 465
    Mike Maru, 458
    Minnesota, 458–459
    Sea Otter, 451
Shrimp, 119, 131, 141, 151, 154, 178
Silicon Forest, the, 358
Silk trade, 459
Silk trains, 439, 459
Silmilkameen River, 302
Silver, 295–296, 298–299, 301–305
Silver fir (Abies amabilis), 205
Simpson Lumber Company, 31
Simpson, Sir George, 11, 13, 417
Sisters, OR, 422–423
Sitka spruce (Picea sitchensis), 106, 191
Ski resorts, 435
Skid Road (in Seattle, WA), 430
Smelterville, ID, 435
Smet, Father Pierre Jean de, 9, 13
Smith Jedidiah, 9
Smolt, 165, 173
Smoltification, 145, 170
Snake River, 49, 55, 59, 85, 430, 432, 434, 436–437, 444, 456, 466, 468, 470, 476

Snake River Plain, 52, 54, 56, 134, 432, 436
Snoqualmie Pass, 476
Snow, 88
Socio-economic specialization, 330, 340–342
Soil, characteristics, 117, 119, 121–122, 124–126, 128–133, 135
    organic matter, 119–120
    structure, 117, 119–120, 124, 131, 134
    texture, 117, 119–120, 124, 126
Soil Conservation Service, 194
Soil degradation, 117, 120, 122, 134–136
Soil erosion, 118, 120–122, 125, 134–135
Soil-forming factors (external), 118
    Biota, 118, 120
    Climate, 117–119, 121–122, 130, 132, 134, 136
    Parent material, 117–119, 124, 128, 131–132
    Relief, 118, 121
    Time, 118, 120, 122
Soil orders, 124, 126, 128, 134, 137
    Alfisols, 126, 128, 132
    Andisols, 129–130
    Aridisols, 129
    Entisols, 130, 132
    Histosols, 130
    Inceptisols, 129–132
    Mollisols, 131
    Spodosols, 131–132
    Ultisols, 132
    Vertisols, 132
Soil taxonomy, 117, 124, 126
Soils, 365–366
Solar, 233, 249
Sole, 121, 141, 143–144, 150
Space heating, 260, 287
Spalding, Reverend Henry H., 13, 26
Special forest products, 205–210
Spiny Dogfish Shark, 124, 144, 147
Spokane House, 11
Spokane Indians, 13
Spokane River (valley), 48, 51
Spokane, WA, 86, 89, 429, 431–433, 436, 442, 477
Sprague, WA, 429
Springfield, OR, 442
Squawfish, 144, 170
St. Anthony, ID, 437

St. Paul and Tacoma Lumber Company, 30
Staple industries, 9
Starbucks coffee, 445
State Agencies (Actors), 228–229
Steelhead, 119, 141, 158, 163, 174–175
Steens Mountain, 53, 59, 61
Steilacoom, WA, 454
Steppe, 111
    Desert-shrub, 111, 113
    Steppe-shrub, 111, 112
Stevens, Major Isaac I., 22
Stevens Pass, 476
Stone Container Corporation, 470
Storm systems, 84
Storms, 94
Strait of Anian, 5
Strait of Juan de Fuca, 438
Sturgeon, 119, 150
    Green, 151
    White, 150
Subalpine fir (Abies lasiocarpa), 107, 205
Sublette Brothers, 9
Subtropical air, 83
Suburbanization, 442–444
Succession, 99, 100, 189–190
Sugar pine (Pinus lambertiana), 107, 191
Sun Valley, 435
Sunbelt Shift, 432
Sunnyside Project, 26
Superimposed River, 48, 51
Surface waters, 216–222
Surimi, 124, 152, 147, 178
Sushi, 178
Swauk Creek, 302
Sword fern (Polystichum munitum), 205, 209
Tacoma Dome, 440
Tacoma Mall, 443
Tacoma, WA, 429, 432, 438–441, 443, 467, 477
Talc, 293, 306
Tanoak (Lithocarpus densiflorus), 110, 191
Taxol, 115, 206
TCK smut, 463
Tektronix, 345, 358–359
Tektronix Corporation, 443
Tertiary, 100–102
Textile Manufacture, 351, 354
Thematic retail district, 404–405
Theme Communities, 404, 421–422

Thermoelectric plants, 254–255
Thompson, David, 7
Threatened species, 333
Tilapia, 141, 177
Timber, 311–313, 320, 326, 328, 330–331, 333, 335, 338
Timber summit, 1993, 435
Timothy hay, 466–468
"Tonquin," the, 10, 11
Toppenish, WA, 474
Tourism (defined), 401–402
Tourist satisfaction, 404
Townsend, John Kirk, 9
Transportation Equipment Manufacture, 359–360
Travel Montana, 401
Treaty of Washington, 1846, 12
Tri-Cities (Richland, Kennewick and Pasco, WA), 435
Triploid oyster, 152
Trout, 141, 175, 177
Tshimikain Mission, 13
Tuna (Albacore), 141, 177–178
Twin Falls, ID, 78, 434–435, 437, 442
Tyson II shopping mall, Virginia, 444
U.S. Army Corps of Engineers, 439
U.S. Bureau of Reclamation, 26
U.S.-Canadian Treaty on the Columbia, 234–235, 245–246
U.S. Fish and Wildlife, 194
U.S. Forest Service, 187, 193–194, 197–198, 203–204, 208–209
U.S. Travel Data Center, 413
Ullman, Edward L., 434
Umpqua Valley, OR, 87
Union Gap, WA, 474
Union Pacific Railroad, 24, 473, 476
University of Puget Sound School of Law, 443
University of Washington Medical Center, 435
University of Washington, Seattle, 459–460, 474
Unruh, John, 14
Upper "level devils," 95
Upper Snake River Valley, 444
Upstream storage, 234–235
Upwelling cold water, 91
Urban economic base theory, 434
Utah and Northern Railroad, 23
Utah, 437
Valdes, Cayetano, 5

Vancouver, BC, 442
Vancouver, Captain George, 5, 6
Vancouver Expedition, 6
Vancouver, WA, 80, 430, 440
Vanport Manufacturing, 469
Vanport, OR, 431
Vegetable farming, 388–389
Vegetation, 99, 104
    zones, 104, 105
Vernal equinox, 84
Villard, Henry, 22, 23
Virginia City, MT, 435, 437
Viticulture, 392–393
Wade, Richard, 18, 430
Waiilatpu Mission, 13, 26
Walla Walla and Columbia Railroad, 22
Walla Walla, WA, 429, 431, 433, 456
Wallace, ID, 435
Wallowa Mountains, 50
Wallula, 456
Warm Springs Indian Reservation, 429
Wasabi, 178
Washington, 118–119, 126, 129–135
Washington Apple Commission, 466
Washington Beef, 474
Washington State Convention Center (in Seattle, WA), 440, 443
Washington Territory, 15, 21–22, 430–431, 437
Washington Tourism, 411
Washtucna, Washington, 134
Water heating, 260–261, 275
Water rights, 228–229
Water use system, 213–215, 231–232, 247
Weiser, ID, 437
Wenatchee Gold Belt, 302
Wenatchee, WA, 466
West Valley Line, 22
Western hemlock (Tsuga heterophylla), 106–107, 191
Western larch (Larix occidentalis), 191
Western red cedar (Thuja plicata), 106, 205
Westlake Mall (in Seattle, WA), 443
Weyerhaeuser Company, 30, 439, 443
Wheat, 455–456, 463, 465

Wheat Marketing Center, Portland, 465
Whidbey Island, WA, 419, 429, 431, 435, 438
Whiting (Pacific), 141, 143–144, 146–147, 178, 180
Whitman, Dr. Marcus, 12–13, 26
Whitman, Massacre, 13
Whitman, Narcissa, 13
Wild and Scenic Rivers, 241, 249, 421
Wild Fish, 173
Wilkes, Lieut. Charles, 9
Willamette River, 393, 461, 465
Willamette River and Valley, 431, 436, 440–442, 444
Willamette Silt, 69
Willamette Trough, 66, 68–69
Willamette University, 20
Willamette Valley, 110, 455, 468
Willapa Hills, 66, 67
Wind, 260, 278
Winthrop, WA, 435
Wisconsin glaciation, 48, 66, 69
  moraines, 57
Wise Use, 338
Wobblies, 34
Wood, 271
Work, John, 9
Work Projects Administration (WPA), 441
World's Fair: Portland (1905); Seattle (1902 and 1962), 431, 439
Worster, Donald, 379
Yakima Canal Company, 378
Yakima Folds, 51–52, 56–57
Yakima Improvement Company, 26
Yakima Indians, 26
Yakima, WA, 429, 466
Yellowstone National Park, 434
Yesler, Henry, 430
Yesler's Mill, 16
Yew (Taxus brevifolia), 206
Young, Brigham, 437
Zenchiku Land and Livestock Company, 474